Practical
Fracture Mechanics
in Design

MECHANICAL ENGINEERING
A Series of Textbooks and Reference Books

Founding Editor

L. L. Faulkner

*Columbus Division, Battelle Memorial Institute
and Department of Mechanical Engineering
The Ohio State University
Columbus, Ohio*

Additional Volumes in Preparation

Mechanical Engineering Software

Spring Design with an IBM PC, Al Dietrich

Mechanical Design Failure Analysis: With Failure Analysis System Software for the IBM PC, David G. Ullman

Practical Fracture Mechanics in Design

Second Edition, Revised and Expanded

Arun Shukla

University of Rhode Island
Kingston, Rhode Island, U.S.A.

CRC Press
Taylor & Francis Group
Boca Raton London New York

CRC Press is an imprint of the
Taylor & Francis Group, an **informa** business

CRC Press
Taylor & Francis Group
6000 Broken Sound Parkway NW, Suite 300
Boca Raton, FL 33487-2742

First issued in paperback 2019

© 2005 by Taylor & Francis Group, LLC
CRC Press is an imprint of Taylor & Francis Group, an Informa business

No claim to original U.S. Government works

ISBN-13: 978-0-8247-5885-1 (hbk)
ISBN-13: 978-0-367-39350-2 (pbk)

Library of Congress Cataloging-in-Publication Data

A catalog record for this book is available from the Library of Congress.

**Visit the Taylor & Francis Web site at
http://www.taylorandfrancis.com**

**and the CRC Press Web site at
http://www.crcpress.com**

Preface to the Second Edition

This is the second edition of a book that was originally published by Alexander Blake. This edition contains revisions in each chapter, more design problems, and four additional new chapters. The underlying theme of the book "Practical Fracture Mechanics" has not been changed and the book is still intended for the design and practicing engineer. The new chapters deal with the fractographic analysis, fracture of composites, dynamic fracture, and experimental methods in fracture mechanics. Presented below is the preface from the first edition with some additions and deletions reflecting the new material included in the book.

For many years fracture mechanics has been developing from a purely theoretical basis of a new science to a more practical alternative of dealing with such matters as component design, control of brittle fracture in service, and development of material specifications. In particular, linear elastic fracture mechanics, generally referred to in the technical literature as LEFM, has now been established for many years as a design methodology that can be employed with confidence under elastic conditions. The pragmatic alternative, however, should be further refined because of so many areas of potential use that are developing in military applications and widespread private industry, particularly since the fundamental principles of fracture mechanics can cover both metallic and nonmetallic materials. Almost any type of brittle fracture—be that in a giant crane hook in heavy construction or a tiny layer of plaque inside a delicate human artery—can well be analyzed with the aid of this branch of engineering science.

Reliable fracture control plans are already required in such areas as offshore drilling rigs, nuclear power plants, space shuttles, ships, airplanes, bridges, pressure vessels, pipelines, lifting gear, and other mechanical and structural systems in which rapid fracturing cannot be tolerated.

These introductory comments are intended to set the tone for this text written for engineering practitioners, designers, and students interested in exploring the applications of fracture mechanics techniques to mechanical components, structures, and control of fractures. It appears that current researchers of fracture phenomena still outnumber practitioners at all levels of academic and industrial "hands-on" experience. And to a similar degree theoretical publications in the field vastly outnumber practice-oriented papers and books. It is therefore high time to emphasize the need for equal time and effort to balance the learning processes of theory and design practice. By the same token, this book is not directed specifically to researchers and experts in fracture mechanics.

One of the prime objectives of preparing this text is to convey a clear, practical message that the appropriate elements of fracture mechanics, materials science, and stress analysis must be closely interconnected in forming a modern approach to engineering design. These elements directly affect the process of assessing the causes of fracture and the development of the methodology for minimizing the frequency and extent of structural failures. It is, however, only proper to recognize that, in spite of the considerable progress in the area of fracture mechanics, our understanding of the basic fracture mechanism is still hampered by various uncertainties, so that, even with the best analytical tools, certain approximations and simplifying assumptions can hardly be avoided.

The calculations and experimental aspects of fracture mechanics discussed in this volume have naturally been considered supplements to conventional stress analysis and materials technology. For this reason, this book is a compilation of fundamentals, definitions, basic formulas, elementary worked examples, and references with a general emphasis on linear elastic fracture mechanics, supported by several case studies and a survey of stress calculations and material selection used for developing fracture control decisions. Pertinent numerical results included in several chapters are given in English and SI units. Furthermore, in the spirit of recognizing the need for simplicity of presentation, the academic rigor of derivations and mathematical details has been reduced to a minimum during the preliminaries of dealing with toughness parameters, defect characterization, and design constraints.

That a practical book such as this is heavily derived from the contributions of others is evident. I have taken utmost care to acknowledge all the sources of information consulted. At the same time no statement is advanced that all up-to-date standards, codes, specifications, and regulatory guides were utilized in the preparation of the book material, since the process of updating standards and improving fracture mechanics technology continues unabated.

The first two chapters provide a short historical sketch outlining in simple terms the development of the concepts of stress, energy, and material behavior and their relationship to the fundamental parameters of fracture mechanics.

Chapters 3-5 are intended to serve as a general primer for calculation, involving fundamental definitions and symbols describing fracture mechanics and materials input. These chapters contain the majority of design formulas and the numerical illustrations consistent with the typical line of questions encountered in design.

Chapter 6 includes a brief review of fracture modes and design methodologies that should be of direct interest to design engineers. It deals with the elements of structural integrity in relation to materials behavior under brittle or ductile conditions and describes specific design approaches and criteria.

Chapter 7 presents a brief discussion on experimental techniques currently used by practitioners to determine stress intensity factors for different loadings in a given geometry.

Chapter 8 deals with different aspects of dynamic fracture mechanics including crack initiation, crack arrest and crack propagation.

Chapter 9 presents a discussion on fracture of fiber reinforced composite materials.

Chapter 10 covers the important topic of fractographic analysis. The fracture surface features have a wealth of information and this chapter attempts in brief to elucidate this.

Many aspects of the book are based on the direct experiences of Alexander Blake, that he accumulated over a long period working on programs of national defense. Chapter 11 is based on his documented experience, and it highlights a number of field problems involving a mix of materials issues, fracture mechanics parameters, and stress analysis techniques at work.

Chapters 12 and 13 deal with the practical aspects of fracture control, selected design formulas, and definitions in fracture mechanics. This material is supplemented with a number of "design comments" on the application of special formulas to various design situations requiring a typical mix of knowledge of elementary fracture mechanics, materials, and stress analysis.

The invaluable contributions of Alexander Blake to this book are gratefully acknowledged. On his behalf, I want to acknowledge the contributions of all the individuals who helped him in the preparation of the first edition. I also wish to acknowledge the help of my graduate students V. Parameswaran, V. Chalivendra, N. Jain, A. Tekalur and V. Srivastava in the preparation of this revised edition. I also thank Anish Shukla for proofreading the original manuscript.

Arun Shukla

Preface to the First Edition

For many years fracture mechanics has been developing from a purely theoretical basis of a new science to a more practical alternative of dealing with such matters as component design, control of brittle fracture in service, and development of material specifications. In particular, linear elastic fracture mechanics, generally referred to in the technical literature as LEFM, is now established as a design methodology that can be employed with confidence under elastic conditions. The pragmatic alternative, however, should be further refined because of so many areas of potential use that are developing in military applications and widespread private industry, particularly since the fundamental principles of fracture mechanics can cover both metallic and nonmetallic materials. Almost any type of brittle fracture, be that in a giant crane hook in heavy construction or a tiny layer of plaque inside a delicate human artery, can well be analyzed with the aid of this branch of engineering science.

Reliable fracture control plans are already required in such areas as offshore drilling rigs, nuclear power plants, space shuttles, ships, airplanes, bridges, pressure vessels, pipelines, lifting gear, and other mechanical and structural systems in which rapid fracturing cannot be tolerated.

These introductory comments are intended to set the tone for this text written by an engineer for engineering practitioners, designers, and students interested in exploring the applications of fracture mechanics techniques to mechanical components, structures, and control of fractures. It appears that current researchers of fracture phenomena still outnumber practitioners at all levels of academic and industrial "hands-on" experience. And to a similar degree theoretical publications in the field vastly outnumber practice-oriented papers and books. It is therefore high time to emphasize the need for equal time and effort to balance the learning processes of theory and design practice. By the same token, this book is not directed specifically to researchers and experts in fracture mechanics.

One of the prime objectives of preparing this text was to convey a clear, practical message that the appropriate elements of fracture mechanics, materials science, and stress analysis must be closely interconnected in forming a modern approach to engineering design. These elements directly affect the process of assessing the causes of fracture and the development of the methodology for minimizing the frequency and extent of structural failures. It is, however, only proper to recognize that, in spite of the considerable progress in the area of fracture mechanics, our understanding of the basic fracture mechanism is still hampered by various uncertainties, so that, even with the best analytical tools, certain approximations and simplifying assumptions can hardly be avoided.

The calculational and experimental aspects of fracture mechanics discussed in this volume have naturally been considered supplements to conventional stress analysis and materials technology. For this reason, this book is a compilation of fundamentals, definitions, basic formulas, elementary worked examples, and references with a general emphasis on linear elastic fracture mechanics, supported by several case studies and a survey of stress calculations and material selection used for developing fracture control decisions. Pertinent numerical results included in several chapters are given in English and SI units. Furthermore, in the spirit of recognizing the need for simplicity of presentation, the academic rigor of derivations and mathematical details has been reduced to a minimum during the preliminaries of dealing with toughness parameters, defect characterization, and design constraints.

That a practical book such as this is heavily derived from the contributions of others is evident. I have taken utmost care to acknowledge all the sources of information consulted. At the same time no statement is advanced that all up-to-date standards, codes, specifications, and regulatory guides were utilized in the preparation of the book material, since the process of updating standards and improving fracture mechanics technology continues unabated.

Several recent textbooks were found to be of unusual value in the author's search for practical design solutions and graphical illustrations of fracture mechanics parameters to assist the creation of this volume. The authors of the selected titles [References 2, 14, 20, 21, 36, 121, 122 and 140] deserve a special credit for giving the engineering profession a pool of scientific and technical wisdom derived from countless papers and years of research in fracture mechanics and related issues.

The opening two chapters provide a short historical sketch outlining in simple terms the development of the concepts of stress, energy, and material behavior and their relationship to the fundamental parameters of fracture mechanics.

Chapters 3–5 are intended to serve as a general primer for calculation, involving fundamental definitions and symbols describing fracture mechanics and materials input. These chapters contain the majority of design formulas and the numerical illustrations consistent with the typical line of questions encountered in design.

Chapter 6 includes a brief review of fracture modes and design methodologies that should be of direct interest to design engineers. It deals with the elements of structural integrity in relation to materials behavior under brittle or ductile conditions and describes specific approaches and criteria.

The task of preparing the book material based on direct experience with fracture control would not have been possible without cooperation of the research laboratories and industrial organizations supporting the programs of national defense over the past 30 years. During that period the mechanical and structural design philosophy was shaped in terms of the principles of practical fracture mechanics at all stages of procurement and fielding of critical hardware components and systems. One of the specific programs was concerned with the engineering aspects of fielding underground experiments, reflected briefly in the case studies and the elements of fracture control in Chapter 7. The entire chapter is based on documented experience, and it highlights a number of field problems involving a mix of materials issues, fracture mechanics parameters, and stress analysis techniques at work.

Chapters 8 and 9 deal with the practical aspects of fracture control, selected design formulas, and definitions. This material is supplemented with a number of "design comments" on the application of special formulas to various design situations requiring a typical mix of knowledge of elementary fracture mechanics, materials, and stress analysis.

I wish to acknowledge the individual contributions and reviews that greatly influenced the intent, scope, and technical presentation of the material.

Mr. Philip R. Landon offered his extensive knowledge of metallurgy, practical use of fracture mechanics, and history of special case studies related to structural failures in support of the presentation of Chapter 7 and other portions of the book. His insight into the process of blending materials science, the fundamentals of fracture control, and basic stress analysis was of special value during the planning and development of the book's concept.

Mr. Anthony M. Davito undertook the painstaking task of a detailed technical review of the entire manuscript with special emphasis on fracture mechanics methodology, selection of working formulas, and the numerical illustrations. His long-standing experience in advanced engineering design and his superior technical knowledge of the structural and mechanical aspects of modern technology have provided a high level of confidence in presentation of the design arguments and the numerical accuracy of the calculations. Last but not least, the reviewer has performed a valuable service to the reader by confirming the relevance of the book's material to design.

Dr. Donald W. Moon conducted a critical review of the manuscript from the point of view of a specialist in materials science and the engineering aspects of fracture mechanics. His varied professional background and interest in this book project have been of special help in clarifying the presentation of scientific concepts, definitions, and pragmatic elements of the two disciplines in relation to a modern approach to fracture control technology. The emphasis of the review was on the scientific and educational aspects of the book and reader-friendly characteristics.

I am open to comments and constructive suggestions for future improvements of this book's substance and style. This work is offered here as a plea for equal time in dealing with the theory and practice of fracture mechanics.

Alexander Blake

Contents

Contents   xiii

1

Historical Developments in Fracture Mechanics and Overview

Several structural failures can be associated with the fracture of one or more of the materials making that structure. When such events occur, they are mostly unexpected, sudden, and unfortunate, and it is natural for us to focus attention on minimizing the undesired consequences when designing and analyzing modern-day structures. The study of crack behavior, prevention and analysis of fracture of materials is known as *fracture mechanics*.

In every discipline, including fracture mechanics, it is of critical importance to examine the historical antecedents. Progress not only depends on revolutionary ideas, but a significant part of it depends on retentiveness as well. People who tend to ignore the past are more prone to repeat mistakes. Although developments in fracture mechanics concepts are quite new, designing structures to avoid fracture is not a new idea. The fact that many ancient structures are still standing is a testimony to this. The stability of some of the ancient structures is quite amazing when we consider the fact that the choice of construction material was limited at that time. Brick and mortar, which were relatively brittle and unreliable for carrying tensile loads, were the primary construction materials. Even though the concept of brittle fracture did not exist, the structures were inadvertently designed against fracture by ensuring that the weaker components of the structure were always in compression. An arch-shaped Roman bridge design, as shown in Fig. 1.1, is an excellent example of a structure where fracture was avoided by virtue of design. The possibility of fracture in the bridge design

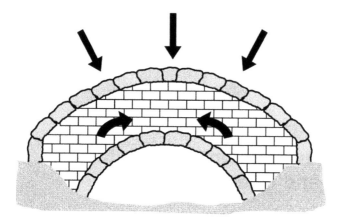

FIGURE 1.1   Schematic of an ancient Roman bridge design.

was avoided as the arch shape of the bridge results in compressive rather than tensile stresses being transmitted through the structure.

EARLY FOUNDATIONS OF STRENGTH OF MATERIAL CONCEPTS

The notes of Leonardo da Vinci (1452–1519) are the earliest records indicating a concept for evaluating the strength of materials. He suggested an experiment (da Vinci's sketch shown in Fig. 1.2), which is believed, was intended to establish a "law" for the influence of length on the strength of all types of materials. Even though it is unknown if the experiments were performed at that time, this was an early indication of the size effects on the strength of material. A longer wire corresponds to a larger sample volume and provides a higher probability of sampling a region containing a flaw.

The science and early evolution of the strength of materials concepts can be attributed to Galileo. In the early 17th century, Galileo turned his attention to structural mechanics while he was under house arrest and was banned from celestial mechanics. In his book *Due Nuove Scienze*, which was published in 1638,[1] Galileo introduced the concept of tensile strength in simple tension (Fig. 1.3), which he referred to as "*absolute resistance to fracture.*" His observation that the strength of the bar is proportional to the cross-sectional area and is independent of the length produced early strength of material concepts. Figure 1.3 illustrates Galileo's method of evaluating tensile strength in a column. It is an interesting fact that even Galileo noted an indication of size effect while he was visiting the Venetian Arsenal. He noticed a greater attention used by workers

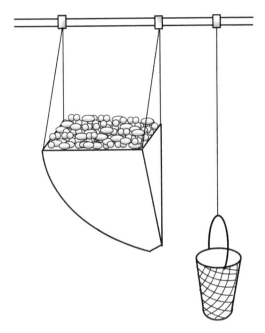

FIGURE 1.2   Da Vinci's concept for measuring the strength of wires.

in the construction of big ships than in small ships. At that time one of the master builders explained to him that the large ships were assumed to be more brittle than the smaller vessels.

DEVELOPMENT OF CONCEPT OF STRESSES

It was Robert Hooke (1635–1702) who broke away from the traditional thinking of his era and introduced the concept of the *true theory of elasticity or springiness* in 1678. Hooke tested wire strings of 20 to 40 ft in length by adding weight and measuring displacements. He made an important observation: that the wire always returned to its original length after several tests on the same wire. Hooke published his research in 1679,[2] outlining the principle that has since been known as Hooke's law. His far-reaching statement *ut tensio sic vis (Latin)* implied that when a mechanical force is applied to a solid object, change in shape (by extension or compression) must take place, and accordingly the solid produces a reaction.

The nature of a relationship between forces and deflections in a solid, under normal engineering conditions, is of course macroscopic. However, Hooke reasoned that when a structure is deflected, the structural material is also

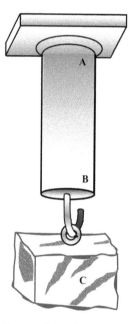

FIGURE 1.3   Galileo's illustration of tensile strength in a column.

deformed internally. This was a remarkable observation, because we know nowadays that the atoms and molecules can move under external forces. The chemical bonds joining the atoms can therefore be stretched or compressed, although on a nanoscopic scale. Hooke continued his work (in spite of experimental difficulties) to prove this point, and he also showed that the deflection of a structure was proportional to the load. Essentially, Hooke arrived at the conclusion that all solids and objects can behave like springs. Gordon provides an excellent assessment of Hooke's mental effort by saying that it is perhaps one of the great intellectual achievements of history.[3] Hooke's law advanced two very important principles:

1. Recovery from elastic deformation.
2. Linear relationship between applied load and elastic deformation.

Although rather simplistic in mathematical terms, this principle has been a significant help to engineering practitioners for more than 300 years. It certainly represents the early history of a conventional stress analysis that deals with relationships between the deflection, geometry, and material parameters of a given structure. It further denotes the science of elasticity concerned with the interactions between forces and displacements.

From the principles of Hooke's law, all subsequent contributions were based on the theory of elasticity. In 1807, Thomas Young published[4] the definition of modulus of elasticity, which is also known as Young's modulus. Young related stress (σ) and strain (ε) by using the modulus of elasticity (E) with a very simple equation

$$\sigma = E\varepsilon \tag{1.1}$$

From this point on in structural mechanics, quantitative methods could be used to design structures without having to constantly resort to testing.

DEVELOPMENT OF MODERN FRACTURE MECHANICS

By the end of the 19th century, the influence of crack on the structural strength was widely appreciated, but its nature and influence was still unknown. In 1913, Inglis published the first significant work in the development of fracture mechanics.[5] The work was an analytical formulation of stresses in a plate in the vicinity of a two-dimensional elliptical hole. The plate was pulled at both ends perpendicular to the ellipse as shown in Fig. 1.4. Inglis observed that the corner of the ellipse (point A) was feeling the most pressure and as the ellipse gets longer and thinner the stresses at A become larger. He examined local stresses at the tip of the ellipse and estimated that the stress concentration was

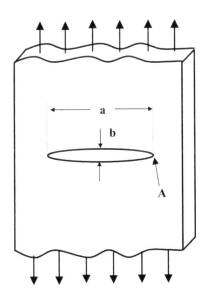

FIGURE 1.4   Elliptical hole in a flat plate.

approximately

$$2\left(\frac{a}{b}\right) \quad \text{or} \quad 2\sqrt{\frac{a}{\rho}} \tag{1.2}$$

where a and b are the semimajor and semiminor axes respectively and ρ is the root radius at the tip of the ellipse. Inglis evaluated various hole geometries and realized it is not really the shape of the hole that matters but the length of hole perpendicular to the load and the curvature at the end of the hole that matters in cracking. He also noticed that pulling in a direction parallel to the hole does not produce a great effect.

The basic ideas leading to the start of modern fracture mechanics can be attributed to a theory of fracture strength of glass, which was published by A.A. Griffith in 1920.[6] Using Inglis' work as a foundation, Griffith proposed an energy balance approach to study the fracture phenomenon in cracked bodies. A great contribution to the ideas about breaking strength of materials emerged when Griffith suggested that the weakening of material by a crack could be treated as an equilibrium problem. He proposed that the reduction in strain energy of a body when the crack propagates could be equated to the increase in surface energy due to the increase in the surface area. The Griffith theory assumed that the fracture strength was limited by the existence of initial cracks and that brittle materials contain elliptical microcracks, which introduce high stress concentrations near their tips. He developed a relationship between crack length (a), surface energy connected with traction-free crack surfaces (2γ), and applied stress, which is given by

$$\sigma^2 = \frac{2\gamma E}{\pi a} \tag{1.3}$$

Plasticity effects in metals limited the theorem and it was not until Irwin's work in 1948, that a modification was made to Griffith's model to make it applicable to metals. Irwin's first major contribution was to extend the Griffith approach to metals by including the energy dissipated by local plastic flow.[7] Orowan independently proposed a similar modification[8] to Griffith's theory in 1949. Orowan limited practical use to brittle materials while Irwin made no such restrictions.

It is an interesting fact and perhaps relevant to point out that the scientific curiosity towards fracture mechanics became a significantly important engineering discipline after the unfortunate failures of Liberty ships during World War II. The Liberty ships were built by the United States to support Britain's war effort and used a new construction method for mass production in which the hull was welded instead of riveted. The Liberty ship program was an astounding success until 1943, when a Liberty ship broke completely in two while sailing in the

North Pacific. Later, hundreds of other vessels sustained fractures. An investigation into Liberty ship failures pointed out poor toughness of steel and transition from ductile to brittle behavior at the service temperatures that ships experienced. It was noticed that the fractures initiated at the square hatched corners on the deck where there was a local stress concentration and the sharp corners acted like starter cracks. Research into this problem was led by George Rankine Irwin at the Naval Research Laboratory in Washington, DC. It was the research during this period that resulted in the development and definition of what we now refer to as linear-elastic fracture mechanics (LEFM). A major breakthrough occurred in the early 1950s when Irwin and Kies[9,10] and Irwin[11] provided the extension of Griffith theory for an arbitrary crack and proposed the criteria for the growth of this crack. The criterion was that the *strain energy release rate* (G) must be larger than the critical work (G_c), which is required to create a new unit crack area. Irwin also related strain energy release rate to the stress field at the crack tip using Westergaard's work.[12] Westergaard had developed a semi-inverse technique for analyzing stresses and displacements ahead of a crack tip. Using Westergaards' method, Irwin showed that the stress field in the area of the crack tip is completely determined by a quantity K called the *stress intensity factor*. Using the method of virtual work, Irwin presented a relationship between the energy release rate and the stress intensity factor as

$$\sigma_{ij} = \frac{K f_{ij}(\theta)}{\sqrt{2\pi r}} \tag{1.4}$$

$$K^2 = EG \tag{1.5}$$

where E is Young's modulus.

Other serious failures that were experienced during that period were those of the de Havilland "Comet" commercial aircraft. The Comet was first manufactured in 1952, and was the first two-jet-engine aircraft to fly at 40,000 ft with a pressurized cabin. After about a year in service, three aircraft failed, resulting in the tragic loss of several lives. In 1955, Wells[13] used fracture mechanics to show that the fuselage failures in several Comet jet aircraft resulted from fatigue cracks reaching a critical size. These cracks were initiated at windows and were caused by insufficient local reinforcement in combination with square corners, which produced higher stress concentrations. It was noticed that the fracture of welded Liberty ships, the pressurized cabin fractures of de Havilland Comet jet airplanes, bursts of several large petroleum storage tanks, and several other unpredicted failures, all seemed understandable in terms of the new fracture-strength points of view. The evaluation method was straightforward, a value of G_c was established from laboratory tests on precracked specimens and the value of the driving force G that tended to extend the starting crack was computed using appropriate stress analysis methods. The comparison showed that the fracture toughness had not been large enough to prevent crack propagation in the failure cases mentioned above.

In 1957, Williams[14] developed an infinite series that defined stress around a crack for any geometry. The use of the optical method "photoelasticity" to examine the stress fields around the tip of a running crack was published by Wells and Post in 1958,[15] and Irwin[16] observed that the photoelastic fringes not only formed closed loops at the crack tip as predicted by singular stress field equations but also showed a tilt as a result of the near specimen boundaries. In 1960, a significant contribution to the development of LEFM was put forth when Paris and his coworkers advanced an idea to apply fracture mechanics principles to fatigue crack growth. Although they provided convincing experimental and theoretical arguments for their approach, the initial resistance to their work was intense and they could not find a peer-reviewed technical journal to publish their manuscript. They finally opted to publish their work in a University of Washington periodical entitled *The Trend in Engineering*.[17] The work by Paris and colleagues was a landmark in the fatigue aspects of fracture mechanics, and yielded the equation

$$\frac{\mathrm{d}a}{\mathrm{d}N} = c(\Delta K)^n \tag{1.6}$$

where c and n are the curve-fitting parameters of experimentally obtained fatigue data.

Linear elastic fracture mechanics is not valid when significant plastic deformation precedes failure. Although earlier theoretical developments were aimed at understanding brittle crack behavior, it became apparent from experiments that except for a few, most materials are ductile and therefore linear elastic analysis should be modified accordingly. Dugdale[18] in 1960 and Barenbelt[19] in 1962 made the first attempts to include cohesive forces in the crack tip region by developing an elaborate model within the limits of elasticity. Later, in 1968, Rice conducted a simplified analysis of complete plastic zone formation, approximated by a circular region ahead of the crack tip.[20,21] The results derived from the energy–momentum tensor concept and applied to elastic cracks were extended to include plastic cracks by defining a path-independent integral termed the J integral. The plastic zone size and the crack opening displacement were found to correlate with the elastic stress intensity factor criterion. The successful experiments in 1971 by Begley and Landes,[22] who were the research engineers at Westinghouse, led to the publication of a standard procedure for J testing of metals in 1981.[23] In 1976, Sih[24] introduced the strain–energy density concept, which was a departure from classical fracture mechanics. He was able to characterize mixed-mode extension problems with this method, which also provided the direction of the crack propagation in addition to the amplitude of the stress field.

Contemporary research and development in fracture mechanics focuses on several interesting areas, such as dynamic fracture mechanics, interface fracture mechanics, shear ruptures in earthquakes, stress corrosion cracking, environmental effects on fatigue crack propagation, fracture of novel materials such as nanocomposites and graded materials. Extremely powerful numerical codes that are able to investigate fracture due to separation of atoms are being developed. Also, experimental techniques have progressed enough to investigate fracture in materials at nanometer length scales and nanosecond time resolution. However, experimental techniques that could provide spatial and temporal resolution simultaneously at the nanolevel are still not available. At this point it is pertinent to point out that in the present age of unprecedented technological growth, we are inclined to believe that technology and knowledge are accelerating in an exponential fashion. However, we must recognize that we are observing an exceedingly tiny period of human development on a historic time scale. Table 1.1 represents a compression of the elapsed time from the Big Bang event to the present day and puts in perspective that on a cosmic scale, our knowledge is still in its infancy. Although the field of fracture mechanics has matured in recent years, there will be a lot more to learn in the future.

BASIC CONCEPTS OF STRESS AND STRAIN

It is a rather strange twist of history that no real progress was made in solving practical problems in stress analysis and elasticity until 120 years after Hooke's

TABLE 1.1 Historical Events on a Cosmic Time Scale.

Events	Time: Compressed and scaled to 1 year
Big Bang	Jan 1st
Origin of Milky Way galaxy	May 1st
Origin of Solar System	Sept 9th
Formation of Earth	Sept 14th
First Humans	Dec 31st, 10:30 p.m.
Euclidian Geometry, Archimedean Physics, Roman Empire, Birth of Christ	Dec 31st, 11:59:56 p.m.
Renaissance in Europe, experimental methods in science	Dec 31st, 11:59:59 p.m.
Major developments in science and technology, power, flight, space, computers and strength of material concepts	Last second of the year (present)

death. In terms of our modern rush to saturate every scientific field with advanced numerical techniques, theoretical solutions, and software tools, 18th century experience appears to be unreal. There were, of course, valid reasons for the lack of progress, some of which have persisted until today. Theoretically inclined engineers and scientists, as well as most philosophers, appear to have limited interest in the problems of design and manufacture of industrial products, and in the numerous technical decisions affecting the integrity and economics of structures and machines. Hooke's accomplishments in the field of mechanics and his long-standing inventions were more than enough for the majority of interests of a scientific and technical nature during the 18th century. Finally, after many years of outstanding intellectual effort on the part of Leonhard Euler,[25] Thomas Young,[4] and an applied mathematician, Augustin Cauchy,[26] the concept of stress and strain was becoming a practical engineering tool, with Cauchy securing the major part of credit for this development in 1822.

Although they are rudimentary in nature, it is well at this point of our trek through history to sum up the concepts of stress and strain without which even the current, sophisticated science could not survive. While these concepts were few and, in modern terms, so obvious, they took centuries to develop. It appears that Galileo himself (1564–1642) almost stumbled upon the idea of stress, but the world needed another 180 years for this concept to mature. The simplicity is bewildering when we say that

$$\sigma = \frac{load}{area} = \frac{W}{A} \tag{1.7}$$

This can be tension or compression at a given point in a material that is acting in the direction of the applied load. In English units the stress (σ) is usually given in pounds (force) per square inch, or psi for short. Scientists prefer using the SI (Système Internationale) units while the Continental countries (generally speaking, Europe and Asia) employ metric units such as kilogram (force) per square centimeter, or kg/cm^2 for short. The basis for the SI, applicable to conventional stress analysis and fracture mechanics, involves the *newton* as a unit of force or weight while the unit of stress is known as the *pascal*, with the relevant symbols of (N) and (Pa). Hence pressure, stress, material strength, or elastic constants in the SI world are denoted by pascals or their more convenient multiples. Unfortunately, in engineering work, the unit of the pascal is far too small for all practical purposes. After struggling with the pascal unit for years it is hard not to promote the use of N/mm^2. The dimensions of countless machine components and structural elements are still expressed in millimeters worldwide. Also, 1 atmosphere, for practical reasons, can be taken as $0.1\ N/mm^2$, the strength of typical mild steel as $250\ N/mm^2$, or the elastic modulus for steel in general as

$2 \times 10^5 \, \text{N/mm}^2$. Of course, the proposed unit relates to the pascal as follows

$$\frac{\text{N}}{\text{mm}^2} = \frac{\text{N}}{\text{m}^2 \times 10^{-6}} = \frac{10^6 \, \text{N}}{\text{m}^2} = 10^6 \, \text{Pa} = 1 \, \text{MPa}$$

Here MPa denotes 1 megapascal. Unfortunately, the literature lacks uniformity because of the open choice of pascal, kilopascal, megapascal, and other potential multiples of the small and cumbersome unit of stress. The traditional unit of stress (psi) is still used by many engineers in English-speaking countries in spite of the efforts to convert industries and the public at large to the SI standard of weights and measures.

In dealing with practical issues of engineering formulas and the meaning of numerical results in various portions of this book, it may be helpful to the reader to have the following brief summary of the basic conversions at hand.

$1 \, \text{lb} = 4.4482 \, \text{N}$
$1 \, \text{psi} = \text{lb/in}^2 = 4.4482 \, \text{N}/(0.0254 \, \text{m})^2 = 6895 \, \text{Pa}$
$1 \, \text{ksi} = 6.895 \, \text{MPa}$
$1 \, \text{in.} = 25.4 \, \text{mm}$
$1 \, \text{psi} = 0.006895 \, \text{N/mm}^2$
$1 \, \text{lb-in.} = 4.4482 \, \text{N} \times 25.4 \, \text{mm} = 112.9842 \, \text{N-mm}$
$1 \, \text{lb/in.} = 4.4482 \, \text{N}/25.4 \, \text{mm} = 0.1751 \, \text{N/mm}$
$1 \, \text{MPa} = 1 \, \text{N/mm}^2 = 145 \, \text{psi}$
$1 \, \text{kg} = 9.8066 \, \text{N}$
$1 \, \text{MN} = 1 \, \text{meganewton} \cong 100 \, \text{long tons force}$

The second elementary but very important formula defines the concept of strain that enters considerations of stress and fracture in engineering materials. For the case of a bar in uniaxial tension or compression, the strain is

$$\varepsilon = \frac{\Delta L}{L} \tag{1.8}$$

where ΔL denotes the increase or decrease of the original length of the bar and L is the bar length. Engineering strain given by Eq. (1.8) is relatively small under elastic conditions and it is convenient to express strains as percentages in order to minimize typographical errors with zeros and decimal points.

It should be added that the original efforts to define and verify the concept of strain in Hooke's days were rather irksome because of experimental difficulties and a certain amount of confusion about whether to deal with the structure as a whole or at any given point within the material. Today, of course, we take a "test piece" from the structure under consideration and the stress–strain diagram obtained from the test is essentially unaffected by the size of the test piece. However, the shape of the diagram is characteristic of any given material, as shown in Fig. 1.5.

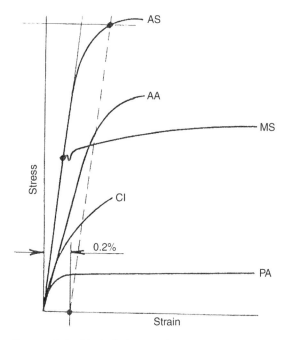

FIGURE 1.5 Typical shapes of stress–strain curves (AS, alloy steel; AA, aluminum alloy; MS, mild steel; CI, cast iron; PA, pure aluminum) (from Ref. 27).

When the major portion of the stress–strain curve is a straight line, it is customary to say that we are dealing with a Hookean material. The slope of the curve indicates the degree of stiffness of a given solid, which leads to the definition of a material constant E known as Young's modulus, also known as the modulus of elasticity or the elastic modulus:

$$E = \frac{\sigma}{\varepsilon} \tag{1.9}$$

The elastic modulus has the same dimensions as stress. This physical property is now regarded as a fundamental concept in materials science and engineering, and it has made some inroads in other science disciplines such as biology. For instance, in the cardiovascular field Young's moduli are measured for plaque and artery materials in order to better understand plaque failure strength and fracture characteristics. The importance of the Young's modulus concept can hardly be disputed, and yet it took the entire first half of the 19th century for scientists and engineers to accept it. The stress analysis and fracture mechanics principles cannot be understood and applied without the full acceptance of this key concept. Table 1.2 shows approximate values of Young's modulus for various materials.

TABLE 1.2   Approximate Young's Moduli.

Material	Young's modulus (E)	
	psi	(MPa) or (N/mm²)
Artery	14.5	0.1
Plaque	145	1
Rubber	1,000	6.9
Plastics	200,000	1,380
Plywood	1,000,000	6,897
Birch	2,070,000	14,280
Fresh bone	3,000,000	20,690
Brick (hard)[a]	3,500,000	24,140
Magnesium	6,000,000	41,380
Granite	7,000,000	48,280
Marble	8,000,000	55,170
Glass	10,000,000	68,970
Aluminum	10,000,000	68,970
Brass (naval)	15,000,000	103,450
Cast iron	15,000,000	103,450
Titanium (alloy)	17,000,000	117,240
Cast iron (malleable)	26,000,000	179,310
Steel	30,000,000	206,900
Chromium	42,000,000	289,700
Tungsten	58,000,000	400,000
Aluminum oxide (sapphire)	60,000,000	413,800
Tungsten carbide	102,000,000	703,400
Diamond	170,000,000	1,172,000

[a]Concrete, not shown here, has an average modulus of 3,000,000 psi. However, its value depends upon the ultimate strength according to the formula $E = 1000 \times$ compressive strength.[28]

Combining Eqs. (1.7) through (1.9) leads to a simple formula for estimating the amount of tension or compression in uniaxial loading:

$$\Delta L = \frac{WL}{AE} \tag{1.10}$$

To clarify some of the basic relationships in uniaxial loading we can look at the test piece in Fig. 1.6, where L denotes the original length of the specimen and Δr is the amount of lateral contraction. The corresponding lateral strain can be defined as

$$u = \frac{\Delta r}{r} \tag{1.11}$$

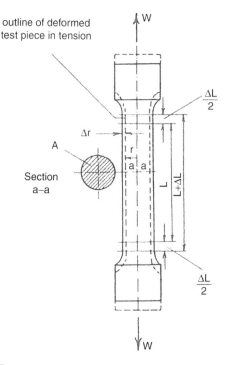

outline of deformed
test piece in tension

A

Section
a–a

FIGURE 1.6 Test specimen symbols (from Ref. 27).

Since the test piece in a standard case is cylindrical, r is also the original radius before the tensile loads are applied.

The relationship given by Eq. (1.11) is identical in form and meaning to Eq. (1.8). Both equations represent strain but in two directions, and the relevant absolute values of strain are applicable to tension and compression. The outline shown in Fig. 1.6 gives elongation as ΔL and contraction as Δr.

At this point it is quite easy to get into an argument as to what relation should exist between ε, of Eq. (1.8), and u, of Eq. (1.11), because it is difficult to measure small changes in axial and radial displacements, even with modern technology. This limitation must have been very acute at the close of the 18th century and required mathematical insight such as that of S.D. Poisson (1781–1840). Poisson determined the ratio (u/ε) analytically by employing the molecular theory of structure of the material. For elastic, isotropic materials Poisson calculated the values of this ratio, which were confirmed experimentally, and, to this day, the ratio

$$\nu = \frac{\text{unit lateral contraction}}{\text{unit axial elongation}} = \frac{u}{\varepsilon} \tag{1.12}$$

within the elastic limit is known as *Poisson's ratio*. This is constant for a given material with the theoretical limits of 0.50 and zero for ductile and brittle materials, respectively.[29]

It follows directly from Eqs. (1.9) and (1.12) that

$$u = \frac{\nu\sigma}{E} \tag{1.13}$$

Hence for a given Poisson's ratio, ν, and Young's modulus, E, one can calculate the axial and lateral strains using Eqs. (1.9) and (1.13). It should be noted that in these calculations we have ignored any sign convention, although strictly speaking one strain such as u is always of opposite sign to ε. Poisson's ratio is never shown in materials properties tables as negative. Several typical values of this ratio are given in Table 1.3.

Theoretically, Poisson's ratio applies to elastic conditions, although we normally use the same definition when the ratio increases with the increase, say, of the stress when the stress–strain curve is no longer a straight line. This characteristic has been proven experimentally,[30] justifying the use of the Poisson ratio term for both elastic and plastic strains. The effects of Poisson's ratio are especially significant in biomechanics, and the theoretical limit of 0.5, so well defended by the elasticians, is likely to be stretched to about 1.0, as shown by a fascinating discussion of biological materials by Gordon.[3] It seems that no matter what twists and turns modern scientific disciplines can take, the archaic but essential concepts of stress and strain survive.

Another view of Poisson's ratio can be acquired by calculating the change in volume due to strain for a bar of uniform circular cross-section subjected to tension,[27] which is given by

$$\frac{\Delta V}{V} = \frac{\Delta L}{L}(1 - 2\nu) \tag{1.14}$$

The practical values of ν vary within a relatively narrow range, say between 0.25 and 0.35. For a material subjected to stresses in the plastic range and with the Poisson ratio reaching the theoretical limit of 0.5, Eq. (1.14) gives $(\Delta V/V) = 0$, so that the volume of the material remains essentially unchanged. For the ideally brittle material behavior, the volume change becomes equal to the linear change in a dimension. It should be noted here that Eq. (1.14), derived for a bar of circular cross-section, also applies to bars with other uniform cross-sections.

The basic relations discussed so far can be extended to the case of volumetric strain of an elemental cube of a material subjected to hydrostatic

TABLE 1.3 Poisson's Ratio for Various Materials.

Upper theoretical limit	0.50
Calcified plaque	0.48
Lead	0.43
Gold	0.42
Platinum	0.39
Silver	0.37
Aluminum (pure)	0.36
Phosphor bronze	0.35
Tantalum	0.35
Copper	0.34
Titanium (pure)	0.34
Aluminum (wrought)	0.33
Titanium (alloy)	0.33
Brass	0.33
Molybdenum	0.32
Stainless steel	0.31
Structural steel	0.30
Fiberglass	0.30
Magnesium (alloy)	0.28
Tungsten	0.28
Granite	0.28
Sandstone	0.28
Plaque	0.27
Artery	0.27
Cast iron (gray)	0.26
Marble	0.26
Glass	0.24
Limestone	0.21
Uranium (D-38)	0.21
Plutonium (alpha phase)	0.18
Concrete (average water content)	0.12
Beryllium (vacuum-pressed powder)	0.027
Lower theoretical limit	0.000

pressure.[31] This gives

$$\frac{\Delta V}{V} = \frac{3\sigma}{E}(1 - 2v) \qquad (1.15)$$

In Eq. (1.15) σ represents hydrostatic pressure acting on all sides of the cube and $(\Delta V/V)$ is the change in volume, which for a fully plastic condition of the material is a negligible quantity. The ratio of the hydrostatic pressure (σ) to the volumetric strain $(\Delta V/V)$ from Eq. (1.15) is called the bulk modulus of the material, denoted here by E_b. This yields

$$E_b = \frac{E}{3(1 - 2\nu)} \tag{1.16}$$

Another useful formula that employs Young's modulus and Poisson's ratio concepts, discussed in this chapter, defines the modulus of rigidity or the shearing modulus of elasticity denoted by E_S

$$E_S = \frac{E}{2(1 + \nu)} \tag{1.17}$$

Note that the bulk and shear moduli in engineering literature are usually denoted by K and G, respectively. These symbols, however, are not used in this book as it may become a little confusing later when dealing with stress intensity factors and energy release rates in fracture analysis.

THEORY OF ELASTICITY

This section introduces the concept of stress and strain in three dimensions and briefly discusses governing equations in the theory of elasticity from which linear elastic fracture mechanics is derived. In concept, stresses at a point are defined with respect to a plane or area passing through that point and can be obtained by shrinking that area to an infinitesimally small size. Conventionally, the stresses at a point are defined in terms of normal stresses (σ), which act perpendicular to the plane passing through the point and in terms of shear stresses (τ), which act along that plane. In a homogeneous body, the stresses at a point will depend on the orientation of the plane passing through the point and will vary from one point to another.

Stresses in a three-dimensional system can be defined by constructing a Cartesian coordinate system at a point and considering the average forces acting on the faces of an infinitesimal cube surrounding that point. The stresses on each face of an infinitesimal cube around a point in a Cartesian coordinate system are shown in Fig. 1.7. As a general convention, the tensile stresses are considered positive normal stresses and they are presented as arrows pointing outwards and along the surface normal for that face and are denoted with the corresponding coordinate as subscript, that is, σ_x, σ_y, σ_z. Shear forces require two subscripts. The first subscript denotes the face on which the shear forces act and the second subscript indicates the direction in which the resultant shear is resolved. All the stresses shown in Fig. 1.7 are positive stresses.

Since the cube is infinitesimally small and the stresses are slowly varying across the cube, the moment equilibrium about the centroid of the cube gives

$$\tau_{xy} = \tau_{yx} \qquad \tau_{xz} = \tau_{zx} \qquad \tau_{yz} = \tau_{zy} \tag{1.18}$$

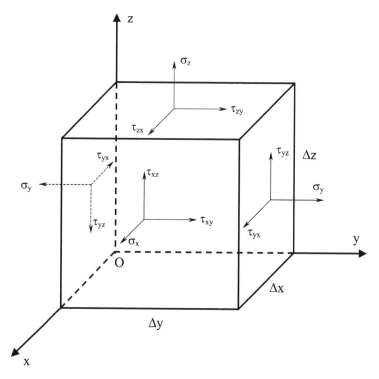

FIGURE 1.7 Stress components in a three-dimensional Cartesian coordinate system.

Equation (1.18) reduces nine stress components to six independent components. The stress components can be represented in an array form as

$$\sigma_{ij} = \begin{pmatrix} \sigma_x & \tau_{xy} & \tau_{xz} \\ \tau_{xy} & \sigma_y & \tau_{yz} \\ \tau_{xz} & \tau_{yz} & \sigma_z \end{pmatrix} \tag{1.19}$$

The above array has the transformation properties of a symmetric second-order tensor and is called the *stress tensor*.[32] If we rotate the infinitesimal element, all the shear stresses vanish at one particular orientation of that element. In essence, in a three-dimensional stress system one can find three mutually perpendicular directions in which only normal stresses σ_1, σ_2, and σ_3 are acting. Under these conditions where no shearing stresses are present, σ_1, σ_2, and σ_3 are defined as principal stresses. The corresponding principal strain in one direction can be stated as

$$\varepsilon = \frac{1}{E}(\sigma_1 - \nu\sigma_2 - \nu\sigma_3) \tag{1.20}$$

This equation indicates the extension of Hooke's law to the triaxial state of stress and it helps to define the degree of a constraint at the corner of a notch or a similar discontinuity. If σ_2 and σ_3 act in the same plane, then the direction of σ_1 must be perpendicular to the σ_2–σ_3 plane. Further ramifications of Eq. (1.20) are relegated to Chapter 2. The stress tensor when the coordinate system is oriented in principal stress directions will be

$$\sigma_{ij} = \begin{pmatrix} \sigma_1 & 0 & 0 \\ 0 & \sigma_2 & 0 \\ 0 & 0 & \sigma_3 \end{pmatrix} \tag{1.21}$$

where by convention, $\sigma_1 > \sigma_2 > \sigma_3$.

The strains are defined in terms of displacements of a point from its undistorted position and its derivatives. In the Cartesian system we can define the displacements in the x, y, and z directions as u, v, and w, respectively. For small displacements in a continuous body, the normal strains (also known as dilatational strain) in terms of displacement are given by

$$\varepsilon_x = \frac{\partial u}{\partial x} \qquad \varepsilon_y = \frac{\partial v}{\partial y} \qquad \varepsilon_z = \frac{\partial w}{\partial z} \tag{1.22}$$

The engineering shear strains (measure of angular distortion) are given by

$$\gamma_{xy} = \frac{\partial v}{\partial x} + \frac{\partial u}{\partial y} \qquad \gamma_{xz} = \frac{\partial w}{\partial x} + \frac{\partial u}{\partial z} \qquad \gamma_{yz} = \frac{\partial w}{\partial y} + \frac{\partial v}{\partial z} \tag{1.23}$$

An array that defines a complete state of strain and turns out to be symmetric is

$$\varepsilon_{ij} = \begin{pmatrix} \varepsilon_x & \dfrac{\gamma_{xy}}{2} & \dfrac{\gamma_{xz}}{2} \\ \dfrac{\gamma_{xy}}{2} & \varepsilon_y & \dfrac{\gamma_{yz}}{2} \\ \dfrac{\gamma_{xz}}{2} & \dfrac{\gamma_{yz}}{2} & \varepsilon_z \end{pmatrix} \tag{1.24}$$

When defining stresses, it is assumed that a material medium exists for the stress to act against some resistance and the medium is continuous so that the derivatives defining strain in Eqs. (1.22) and (1.23) are meaningful. The equations relating the state of stress with the state of strain are called the *constitutive equations*. While introducing the basic theory of elasticity, we will assume isotropic, homogeneous, and elastic material medium. In addition, strains are assumed to be sufficiently small and it is also assumed that the normal and shear modes of deformation are uncoupled. A set of constitutive equations that can be applied to linear, elastic, and isotropic materials are referred to as "generalized Hooke's law." In a three-dimensional Cartesian coordinate system, the strain is related to stress by the following relations:

$$
\begin{aligned}
\varepsilon_x &= \frac{1}{E}[\sigma_x - \nu(\sigma_y + \sigma_z)] \\
\varepsilon_y &= \frac{1}{E}[\sigma_y - \nu(\sigma_x + \sigma_z)] \\
\varepsilon_z &= \frac{1}{E}[\sigma_z - \nu(\sigma_x + \sigma_y)] \\
\gamma_{xy} &= \frac{1}{E_S}\tau_{xy} \\
\gamma_{xz} &= \frac{1}{E_S}\tau_{xz} \\
\gamma_{yz} &= \frac{1}{E_S}\tau_{yz}
\end{aligned}
\tag{1.25}
$$

where E is the elastic modulus, ν is Poisson's ratio, and E_S is the shear modulus.

If $\nu \neq \frac{1}{2}$ then Eqs. (1.25) can be inverted to express stress in terms of strain components:

$$
\begin{aligned}
\sigma_x &= \frac{E}{1+\nu}\varepsilon_x + \frac{\nu E}{(1+\nu)(1-2\nu)}(\varepsilon_x + \varepsilon_y + \varepsilon_z) \\
\sigma_y &= \frac{E}{1+\nu}\varepsilon_y + \frac{\nu E}{(1+\nu)(1-2\nu)}(\varepsilon_x + \varepsilon_y + \varepsilon_z) \\
\sigma_z &= \frac{E}{1+\nu}\varepsilon_z + \frac{\nu E}{(1+\nu)(1-2\nu)}(\varepsilon_x + \varepsilon_y + \varepsilon_z) \\
\tau_{xy} &= E_S\gamma_{xy} \\
\tau_{xz} &= E_S\gamma_{xz} \\
\tau_{yz} &= E_S\gamma_{yz}
\end{aligned}
\tag{1.26}
$$

The governing equations of elasticity can be simplified and formulated in two dimensions for the two special cases of plain strain and plain stress. If a body is in a state of plain strain such that all its strain components in the z-direction are zero, then

$$
\varepsilon_z = 0 \qquad \gamma_{xz} = 0 \qquad \gamma_{yz} = 0
\tag{1.27}
$$

and we obtain the plane-strain form of Hooke's law as

$$\varepsilon_x = \frac{1+v}{E}[(1-v)\sigma_x - v\sigma_y]$$

$$\varepsilon_y = \frac{1+v}{E}[(1-v)\sigma_y - v\sigma_x]$$

$$\sigma_x = \frac{vE}{(1+v)(1-2v)}[(1-v)\varepsilon_x + v\varepsilon_y] \qquad (1.28)$$

$$\sigma_y = \frac{vE}{(1+v)(1-2v)}[(1-v)\varepsilon_y - v\varepsilon_x]$$

$$\sigma_z = v(\sigma_x + \sigma_y)$$

$$\tau_{xy} = E_S\gamma_{xy}$$

If a body is in a state of plane stress such that all the stresses in the z-direction are zero, then

$$\sigma_z = 0 \qquad \tau_{xz} = 0 \qquad \tau_{yz} = 0 \qquad (1.29)$$

and

$$\varepsilon_x = \frac{1}{E}(\sigma_x - v\sigma_y)$$

$$\varepsilon_y = \frac{1}{E}(\sigma_y - v\sigma_x)$$

$$\varepsilon_z = -\frac{v}{E}(\sigma_x + \sigma_y) \qquad (1.30)$$

$$\sigma_x = \frac{E}{1-v^2}(\varepsilon_x + v\varepsilon_y)$$

$$\sigma_y = \frac{E}{1-v^2}(\varepsilon_y + v\varepsilon_x)$$

$$\tau_{xy} = E_S\gamma_{xy}$$

The above section summarizes some of the basic equations that are required in the theory of linear elastic fracture mechanics. A textbook on the theory of elasticity should be referred to for a detailed study of this section.

STRENGTH THEORIES AND DESIGN

The first half of this century exhibited considerable interest in developing practical techniques for dealing with more classical behavior of ductile and brittle materials and the justification of the various strength theories in design.[33] Ductile material, where the plastic deformation region on the stress–strain curve is well defined, was considered to have failed when the last point on the elastic portion of the curve was reached, that is, plastic deformation began. A brittle material, on the other hand, was not considered completely failed until it had broken through a tensile fracture at ultimate strength. In compression the failure of a brittle material appears to be a shear fracture. The elongation of 5% was used as the arbitrary dividing line between ductile and brittle materials. However, under special circumstances involving low temperature, high strain rate, combined loading, residual stress, stress raisers, large size, or hydrogen absorption, ductile steel may show a brittle response.

The following four theories of elastic failure received probably the widest acceptance:

Maximum stress theory: Elastic failure occurs when the maximum working stress equals the yield value σ_y.

Maximum strain theory: Elastic failure occurs when the maximum tensile strain reaches (σ_y/E).

Maximum shear theory: Elastic failure occurs when the maximum shear stress becomes equal to $(0.5\sigma_y)$.

Distortion energy theory: Elastic failure occurs when the principal stresses σ_1, σ_2, σ_3 satisfy the following relation:

$$(\sigma_1 - \sigma_2)^2 + (\sigma_2 - \sigma_3)^2 + (\sigma_3 - \sigma_1)^2 = 2\sigma_y^2 \tag{1.31}$$

For the case of a two-dimensional stress the foregoing equation simplifies to

$$\sigma_1^2 - \sigma_1\sigma_2 + \sigma_2^2 = \sigma_y^2 \tag{1.32}$$

Experiments indicate that brittle materials such as glass and Bakelite®[34] fracture in general agreement with the maximum stress and strain theories in tension. The conditions of yielding given by Eqs. (1.31) and (1.32) are usually accepted as valid for ductile materials. Similar comment can be made about the maximum shear theory, which is in good agreement with experiments on ductile materials and is rather simple to apply:

$$\sigma_y = \sigma_1 - \sigma_3 \tag{1.33}$$

This brief chapter on the issues of stress and strain shows how the basic elements of practical stress analysis have developed prior to more modern techniques of design reliability and safety. In the second half of the 19th century, British and American engineers in particular relied on calculated tensile stresses in structures, with factors of safety between 3 and 7, as certification of the tensile strength of a material. There was no real pressure to trim the weight and cost of the structures and, for all practical purposes, the discrepancies between the theoretical and the actual strengths of the materials used in construction projects were not alarming. Certainly the "factor of safety" or "ignorance" in vogue at the time was an order of magnitude greater than a natural few percent variation in strength.

The design of ships, boilers, bridges, support beams, parts of locomotives, and various structural members was based essentially on tensile stresses using relatively safe materials such as wrought iron or mild steel. However, in spite of large factors of safety, some accidents continued to occur. The demand for speed and lower weight, particularly in the shipbuilding industry, has gradually eroded the level of design confidence and opened new areas of experimental and analytical scrutiny. The problems of structural failure — whether from simple overload, insidious stress concentration, or crack propagation — were equally distressing.

SYMBOLS

A	Cross-sectional area, in^2 (mm^2)
E	Modulus of elasticity, psi (N/mm^2)
E_b	Bulk modulus, psi (N/mm^2)
E_s	Shearing modulus of elasticity, psi (N/mm^2)
L	Length, in. (mm)
MPa	Megapascal, 10^6 Pa (N/mm^2)
N	Newton
Pa	Pascal
r	Radius of solid bar, in. (mm)
u	Lateral strain, in./in. (mm/mm)
V	Volume of stressed material, $in.^3$ (mm^3)
W	External load, lb (N)
ΔL	Change in length, in. (mm)
Δr	Change in radius, in. (mm)
ΔV	Change in volume, $in.^3$ (mm^3)
ε	Engineering strain, in./in. (mm/mm)
ν	Poisson's ratio
σ	General symbol for normal stress, psi (N/mm^2)
σ_y	Yield strength, psi (N/mm^2)
$\sigma_1, \sigma_2, \sigma_3$	Principal stresses, psi (N/mm^2)
τ	General symbol for shear stress, psi (N/mm^2)

REFERENCES

1. Galileo. 1638. *Two New Sciences*; Crew, H., de Salvio, A., Eds.; Macmillan Co.: New York, 1933.

2. Hooke, R. *De Potentia restitutiva*; London. Printed by John Martyn, printer to the Royal Society, 1678.

3. Gordon, J.E. *Structures, or Why Things Don't Fall Down*; Plenum Press: New York, 1978.

4. Young, T. *A Course of Lectures on Natural Philosophy and the Mechanical Arts*; Johnson Reprint Corporation: London, 1807.

5. Inglis, C.E. Stresses in a plate due to the presence of cracks and sharp corners. Transactions of the Institute of Naval Architects (London) **1913**, *55*, 219–241.

6. Griffith, A.A. The phenomena of rupture and flow in solids. Phil. Trans. Royal Society **1920**, *221*, 163–198.

7. Irwin, G.R. Fracture dynamics. In *Fracturing of Metals*; American Society of Metals: Cleveland, 1948.

8. Orowan, E. Fracture strength of solids. In *Report on Progress in Physics*; Physical Society of London: London, 1949; Vol. 12, 185–232.

9. Irwin, G.R.; Kies, J.A. Fracturing and fracture dynamics. Welding Journal **1952**, *31* (Research Supplement), 95s–100s.

10. Irwin, G.R.; Kies, J.A. Critical energy rate analysis of fracture strength of large welded structures. Welding Journal **1954**, *33*, 193s–198s.

11. Irwin, G.R. Onset of fast crack propogation in high strength steel and aluminum alloys. Sagamore Research Conference Proceedings **1956**, *2*, 298–305.

12. Westergaard, H.M. Bearing pressures and cracks. Trans. ASME J. Appl. Mech. **1939**, *6*, 49–53.

13. Wells, A.A. The condition of fast fracture in aluminum alloys with particular reference to comet failures. British Welding Research Association Report, April 1955.

14. Williams, M.L. On the stress distribution at the base of a stationary crack. J. Appl. Mech. **1957**, *24*, 109–114.

15. Wells, A.A.; Post, D. The dynamic stress distribution surrounding a running crack — A photoelastic analysis. Proceedings of the Society of Experimental Stress Analysis **1958**, *16*, 69–92.

16. Irwin, G.R. Discussion of: The dynamic stress distribution surrounding a running crack — A photoelastic analysis. Proceedings of the Society of Experimental Stress Analysis **1958**, *16*, 93–96.

17. Paris, P.C.; Gomez, M.P.; Anderson, W.P. A rational analytic theory of fatigue. Trend Engng **1961**, *13*, 9–14.

18. Dugdale, D.S. Yielding of steel sheets containing slits. J. Mech. Phys. Solids **1960**, *8*, 100–104.

19. Barenblatt, G.I. The mathematical theory of equilibrium cracks in brittle fracture. Adv. Appl. Mech. **1962**, VII, 55–129.

20. Rice, J.R. A path independent integral and the approximate analysis of strain concentration by notches and cracks. J. Appl. Mech. **1968**, *35*, 379–386.

21. Rice, J.R. *Mathematical Aspects of Fracture*; Academic Press: New York, 1968; Vol. 2.

22. Begley, J.A.; Landes, J.D. The *J* integral as a fracture criterion. In *Stress Analysis and Growth of Cracks*, ASTM STP 514; American Society for Testing and Materials: Philadelphia, 1972; 1–20.

23. E813-81. Standard test method for J_{IC}, a measure of fracture toughness. American Society for Testing and Materials: Philadelphia, 1981.

24. Sih, G.C. *Handbook of Stress Intensity Factors*; Institute of Fracture and Solid Mechanics: Lehigh University, Bethlehem, PA, 1973.

25. Euler, L. *Histoire de l'Academie*; Berlin, 1757; 13.

26. Cauchy, A.L. French Academy of Sciences Paper, Paris, 1822.

27. Blake, A. *Practical Stress Analysis in Engineering Design*, 2nd ed.; Marcel Dekker: New York, 1990.

28. Large, G.E. *Basic Reinforced Concrete Design: Elastic and Creep*, 2nd ed.; Ronald Press: New York, 1957.

29. Timoshenko, S. *History of Strength of Materials*; McGraw-Hill: New York, 1953.

30. Stang, A.H.; Greenspan, M.; Newman, S.B. Poisson's ratio of some structural alloys for large strains. J. Res. Natl. Bur. Stand. **1946**, *37* (4).

31. Case, J.; Chilver, A.H. *Strength of Materials*; Edward Arnold: London, 1959.

32. Boresi, A.P. Chong, K.P. *Elasticity in Engineering Mechanics*; Elsevier Science Pub. Co., Inc.: New York, 1987.

33. Roark, R.J. *Formulas for Stress and Strain*, 4th ed.; McGraw-Hill: New York, 1965.

34. Weibull, W. Investigations into Strength Properties of Brittle Materials. Proc. R. Swedish Inst. Eng. Res. **1938**, 149.

2

Stress and Energy Criteria and Fracture

EFFECTS OF GEOMETRY

For many years the assumption of uniform distribution of normal stresses over a cross-section of a nonprismatic bar gave satisfactory results as long as no abrupt changes in cross-section along the bar axis were involved. The general design reliability was even better when, over many years, engineers utilized high factors of safety. Although the presence of higher factors obscured some of the more questionable design details, the notion that smooth structural surfaces and limited changes of shape provided a certain amount of reliability and safety was hard to dispute.

In most early designs geometrical features such as holes, cracks, and sharp corners had been known in advance, and some of them were utilized for a specific purpose such as, for instance, grooves in slabs of chocolate or perforations in postage stamps. Although these particular geometric discontinuities were convenient, they were not engineered properly by calculating the stress concentration effects to, at least, indicate the ratio of the elevation of the local stress to the nominal stress. This practice existed in spite of the fact that there was some understanding of the general problem of perturbation in the stress field due to the presence of a hole or groove in a continuous solid. An example of a sharp groove effect is illustrated in Fig. 2.1.

The "trajectories" (Fig. 2.1) are simply "pathways" of stress that go around a particular irregularity such as a groove because the tensile forces applied to the

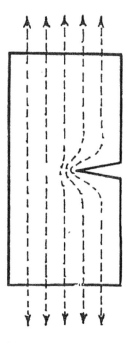

FIGURE 2.1   Typical trajectories of stress in a grooved solid.

solid, as in Fig. 2.1, must be balanced in some way.[1] The stress trajectories are crowded together near the bottom of the groove where the force per unit area is higher than at any other location within the boundaries of the solid. The degree of "crowding together" of the trajectories depends upon the shape of the discontinuity, and, indeed, around a sharp corner this crowding can be rather severe. Where there is no groove, as in the left side of the solid in Fig. 2.1, our imaginary flow lines are straight and equally spaced, indicating a uniform stress field. Another important feature of this stress concentration mechanism is that the crowding effect is very local. In a practical way this feature can be verified by pushing a solid wedge against a rubber hose. These flow lines (trajectories) show the direction of the local tensile stress. Furthermore, in the vicinity of the groove the local stress can have vertical and horizontal components that, in the language of the theoretical elasticity, constitute a biaxial stress field. Hence the groove causes stress concentration and lateral stresses.[2]

INGLIS THEORY OF STRESS

At the turn of this century the practice of using high factors of safety (up to 18 in locomotive design) was not always successful in preventing structural failures, particularly in the areas of large and complex systems such as in the shipbuilding industry. In 1901 the fastest ship in the world (*H.M.S. Cobra*) suddenly broke in two and sank with loss of life in the North Sea during ordinary weather.[1] Subsequent experiments on full-scale structures and verification of engineering calculations using a factor of safety higher than 5 did not provide sufficient explanation of the fracture mechanism responsible for the North Sea disaster.

One of the first investigations into the general area of modeling geometric irregularities and defects was conducted several years later by Professor Inglis of Cambridge University.[3] His theoretical analysis resulted in a design formula for an elliptical hole that also applied to openings such as portholes, doors, and hatchways with reasonable accuracy.

$$\sigma_{max} = \sigma[1 + 2(h/\rho)^{1/2}] \tag{2.1}$$

where σ_{max} = maximum elastic stress at the tip of hole, σ = nominal stress away from the stress concentration, h = major semiaxis of the ellipse, and ρ = root radius of an ellipse, and

$$\rho = b^2/h \tag{2.2}$$

where b is the minor semiaxis of the ellipse. The dimension ρ can also be described as the local radius of curvature at the tip of the ellipse (Krummungsradius), which can be derived from the general parametric equations of the ellipse.[4] The notation for the elliptical hole in the Inglis formula is given in Fig. 2.2.

An alternative form of the Inglis equation can be obtained by substituting the tip radius, ρ, in Eq. (2.1). This should yield

$$\sigma_{max} = \sigma\left(1 + 2\frac{h}{b}\right) \tag{2.3}$$

When $h = b$, Eq. (2.3) gives $(\sigma_{max}/\sigma) = 3$, which is the conventional stress concentration factor for a small circular hole or a semicircular notch. This is indeed a remarkable result considering that Eq. (2.1) was proposed many years ago. At about the same time, Kirsch in Germany (1898) and Kolosoff in Russia (1910) derived similar equations, and it was generally disappointing that little notice was taken of this development in shipbuilding and other industries.

When b tends to a small value in comparison with the dimension h, the stress concentration factor increases markedly, as illustrated in Fig. 2.3. This suggests that a rather narrow opening perpendicular to the direction of nominal tension can produce a very high stress concentration, which may account for unexpected fractures even under moderate applied stresses. Under these conditions, however, dominated by high (h/b) ratios, we are entering a rather

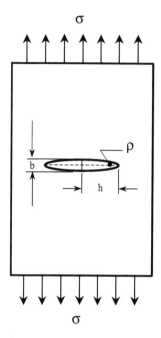

FIGURE 2.2   Notation for elliptical hole.

different approach to judging the degree of severity of a particular discontinuity. Here we no longer deal with a conventional notch geometry but with a deep and sharp crack, with the tip-of-the-crack radius having, perhaps, molecular dimensions. Hence the conventional definition of stress concentration factor cannot be applied.

One should not take the results based on the Inglis formula entirely at their face value. This would only lead to a conclusion that it is impossible to design any structure to carry tensile loads because all structures and materials are scarcely free of discontinuities and cracks. In real-life situations bridges, machinery members, ships, and airplanes may well be infested with stress concentrations caused by holes, notches, and cracks, and yet such irregularities are seldom dangerous. Certainly, since the appearance of Inglis's paper,[3] a wealth of information on classical stress concentration methodology has developed,[5-7] so that almost any geometrical transition can be handled by calculation to enhance product safety. However, we should be careful with designing around a particular geometrical weakness by adding extra material, such as gussets or webs, so that "strengthening" does not produce some other form of weakness. It may not be easy to assure a proper design balance, because only nature is really good in mitigating undue stress effects.

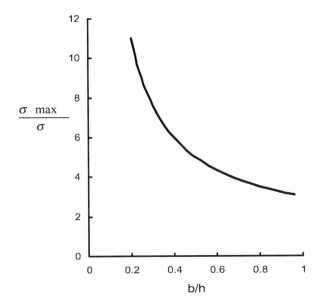

FIGURE 2.3 Variation of stress factor.

ADVENT OF THE ENERGY CONCEPT

While the Inglis formula planted certain questions in the minds of practical engineers and some startling results could be predicted, the design profession as a whole was, for a long time, eager to dismiss Inglis's implications by invoking the ductility of metals and plastic flow around the tip of a crack or a geometrical discontinuity. In effect, local plastic action was regarded as a "rounding off" mechanism for blunting the sharp tip.

In the meantime, additional structural failures continued to crop up and persisted until modern times, with some spectacular incidents involving ships, bridges, and oil rigs. It has become painfully obvious that the classical concepts of stress and strain — developed by Hooke, Young, Cauchy, and others — were not really enough, by themselves, to predict structural failures. After all, until quite recently, elasticity was taught in terms of forces and distances, and even now we seldom think of the stress–strain curve as a symbol of energy and the measure of conservation of energy. Yet the quantity of energy required to break a given material or structure defines the toughness, sometimes called fracture energy or work of fracture.

Although energy can exist in different forms — such as, for instance, electrical, chemical, heat, and potential energy — in the field of mechanical engineering and biomechanics, the concept of strain energy is more widespread because every elastic material under stress contains strain energy. In its simplest definition, the area under the stress–strain curve, shown in Fig. 2.4, represents strain energy, where the stress can be either tensile or compressive. Hence the strain energy per unit volume of the material, in line with Fig. 2.4, is

$$U = \frac{\sigma\varepsilon}{2} \tag{2.4}$$

or using Eq. (1.3), we can directly obtain

$$U = \frac{\sigma^2}{2E} \tag{2.5}$$

The basic physical rule is that in any manner of transformation of energy, we cannot get something for nothing. Also, energy can neither be created nor destroyed, which is known as the principle of conservation of energy. However, this principle was not generally accepted until quite late in the 19th century.

As far as the units of energy are concerned, there is little uniformity. In mechanical engineering the tradition is still to use foot-pounds, while the SI unit of energy is the joule. It represents the work done when 1 newton (N) acts through 1 meter (m), or in short

$$1 \text{ joule (J)} = N \times m$$

Other equivalents are

$$1 \text{ joule} = 10^7 \text{ ergs} = 0.734 \text{ ft-lb} = 0.239 \text{ calories}$$

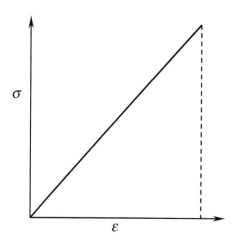

FIGURE 2.4   Strain energy diagram.

While the concept of force times distance is not too difficult to grasp, the measure of a joule is not that easy to comprehend. Gordon[1] suggests that the energy of 1 joule is roughly equal to the energy of an apple hitting the floor after falling off a regular height table.

GRIFFITH THEORY OF FRACTURE

It was shown by Eq. (2.3) and Fig. 2.3 that as the elliptical shape of the opening within a loaded body degenerates into a crack, the theoretical stresses at the end of the major axis tend to infinity. It becomes clear that while the stress concentration factor indicates the degree of the elevation of the local stress, the factor by itself is not a criterion of failure and it does not explain why a distinctly sharp crack does not produce a structural failure. Griffith[8] was first to put forth a rational theory of fracture mechanics concerned with the specific conditions under which a small, sharp crack in a stressed body becomes unstable. Since Griffith's approach to this problem was by way of energy, rather than the traditional force and stress, the entire idea was rather foreign at the time. Griffith regarded Inglis's stress concentration as a mechanism for converting strain energy into fracture energy. Such a mechanism, of course, can only work under a continuous supply of energy. If such a supply dries up, then the fracture process must stop.

Griffith assumed that incipient fracture in ideally brittle materials takes place when the elastic energy supplied at the crack tip is equal to or greater than the energy required to create new crack surfaces.[9,10] His analysis was based on a model in the form of an elliptical cutout of length $2h$, where for a very small dimension b (minor half-axis) and sharp corner radius, the cutout resembles a typical crack geometry, as shown in Fig. 2.5. At this point in our discussion the symbol h, standing for major half-axis of the ellipse, is changed to a, denoting crack length.

Additional conditions required in analyzing crack extension include:

- The stresses ahead of the crack tip must reach a critical magnitude.
- The total energy of the system must be reduced during crack extension.

As stated previously in connection with stress trajectories, the state of stress in the $y-x$ plane (Fig. 2.5) at the tip of the crack is expected to be at least biaxial. If the tip of the crack has a finite radius, it is also a free surface and the stress along the x-axis at $x = a$ must be zero, while the stress along the y-axis at the same point attains a maximum value. The free faces of the plate carry no stresses. The term *plane stress*, as used in the science of fracture mechanics, defines a state of stress in which one of the principal stresses is zero. This condition may be found in those applications where the thickness of a machine member or a structural element is small compared to other dimensions.

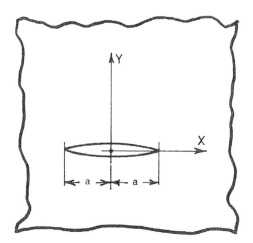

FIGURE 2.5   Griffith crack form.

The term *plane strain*, used in conjunction with the various definitions and criteria of fracture mechanics, refers to the state of a constraint in the vicinity of the crack tip. This situation develops when the surrounding material prevents, say, contraction so that a high tension develops in the thickness direction. In essence we have a triaxial state of stress, and for a complete constraint, the strain in the z-direction (normal to the $x-y$ plane in Fig. 2.5) is zero.

The foregoing description of stress and strain conditions may be summarized in simple mathematical terms.

Plane stress:

$$\varepsilon_z = -\nu \frac{\sigma_x}{E} - \nu \frac{\sigma_y}{E} \tag{2.6}$$

Plane strain:

$$\varepsilon_z = \frac{\sigma_z}{E} - \nu \frac{\sigma_x}{E} - \nu \frac{\sigma_y}{E} = 0 \tag{2.7}$$

from which

$$\sigma_z = \nu(\sigma_x + \sigma_y) \tag{2.8}$$

The stress given by Eq. (2.8) cannot exist at the free surface although it can build up rather quickly going inward through the thickness of the material. Broek notes[2] that in the absence of σ_z and presence of ε_z at the surface, a small dimple can develop. Since in practice a complete constraint is unlikely, a triaxial state of stress rather than a plane-strain condition should exist. However, the state of stress may not always be totally dictated by thickness.

It was recently shown in a couple of examples[1,9] that the mathematical process by which Griffith obtained his solution could be simplified. The idea here is to calculate the energy stored in a remotely clamped and uniformly stressed plate in the absence of a crack and then to approximate the strain energy released by a Griffith-type crack (Fig. 2.5) of length $2a$ introduced into the plate. The variation of the two energies as a function of crack length is sketched roughly in Fig. 2.6. Curve (A) represents positive energy of input while curve (B) is the negative quantity of release energy.

The positive input energy, which changes linearly with the crack length, is required to break atomic bonds ahead of the crack and in this manner to form new crack surfaces. The strain energy release as a negative quantity is assumed to vary as the square of crack length. By denoting G and σ as the elastic energy release rate and the uniform stress field, respectively, Parker[9] provides the basic equations for the elastic stress fields, in good agreement with Griffith's solution. The left-hand side of the equations contains material properties while the right-hand side includes, essentially, geometrical and loading parameters. Here we have a splendid example of a practical and relatively simple approach to the somewhat alarming complexity of a theoretical problem. The energy release rate G is also called "crack extension force" or "crack-driving force." The Griffith equations for a wide plate are given here for plane stress and plane strain conditions.

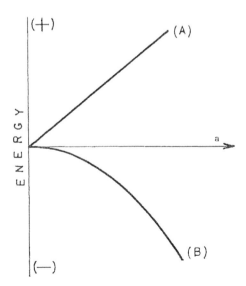

FIGURE 2.6 Energy as a function of crack length, a [(A) input energy; (B) release energy].

Plane stress

$$(GE)^{1/2} = \sigma(\pi a)^{1/2} \tag{2.9}$$

Plane strain

$$(GE)^{1/2} = \sigma(\pi a)^{1/2}(1 - \nu^2)^{1/2} \tag{2.10}$$

It follows therefore that there is a way of estimating the length of a critical crack, and that a crack shorter than this should be stable under normal conditions. Also, a crack longer than this is likely to be self-propagating. Although Griffith's theory may not be the answer to all design problems it has served us well in clarifying the various structural situations beyond the conventional methodology of stress concentration.

CONCEPT OF STRESS INTENSITY

Many years after the Griffith fracture criterion for ideally brittle materials was established, Irwin[10] and Orowan[11] suggested a modification that would extend the Griffith theory to metals exhibiting plastic deformation. This modification was based on the idea that the resistance to crack extension was due to the combined effect of the elastic surface energy and the plastic-strain work.[12] Since the term $\sigma(\pi a)^{1/2}$ entering Eqs. (2.9) and (2.10) also represents the intensity of the stress field at the tip of a through-thickness Griffith crack of length $2a$, there must be a direct relationship between the stress intensity parameter K and the relevant material properties. The symbol K (with the appropriate subscripts) is widely used in the literature dealing with a multitude of theoretical and experimental studies of fracture phenomena and materials science in general. It does refer to a specific zone near the crack tip, as shown in Fig. 2.7. In this zone, the stress field is completely described by the stress intensity factor, K, and the stresses are given by the following equations:

$$\sigma_x = \frac{K}{\sqrt{2\pi r}} \cos\frac{\theta}{2}\left[1 - \sin\frac{\theta}{2}\sin\frac{3\theta}{2}\right] \tag{2.11}$$

$$\sigma_y = \frac{K}{\sqrt{2\pi r}} \cos\frac{\theta}{2}\left[1 + \sin\frac{\theta}{2}\sin\frac{3\theta}{2}\right] \tag{2.12}$$

$$\tau_{xy} = \frac{K}{\sqrt{2\pi r}} \cos\frac{\theta}{2}\sin\frac{\theta}{2}\cos\frac{3\theta}{2} \tag{2.13}$$

where, r and θ are polar coordinates shown in Fig. 2.7.

The subscripts of K are usually given in roman numerals, I, II, and III, which refer to the modes of loading.[2] Hence K_I describes the opening (tensile) mode where the displacement of the crack surface is perpendicular to the crack plane. The K_{II} parameter is applicable to the sliding, or in-plane, shearing

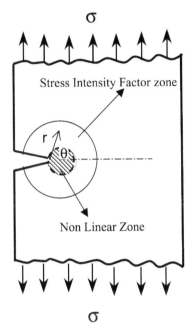

Stress Intensity Factor zone

Non Linear Zone

FIGURE 2.7 Stress intensity zone (SI).

mode, where the crack surface moves in the plane of the crack and, at the same time is normal to the leading edge of the crack. Finally, K_{III} refers to the tearing mode of external loading caused by out-of-plane shear. For practical reasons, Mode I is the most important, and therefore only the K_I parameter is considered in this book. The major (tensile) mode of loading is illustrated in Fig. 2.8, in the vertical plane. Other modes are shown for comparison.

Experience shows that the great majority of cracks result from the opening (tensile) mode while the other two modes (II and III) are rare and occur in a combined fashion only. It appears that the majority of such combinations are converted to Mode I by nature itself, unless there is a preferred direction of crack growth in a particular material. It should also be added that analytically the combined modes are more difficult to handle, and such problems become largely academic. Broek[2] quotes interesting statistics, according to which 90% of the engineering problems involving fracture mechanics are of the Mode I type, another 8% of the combined-mode type, which, immediately upon initiation of loading, transform into Mode I crack behavior. Armed with the preferred mode of loading for eminently practical reasons, the magnitudes of the crack tip stress intensity factor are given as follows.

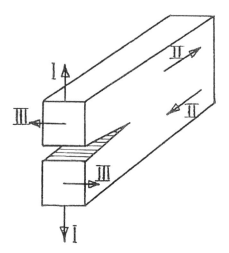

FIGURE 2.8 Primary opening mode (I) and other effects (II) and (III).

Plane stress

$$K_I = (GE)^{1/2} \tag{2.14}$$

Plane strain

$$K_I = \left(\frac{GE}{1 - v^2}\right)^{1/2} \tag{2.15}$$

Also, the stress intensity factor K_I can be related to the applied stress field in a more general way as

$$K_I = \sigma(a)^{1/2} f(g) \tag{2.16}$$

where f(g) is a geometric parameter for various crack shapes and the manner in which the external loads are applied. Parker[9] points out that it would be incorrect to assume that f(g) is always a function of geometry alone.

Some of the formulas such as Eqs. (2.10) and (2.15) involve the Poisson ratio effect on the plane-strain state of stress. The magnitude of this correction is relatively small, as shown in Fig. 2.9, even for the extreme theoretical values of the Poisson ratio. Other effects on the differences between plane-stress and plane-strain conditions should prove to be much more significant. It is quite reasonable, for instance, to anticipate that the toughness should be higher under plane stress rather than plane strain.

It is well to emphasize at this stage of looking at the fundamentals of fracture mechanics that the K parameter, representing crack driving force, can be

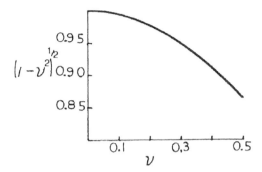

FIGURE 2.9   Effect of Poisson's ratio.

obtained by analytical methods[9] and by special experimental techniques.[13] This puts a particularly heavy burden on a designer seeking a balance between the effects of time, cost, and accuracy in selecting the proper stress intensity level for a case at hand. It is fortunate, however, that many formulas for stress intensity have already been determined[13–16] to help with the design process. Analytical solutions satisfy boundary conditions but still apply only to simple geometries, as is the case with many other areas of solid mechanics. Hence it is of utmost importance to utilize experimental knowledge, a conservative philosophy of estimating, and practical methodology in coping with the design problems, and certification of new products.

PLANE-STRAIN FRACTURE TOUGHNESS

One of the most important parameters in fracture mechanics is the plane-strain fracture toughness, denoted by K_{Ic} This is the critical value of stress intensity K_I at which fracture takes place. This situation can be compared with the case of conventional stress analysis where the working stress σ reaches the yield point of the material S_y. In a similar manner, one can think of the state of plane-stress and the transitional conditions where toughness has a symbol of K_c. There is, of course, some room for confusion with the symbols because plane-stress fracture toughness exists also in the Mode I stress intensity. The variation of fracture toughness with thickness shown schematically in Fig. 2.10 suggests that there should be only one symbol to denote fracture toughness since there is only one curve covering, in a continuous manner, the three regions of constraint ahead of a sharp crack. Broek[2] provides a strong argument for using the K_{Ic} symbol, regardless of the state of stress.

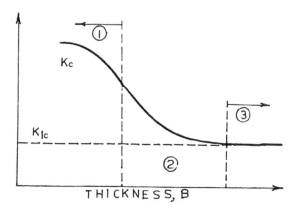

Figure 2.10 Toughness as a function of thickness, B (Arrow (1), direction of plane stress; region (2), denotes transition; arrow (3), direction of plane strain).

With the correct fracture toughness symbol in place, one of the simplest formulas of fracture mechanics can be written as

$$K_{Ic} = \sigma(\pi a)^{1/2} \tag{2.17}$$

This equation is intended for an infinite plate under uniform tensile stress where the length of a through-thickness crack is $2a$, as indicated in Fig. 2.11. This is essentially the Griffith crack.

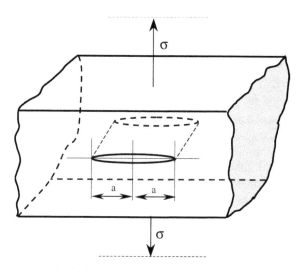

Figure 2.11 Infinitely wide plate with a typical through-thickness crack.

Although Eq. (2.14) is rather elementary and assumes no geometrical correction, it contains three important parameters reflecting the fundamental principles of a quantitative evaluation of structural integrity of mechanical and structural components in the face of a potential failure due to cracks. Here the nominal stress applied to the structural member is denoted by σ. The design parameter a is the half-length of a through-thickness crack (or a similar flaw) in a wide plate. Finally K_{Ic} represents the fracture toughness of the material for static-loading and plane-strain conditions of the maximum constraint. This is a material property that depends on ductile or brittle behavior as the case may be.

It is necessary to emphasize that the K_{Ic} parameter can only be determined from tests. ASTM has standardized the testing procedures and specimen geometries for measuring the plane-strain fracture toughness of metallic materials (ASTM standard E 399). The procedure is that a crack-notched specimen of suitable dimensions is progressively loaded until the crack becomes unstable, causing abrupt extension. Hence the K_{Ic} value consistent with such abrupt extension becomes the K_{Ic} value known as the plane-strain fracture toughness, and this material's property is only a function of strain rate and temperature. Therefore, knowing the appropriate K_{Ic} level of a material containing a crack a, it is possible to estimate the corresponding applied nominal stress from a formula such as that given by Eq. (2.14), intended for a wide plate in tension such as illustrated in Fig. 2.11. If the crack size a in Eq. (2.14) is changed, the applied stress σ must also change to be consistent with the K_{Ic} value for a given material. Also, if the load on a structural member and the existing crack size are known, the required toughness K_{Ic} can be calculated from Eq. (2.14) or another appropriate formula consistent with loading and plate conditions. However, the required toughness level K_{Ic} obtained from such calculation may or may not be equal to the valid K_{Ic} parameter. In other words, Eq. (2.14), or a similar formula, can be used to predict the size of the crack the structural member can tolerate if the valid K_{Ic} and the nominal, applied stress σ are known. Hence the basic formulas can be used with K_I, K_{Ic}, and other terms depending on the nature of the design problem at hand.

The K_{Ic} and K_I parameters have distinct differences, similar to those of strength and stress in a conventional stress analysis. Material toughness in general can be described as the ability to carry the loads or to deform plastically in the presence of a structural discontinuity such as a sharp crack or a notch. The K parameters (K_I, K_{Ic}, K_c, and others), depicting stress intensity under various conditions, have unique dimensions, which can be expressed in units of

$$(stress) \times (length)^{1/2}$$

The more frequently used units are

(English)   ksi (in.)$^{1/2}$
(SI)   MPa·m$^{1/2}$

The conversion from English to SI units may be given as

$$1 \,\text{ksi (in.)}^{1/2} = 6.895 \,\text{MPa} \,(0.02540 \,\text{m})^{1/2} = 1.099 \,\text{MPa} \,(\text{m})^{1/2}$$

or

$$1 \,\text{MPa} \,(\text{m})^{1/2} = 0.91 \,\text{ksi (in.)}^{1/2}$$

VARIATION OF BASIC PARAMETERS

The core of the entire contribution of Griffith and Irwin is that the length and shape of the existing crack, and a similar flaw, could be a design parameter as long as the nominal applied stress can be calculated using the available stress analysis techniques, and provided material toughness can be determined. At last it is possible to have some idea of the "critical" crack length with a reasonable probability that cracks shorter than "critical" should be stable and should not grow catastrophically under normal operating conditions. This should be true even if the theoretical, local stress, estimated by conventional means, is high or even higher than the tensile strength of the material. This may well be the reason why we can live with the geometrical discontinuities and flaws unless the material is unduly sensitive to fracture. It is at this point of a design that error can be introduced, particularly with larger structures, where a material of higher tensile strength is specified as a way of enhancing the design safety. Although it is difficult to obtain consistent data on the relationship between the work of fracture (or toughness) and the tensile strength of the material, it is generally accepted that the toughness of most metals decreases as tensile strength increases. This reduction in toughness can be very drastic. Gordon[1] suggests, for example, that if we double the strength of mild steel (say, by upping the carbon content), the work of fracture may be reduced by at least an order of magnitude. The real message here is that a significant reduction in the "work of fracture" will seriously change the critical crack length. According to Gordon, the critical crack length can be described as

$$a_{CR} = \frac{1}{\pi} \times \frac{\text{work of fracture per unit area of crack surface}}{\text{strain energy stored per unit volume of material}}$$

which in mathematical terms is

$$a_{CR} = \frac{2GE}{\pi\sigma^2} \tag{2.18}$$

The strain energy component is given by Eq. (2.5). The stress symbol σ defines the average tensile stress near the crack with no account of stress concentration. The work of fracture (also known as "crack extension force," "crack driving force," or "elastic energy release rate," to mention a few) can be expressed in terms of fracture toughness as

$$G = K_{\text{Ic}}^2/E \tag{2.19}$$

Hence, combining Eqs. (2.18) and (2.19) gives

$$a_{\text{CR}} = \frac{2}{\pi}\left(\frac{K_{\text{Ic}}}{\sigma}\right)^2 \tag{2.20}$$

In the definition of a Griffith crack, $a_{\text{CR}} = 2a$, which on substitution in Eq. (2.20) reduces to the classical formula given by Eq. (2.17). The foregoing review points to the need for special care in extracting design information from the published sources, with special regard to definitions and units.

For the basic case illustrated in Fig. 2.11 and the nominal design stress of 40 ksi (for instance, alloy steels), Fig. 2.12 shows the variation of the half-length of the critical crack with fracture toughness. The K_{Ic} limits selected for this illustration have been obtained from a survey of room temperature data for metals.[17] These are the typical minimum values. It is not implied, however, that the alloy steels could not be better or worse under extreme processing, environmental, or loading conditions. The nominal design stress of 40 ksi is, of course, low and implies a rather high factor of safety in dealing with the alloy steel of the type

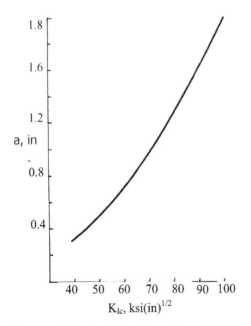

FIGURE 2.12 Variation of crack length with toughness.

used in the survey, which included maraging class of the materials. This brings us also to the question as to what happens to the size of a critical crack when, for a given type of the material and a constant value of fracture toughness, the nominal tensile stress is poorly defined. Although modern fracture mechanics is less concerned with stresses than with how the strain energy is turned into fracture energy, the practical question of the nominal stress does not go away. The classical definition says that nominal stress should be calculated on the net cross-section using the elastic theory without taking into account the effect of dis-continuities, such as holes, fillets, grooves, and cracks. Unfortunately, in general practice the nominal stress is not always defined in the same fashion because there is a choice between the net section (through the notch) and the full section (away from the notch). Furthermore, there may be a nominal stress with or without the effect of any residual stress patterns, which may, or may not, be possible to determine by conventional stress analysis techniques.

To throw some light on the relation between the crack size and the nominal stress, we can take the average value of K_{Ic} to be 70 ksi (in.)$^{1/2}$. The relevant curve is given in Fig. 2.13. This illustration applies to alloy steels having very high yield strength where, with a conventional factor of safety of 2, the design stress level would probably be on the order of 100 ksi. This would correspond

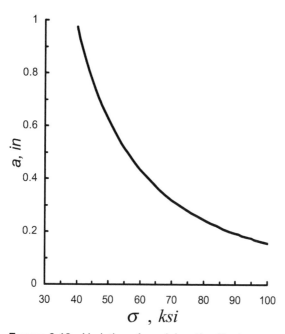

FIGURE 2.13   Variation of crack length with stress.

to a crack length of 0.16 in. Small imperfections and cracks of this size fall in the area of satisfactory visual detection but with perhaps a lower degree of reliability in some NDT (nondestructive testing) techniques.

The important parameter K_{Ic} represents a conservative criterion that gives some idea of the applied stress where crack growth begins. In those areas of industry where fracture toughness is accepted as an evaluation parameter, efforts have been made to assemble information on the dependence of fracture toughness on the yield strength of high-strength alloy steels. One of the original collections of such data is indicated schematically in Fig. 2.14.[18] The existence of scatter bands in the K_{Ic} data should be considered as unavoidable for all practical purposes. This is therefore a normal way of life when we realize how complex is the mechanism of brittle fracture. Another example of correlating K_{Ic} with the tensile strength[18] is illustrated by the design curve in Fig. 2.15. In this case we have typical fracture toughness data for sheet and plate samples fabricated according to the AISI 4340 material specification existing at the time of experimentation. It should be added that the curve shown in Fig. 2.15 was fitted into a scatter band of test points from several heats of the material. The reader is advised to use these results with some caution because of a rapid evolution of fracture mechanics methodology. However, the general trend of decreasing K_{Ic} values with the increase of tensile strength of steel should be valid.

It is normally agreed in the area of practical stress analysis that designs based on a linear theory of elasticity err generally on the conservative side.

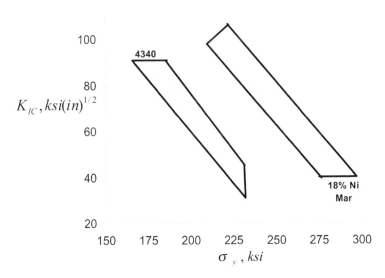

FIGURE 2.14 Approximate K_{Ic} scatter bands for high-strength steels.

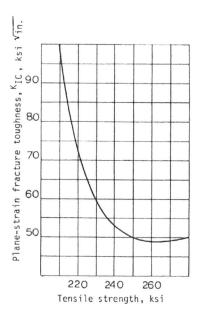

FIGURE 2.15 Variation of K_{Ic} with tensile strength of 4340 steel (from Ref. 19).

Similarly in the closely related discipline of linear elastic fracture mechanics (known as LEFM in the technical literature), the concept of plane-strain fracture toughness denoted by K_{Ic} is a conservative criterion even under conditions of combined plane strain and plane stress, where a slower crack growth can be expected. It is also remarkable how useful is the most elementary formula of fracture mechanics given by Eq. (2.14). It has also been said in the circles of design practitioners how much can be accomplished with a handful of simple formulas of stress and strain in sizing the various mechanical and structural components, and how often the proverbial P/A (force per area) and (M/Z) (moment in bending divided by section modulus) approach has helped to provide a check on a more complex and time-consuming solution. There is absolutely no reason to assume that similar elementary formulas of practical fracture mechanics cannot be used to monitor a design or help to certify a new technological structure containing openings, cracks, or other defects. With this proviso the major portions of the next chapter are devoted to the formulas and cases of immediate practical concern.

SYMBOLS

A	Area of cross-section, in.^{2} (mm^2)
a	Length of crack (flaw), in. (mm)

a_{CR}	Critical flaw size, in. (mm)
b	Minor semiaxis of ellipse, in. (mm)
E	Young's modulus, psi (N/mm^2)
$f(g)$	Geometrical parameter
G	Elastic energy release rate, lb/in. (N/mm)
h	Major semiaxis of ellipse, in. (mm)
J	Joule (unit of energy), lb-in. (N-mm)
K	Stress intensity factor, ksi $(in.)^{1/2}$ $[MPa (m)^{1/2}]$
K_I	Tensile mode stress intensity, ksi $(in.)^{1/2}$ $[MPa (m)^{1/2}]$
K_{II}, K_{III}	Shearing or tearing modes of stress intensity, ksi $(in.)^{1/2}$ $[MPa (m)^{1/2}]$
K_{Ic}	Plane-strain fracture toughness, ksi $(in.)^{1/2}$ $[MPa (m)^{1/2}]$
K_c	Plane-stress fracture toughness, ksi $(in.)^{1/2}$ $[MPa (m)^{1/2}]$
M	Bending moments, lb-in. (N-mm)
P	Concentrated load, lb (N)
S_y	Yield strength, ksi (MPa)
U	Elastic strain energy, lb-in. (N-mm)
Z	Section modulus, $in.^3$ (mm^3)
ε	Strain, in./in. (mm/mm)
ε_z	Strain in z-direction, in./in. (mm/mm)
ν	Poisson's ratio
ρ	Root radius, in. (mm)
σ	Applied (reference) stress, ksi (MPa)
σ_{max}	Maximum elastic stress, ksi (MPa)
$\sigma_x, \sigma_y, \sigma_z$	Three-dimensional stresses, ksi (MPa)

REFERENCES

1. Gordon, J.E. *Structures, or Why Things Don't Fall Down*; Plenum Press: New York, 1978.
2. Broek, D. *The Practical Use of Fracture Mechanics*; Kluwer Academic Publishers: Dordrecht, 1988.
3. Inglis, C.E. Stresses in a plate due to the presence of cracks and sharp corners. Trans. Inst. Nav. Architects (London) **1913**, 60.
4. Sass, F.; Bouche, Ch.; Leitner, A. *Dubbels Taschenbuch für den Maschinenbau*, 12th ed.; Springer-Verlag: Berlin, 1966.
5. Roark, R.J. *Formulas for Stress and Strain*, 4th ed.; McGraw-Hill: New York, 1965.
6. Timoshenko, S. *Strength of Materials*; Van Nostrand: New York, 1956.
7. Peterson, R.E. *Stress Concentration Design Factors*; John Wiley: New York, 1953.
8. Griffith, A.A. The phenomena of rupture and flow in solids. Trans. R. Soc. Lond. **1920**, 221.
9. Parker, A.P. *The Mechanics of Fracture and Fatigue*; E. & F.N. Spon: London, 1981.

10. Irwin, G.R. Fracture Dynamics. In *Fracturing of Metals*; American Society of Metals: Cleveland, 1948.

11. Orowan, E. Fracture strength of solids. In *Report on Progress in Physics*; Physical Society of London: London, 1949; Vol. 12.

12. Barsom, J.M.; Rolfe, S.T. *Fracture and Fatigue Control in Structures*, 2nd ed.; Prentice-Hall: Englewood Cliffs, NJ, 1987.

13. Rooke, D.P.; Cartwright, D.J. *Compendium of Stress Intensity Factors*; Her Majesty's Stationery Office: London, 1976.

14. Tada, H., Paris, P.C.; Irwin, G.R. *The Stress Analysis of Cracks Handbook*; University of St. Louis: St. Louis, 1973.

15. Sih, G.C. *Handbook of Stress Intensity Factors*; Institute of Fracture and Solid Mechanics: Lehigh University, Bethlehem, PA, 1973.

16. Zahoor, A. *Circumferential Through Wall Cracks*, EPRI NP-6301-D, June 1989; Vol. 1. Ductile Fracture Handbook: Palo Alto, CA.

17. Blake, A., Ed. *Handbook of Mechanics, Materials and Structures*; John Wiley: New York, 1985.

18. Steigerwald, E. What you should know about fracture mechanics. Met. Progr. **1967**, November.

19. Blake, A. *Practical Stress Analysis in Engineering Design*, 2nd ed.; Marcel Dekker: New York, 1990.

20. Streit, R.D. *Design Guidance for Fracture-Critical Components at Lawrence Livermore National Laboratory*; Lawrence Livermore Laboratory, UCRL-53254, March 3, 1982.

3

Calculation of Stress Intensity

The basic practical problem facing a designer is to make a decision as to the method for determining stress intensities. It is not easy to strike a balance between the accuracy of the method, time required to get a solution, and cost. Numerous equations for stress intensity factors are available in the literature.[1-4] These factors represent various geometries and loading conditions of fundamental importance in the prediction of structural failure of cracked bodies. In all there are probably more than 600 formulas for calculating K values for different crack configurations, body geometries, and loading situations. However, it appears that the bulk of fracture mechanics work to date has been limited to a single-mode loading largely because little is known about mixed-mode phenomena.[5] For this and other practical reasons, this chapter is restricted to pure Mode I (tensile) loading and plane-strain behavior.

CENTER CRACK

The first case to be considered is uniform tension applied to a panel of finite width with a through-thickness crack of length $2a$, shown in Fig. 3.1. The expression for the stress intensity factor is

$$K_I = \sigma(\pi a)^{1/2} f(b) \tag{3.1}$$

The applied stress is σ and $f(b)$ is the correction factor.

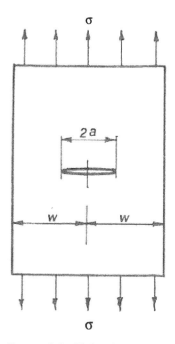

FIGURE 3.1 Finite plate with central crack.

The stress intensity correction factor $f(b)$ is a function of the a/w ratio. The analytical expressions for $f(b)$ can be of the tangent or secant type.[5,6] The theoretical limits of the applicable a/w ratio can be defined in simple terms as follows. Take the length ratio as

$$m = a/w$$

and the total length of two ligaments as

$$n = 2(w - a)$$

then

$$n = 2(w - mw)$$
$$= 2w(1 - m)$$

Hence for $m = 0$, $n = 2w$, and for $m = 1$, $n = 0$. The theoretical limits, while they exist, have no practical bearing because either the crack length or the ligament length vanishes. The size of the ligament comes into play when the concepts and conditions for "residual strength"[5] analysis with LEFM are discussed. However, the term "residual" should not be confused with the definition of "residual stresses" in stress analysis. No connection is intended between the two definitions.

It is not too obvious how to choose between the tangent and secant expressions[5,6] for estimating the value of parameter $f(b)$, although the "secant" version appears to be a bit more straightforward:

$$f(b) = \left(\sec \frac{\pi a}{2w}\right)^{1/2} \tag{3.2}$$

It is necessary to remember that the term $\pi a/2w$ is given in radians. The approximate design curve for $f(b)$ is depicted in Fig. 3.2. Although the design curve in Fig. 3.2 extends to a/w ratios as high as 0.7, in the great majority of design cases, the actual crack length may be rather small in comparison with the width of the plate. Making $K_I = K_{Ic}$ and solving Eq. (3.1) for a, gives the half-length of the crack as

$$a = \frac{1}{\pi} \left(\frac{K_{Ic}}{\sigma}\right)^2 \times \frac{1}{f(b)^2} \tag{3.3}$$

Design Problem 3.1

Determine the critical crack size for a central crack in a plate where the applied stress is 22 ksi in tension and the plane-strain fracture toughness is 55 ksi (in.)$^{1/2}$. Assume a ratio of crack length to plate width of 0.2 and check the plate width is compatible with the correction parameter $f(b)$.

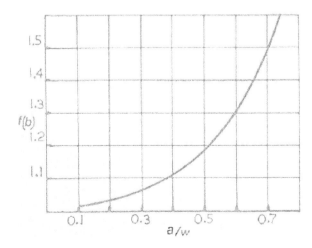

FIGURE 3.2 Secant correction factor for a central crack in a finite plate.

Solution

From Fig. 3.2 the approximate value of $f(b)$ is 1.02. Hence, using Eq. (3.3)

$$a = \frac{1}{\pi}\left(\frac{55}{22}\right)^2 \times \frac{1}{(1.02)^2}$$

$$= 1.91 \text{ in. } (48.5 \text{ mm})$$

$$\frac{a}{w} = 0.2$$

then

$$w = 5 \times 1.91$$

$$= 9.55 \text{ in.}$$

or the total plate width is

$$2w = 19.1 \text{ in. } (485 \text{ mm})$$

The basic difficulty with using Eq. (3.3) is that, in a strict sense, the exact solution can only be obtained by iteration because a is involved in the correction factor $f(b)$. In our case $f(b)$ is given in a graphical form so that there is always a small error in reading the chart (Fig. 3.2). And there is also a practical question. How exact is the theory leading to Eq. (3.3), and what accuracy can be assigned to stress calculation and the strength properties of the materials? Experience easily shows that one has to be fortunate indeed to have the accuracy as small as 10–15% on the material's properties alone. This comment applies to conventional stress analysis as well as fracture mechanics. It is well known, for instance, that a scatter in fracture toughness data of the order of 15% is not unusual.[5]

Design Problem 3.2

A 7.5 in. wide sheet has a 3 in. through thickness crack in the center. If the sheet is subjected to a tensile stress of 60 ksi, determine the level of plane-strain fracture toughness to tolerate the through-thickness crack.

Solution

The ratio $a/w = 3/7.5 = 0.4$. Using $a/w = 0.4$, the approximate reading from Fig. 3.2 gives

$$f(b) = 1.11$$

Rearranging Eq. (3.3) yields

$$K_{\text{Ic}} = \sigma f(b)(\pi a)^{1/2}$$

Hence for $a = 1.5$

$$K_{Ic} = 60 \times 1.11 \, (1.5\pi)^{1/2}$$
$$= 145 \, \text{ksi (in.)}^{1/2}$$
$$= 159 \, \text{MPa (m)}^{1/2}$$

Design Problem 3.3

A 20 in. wide and 0.050 in. thick sheet is constrained at one end and is loaded in tension with a mass of 36,000 lbs on the other end. It is made of maraging steel, which has a fracture toughness of $K_{Ic} = 50 \, \text{ksi (in.)}^{1/2}$. If the sheet contains a small slit in the center, what will be the maximum allowable slit length (critical crack size) before fracture occurs?

Solution

Tensile stress acting on the crack

$$\sigma = \frac{36,000}{20 \times 0.050} = 36 \, \text{ksi}$$

According to Eq. (3.3) and $f(b) = 1$, the crack length is

$$a = \frac{1}{\pi} \left(\frac{K_{Ic}}{\sigma} \right)^2$$

Hence

$$a = \frac{1}{\pi} \left(\frac{50}{36} \right)^2$$
$$= 0.614 \, \text{in.}$$

and

$$a/w = 0.614/10$$
$$= 0.06$$

From Fig. 3.2, $f(b) \cong 1.00$. Then

$$a = \frac{1}{\pi} \left(\frac{50}{1.01 \times 36} \right)^2$$
$$= 0.60 \, \text{in. (15.2 mm)}$$

Since $f(b)$ is very close to unity, there is no need for further iteration, and the total critical length of the crack is

$$2a = 1.20 \text{ in. } (30.4 \text{ mm})$$

The foregoing calculations and the use of Fig. 3.2 imply that in the majority of design cases involving cracked panels or plates, the cracks are relatively small, so that estimates based on the LEFM may be sufficient. Unfortunately, as the a/w ratio in Fig. 3.2 increases, say beyond 0.8, there is a marked increase in the $f(b)$ parameter, which enters Eq. (3.3) as a square. All values given in Fig. 3.2 are directly applicable as long as the reference stress is uniform. However, in the case of a nonuniform stress distribution, such as that due to a bending moment, the designer should consult special handbooks.[1-3] It may be recalled that the "reference stress" is the nominal stress away from the crack. The reader should also be cautioned that in the case of a bending moment, the reference can be based either on the maximum or on intermediate stress, as shown in Fig. 3.3. It is always a good practice to include the definition of the reference stress whenever LEFM calculations are presented.

DOUBLE-EDGE CRACK

The next logical and fundamental design case is concerned with a double-edge crack in a finite-width panel, indicated in Fig. 3.4. The basic formula for the

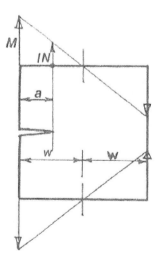

FIGURE 3.3   Bending stress reference (M, maximum; IN, intermediate).

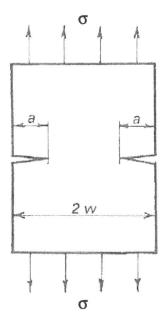

σ

a   a

2 w

σ

FIGURE 3.4   Double-edge crack in a finite-width panel.

plane-strain fracture toughness for a double-edge crack in a panel of infinite width is

$$K_{\mathrm{Ic}} = 1.12\sigma(\pi a)^{1/2} \tag{3.4}$$

When this panel is of finite width, the K_{Ic} value can be estimated by multiplying the result from Eq. (3.4) by the factor $f(b)$ from Eq. (3.2) or Fig. 3.2. The constant 1.12 is a free-surface correction factor for edge cracks or notches that are normal to the applied tensile stress. For all practical purposes the design chart given in Fig. 3.2 should provide a reasonable correction for a finite-width effect. However, for research work where a more precise solution may be required, the reader is directed to the literature,[2–4] where length-to-width ratio and the attenuation of the free-surface correction are taken into account.

It should be noted that in all basic formulas [such as Eqs. (2.13), (2.14), (2.17), (3.3), and (3.4)] where fracture is considered, the toughness can be described as K_{Ic} or K_{C}, whichever is applicable. The symbol K_{C} generally denotes plane stress or transitional conditions and it is inserted in the formula instead of K_{Ic}.

SINGLE-EDGE CRACK

The stress intensity factor for a single-edge crack or notch for the plane-strain case uses the same constant of 1.12 but requires an additional correction for the lack of symmetry. This implies a degree of bending in line with crack opening, as shown in Fig. 3.5.

Combining the free-surface correction 1.12 with the "lack of symmetry" correction factors, tabulated by Barsom and Rolfe,[6] results in a simple design chart, illustrated in Fig. 3.6.

COLLAPSE STRESS

The result of combining the two effects can be substantial when the length of a through-thickness crack (or notch depth) is greater than one-half the width of the plate. This condition represents $a/w > 1.0$ for a single-edge crack (Fig. 3.5). However, for a double-edge crack the ratio a/w approaching 1.0 is not realistic. For certain crack lengths the stress in the net ligament might be sufficiently high to cause yielding. The process in which the complete ligament is yielding is called the collapse process and the stress that causes this is called the collapse stress.

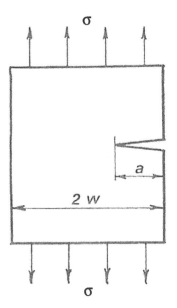

FIGURE 3.5 Single-edge crack in a finite-width panel.

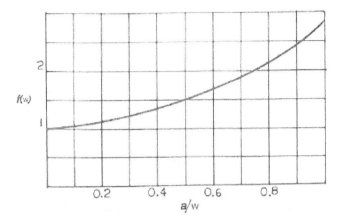

FIGURE 3.6 Correction for single-edge crack in a finite-width panel (Fig. 3.5).

Design Problem 3.4

Determine the fracture and collapse stresses in a double-edge crack for a panel (Fig. 3.4) assuming $a/w = 0.4$, panel width $2w = 8$ in., and plane-strain fracture toughness of 65 ksi $(\text{in.})^{1/2}$. The yield strength of the panel material is 36 ksi.

Solution

$a = 0.4\,w$ and $w = 4$ in., so

$$a = 0.4 \times 4$$
$$= 1.6\,\text{in.}$$

Rearrange Eq. (3.4)

$$\sigma = \frac{K_{\text{Ic}}}{1.12(\pi a)^{1/2}}$$

For $a/w = 0.4$, Fig. 3.2 gives

$$f(b) = 1.11$$

Hence the fracture stress is

$$\sigma = \frac{65}{1.12 \times 1.11(1.6\pi)^{1/2}}$$
$$= 23.3\,\text{ksi}\ (160.7\,\text{MPa})$$

Let the nominal stress at collapse be directly proportional to the yield strength of the material and the ratio of the net section to the total cross-section of the panel.

$$\sigma_{col} = \frac{(w - a)S_y}{w}$$

where σ_{col} = collapse stress (ksi) and S_y = yield strength (ksi). Hence

$$\sigma_{col} = \frac{(4 - 1.6)36}{4}$$
$$= 21.6 \, ksi \, (149 \, MPa)$$

It appears that the two stresses in this particular case are rather close. However, strictly speaking, the two failure modes under consideration cannot be directly compared. The K_{Ic} magnitude on which the fracture stress depends is a local semiplastic effect, while during the collapse process the entire cross-section (or ligament) is yielding. When the material's toughness is high, the fracture stress is also high and it is often higher than the collapse stress. If, on the other hand, the collapse prevails, other conditions must be involved. Broek[5] lists three basic characteristics augmenting the mechanism of collapse:

- very high toughness;
- very small crack;
- very limited panel width.

Design Problem 3.5

A 12 in. wide and 1.0 in. thick flat panel connects a 1860 lb carriage cart with a driving cart. The flat panel has a single-edge crack that is 2.5 in. deep. If the driving cart accelerates with an acceleration of $1g$, estimate the minimum level of plane-strain fracture toughness to assure that the existing crack will not propagate. Use a design chart for a combined correction factor in Fig. 3.6.

Solution

Uniform tensile stress due to acceleration force

$$\sigma = \frac{ma}{Area} = \frac{1860g}{12 \times 1.0} = \frac{1860 \times 386.2}{12} = 59.9 \, ksi$$
$$2w = 12$$
$$w = 6$$
$$a/w = 2.5/6$$
$$= 0.42$$

From Fig. 3.6

$$f(w) \cong 1.3$$

Therefore

$$K_{Ic} = \sigma f(w)(\pi a)^{1/2}$$
$$= 59.9 \times 1.3(2.5\pi)^{1/2}$$
$$= 218.2 \, \text{ksi (in.)}^{1/2}$$
$$= 239.8 \, \text{MPa (m)}^{1/2}$$

This very brief calculation has an important message for design. The required K_{Ic} is very high. It is applicable only to static loading and thick structural members because the plane-strain condition demands a maximum constraint. If we have to live with the level of the applied stress and the presence of the crack, the designer may be faced with the likely cost of repair and certification of the product. In the majority of practical cases of this type, this is not a mundane technical and economic decision. Other input parameters, such as the applied stress, would have to be scrutinized with respect to the factor of safety and stress analysis methodology used.

Design Problem 3.6

A platelike member to be made from a low-carbon structural steel has a width of 8 in. and is designed to carry uniform tensile stress of 32 ksi. Estimate the minimum fracture toughness to tolerate an edge-type, through-thickness crack of 2 mm.

Solution

$$2w = 8$$
$$w = 4$$
$$a/w = 0.08/4$$
$$= 0.02$$

For this small ratio, the correction is also small and of the order of $f(w) = 1.1$. (The reading error in this area is not significant because below $a/w = 0.02$, all values of $f(w)$ tend to 1.0.) And

$$K_{Ic} = 1.1 \times 32(0.08\pi)^{1/2}$$
$$= 17.6 \, \text{ksi (in.)}^{1/2}$$
$$= 19.4 \, \text{MPa (m)}^{1/2}$$

In this sample problem we have a rather low requirement of toughness that is probably unrealistic even for a poor-quality structural steel. This level of K_{Ic} would be more likely appropriate for aluminum. The toughness requirement for this problem is plotted in Fig. 3.7 as a function of crack size. The parameter a enters the calculation in two places, as the square root and as the correction factor read from Fig. 3.6. It may not be obvious how to select the level of K_{Ic} as the acceptable minimum for a common "garden variety" steel. This topic comes up again in this book when determining the K_{Ic} value as a conservative design criterion for unqualified steel.

ROUND HOLE WITH CRACKS

The presence of a circular hole in a platelike structural member is a common occurrence and the conventional stress concentration factor is well known. A hole, unlike a crack, however, does not deform and it remains as a stress raiser in tension or compression. When the hole is filled with a bolt or a pin, involved in some interference, small cracks can be expected; however, these are often below the detection limit of NDT (nondestructive test) instrumentation. If a small crack is present, its behavior is governed by the loading and the ratio between the magnitudes of ligament and hole diameter. Fortunately when the holes are filled, the small-crack problem represents limited technical relevance.

The first classical case of a hole–crack combination is the symmetric arrangement of two small cracks at a hole in a wide panel (no effect of width), as shown in Fig. 3.8. The general stress intensity solution (if neither plane strain nor plane stress is specified) can be stated as

$$K_I = f(a/r)\sigma(\pi a)^{1/2} \tag{3.5}$$

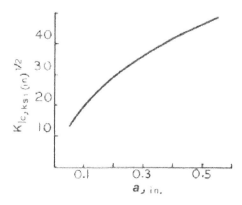

FIGURE 3.7 Toughness vs. crack size (Design Problem 3.6).

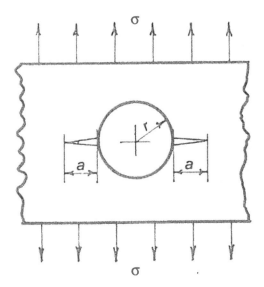

FIGURE 3.8 Circular hole with two symmetrical cracks.

The correction factor $= f(a/r)$ is given as a design curve in Fig. 3.9 based on tabulated data.[6] The maximum value of $= f(a/r)$ is obtained as

$$f(a/r) = 3 \times 1.12 = 3.36$$

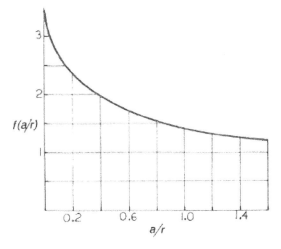

FIGURE 3.9 Correction for a double crack at a circular hole (Fig. 3.8) (from Ref. 6).

where the value 3 is the classical stress concentration factor at a round hole, and 1.12 is the free-surface correction used earlier in other formulas dealing with single- and double-edge cracks in plates. The conventional stress concentration factor is denoted throughout this text by k.

For a large ratio a/r, the correction factor in Fig. 3.9 tends to 1.0. When very short cracks emanate from round or similar openings, with the local stress governed by the conventional principles of stress gradients, the approximate stress intensity factor can be represented by

$$K_I \cong k\sigma(\pi a)^{1/2} \tag{3.6}$$

For the case of a round hole, Eq. (3.6) becomes

$$K_I \cong 3\sigma(\pi a)^{1/2} \tag{3.7}$$

Various methods have been tried at different times to obtain close solutions for the correction factors, often with great effort and cost. Several areas of fracture mechanics are prone to sophisticated approaches in spite of the fact that other parameters — such as material properties, complex geometry, or even traditional stresses — are hard to pin down accurately. Simple solutions must never be underrated, not only because they can approximate more exact methods but also because simple procedures can throw some light on complex issues for which semirigorous answers are still unavailable. Broek[5] makes a number of valuable comments in this regard in his discussion of geometry factors.

Whether we consider small or large cracks at the hole boundary, it appears that the crack behaves as if the hole were an extension of the crack. Hence the effective length of the crack can be stated as

$$a_e = 2(a + r) \tag{3.8}$$

See Fig. 3.8 for an illustration consistent with this effect.

When only one crack is growing from a round hole in a wide panel, the stress intensity can be obtained from Eq. (3.5) using a design curve similar to that shown in Fig. 3.9. The differences between the two curves for the correction factors are minimal, as illustrated in Fig. 3.10. Both curves have been plotted using the tabulated data given by Barsom and Rolfe.[6] These design curves start from the maximum value of 3.36 when a/r approaches zero. For very long cracks in relation to the hole radius, the double-crack curve tends to 1.0, while the single-crack curve has an asymptote of about 0.71.[6]

ELLIPTICAL NOTCHES WITH CRACKS

Recalling the Inglis formula, Eq. (2.1), for an elliptical hole, we know that the conventional stress concentration factor was shown to be 3 when the elliptical

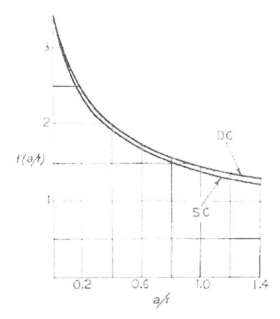

FIGURE 3.10 Comparison of correction factors for double and single cracks at round holes (DC, double crack; SC, single crack) (from Ref. 6).

shape widened and approached a round hole. The other extreme geometrical condition was found when the ratio of semiaxes, h/b, Fig. 2.2, increased to a high value, creating the long and narrow elliptical notch characteristic of the Griffith crack. For a long elliptical notch, the term $2(h/\rho)^{1/2}$ in Eq. (2.1) can be significantly higher than 1.0, so that the Inglis formula simplifies to

$$\sigma_{max} = 2\sigma(h/\rho)^{1/2} \tag{3.9}$$

Since by definition the stress concentration factor is

$$k = \sigma_{max}/\sigma \tag{3.10}$$

the theoretical stress concentration for an elliptical notch in an infinite plate can be taken (in this particular case) as

$$k = 2(h/\rho)^{1/2} \tag{3.11}$$

Now the stress intensity for a short crack emanating from the tip of an elliptical notch is

$$K_{\mathrm{I}} = k\sigma(\pi a)^{1/2} \tag{3.12}$$

Basic geometry and notation are shown in Fig. 3.11.

Although the formula given by Eq. (3.12) is an elegant and practical tool for estimating the stress intensity for the case shown in Fig. 3.11, the general and analytical problem quickly gains in complexity. Barsom and Rolfe[6] stress the importance of a natural limit for design purposes beyond which Eq. (3.12) does not apply. This limit is stated as

$$a(h\rho)^{1/2} = 0.25 \tag{3.13}$$

All symbols used in Eq. (3.13) are noted in Fig. 3.11. When this parameter is at least equal to or greater than 0.25, the stress concentration effect of an elliptical cutout can be neglected. This is consistent with the published data for the ellipse aspect ratio b/h equal to 0.5, intended for an elliptical notch in the center and for a semielliptical notch at the edge of the plate.[6] The geometrical details for the two cases are illustrated in Figs. 3.11 and 3.12.

Since the two Barsom and Rolfe[6] curves appear to have only slight differences in shape and numerical quantities, an approximate design chart

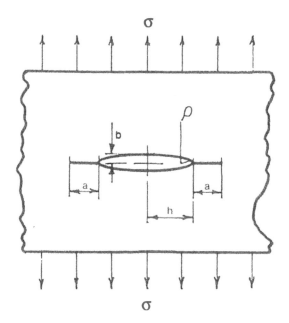

FIGURE 3.11 Geometry and notation for elliptical notch with cracks.

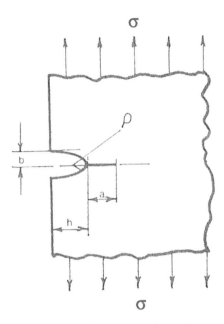

σ

σ

FIGURE 3.12   Geometry and notation for semielliptical edge notch.

can be drawn for the preliminary estimates of stress intensity factors applicable
to both types of elliptical notches. This approximation is shown in Fig. 3.13,
where the equivalent parameter $f(e)$ is plotted against a/b. The origin of the
design curve is unchanged at the coordinates of 0.2 and 0. The curve is also
assumed to pass the point at 1.0 and 0.25 coordinates, roughly halfway
between the Barsom and Rolfe curves. Hence the stress intensity factor for
the tip of a crack emanating from a center notch in an infinite panel
(Fig. 3.11) is

$$K_1 = \sigma f(e)[\pi(h + a)]^{1/2} \qquad (3.14)$$

Similarly, for the case of a crack starting from a notch at the edge of an infinite
plate (Fig. 3.12), we have

$$K_1 = \sigma f(e)f(w)[\pi(h + a)]^{1/2} \qquad (3.15)$$

When the plate width is rather large (consistent with the definition of infinite) the ratio a/w in Fig. 3.6 must be very small as the parameter $f(w)$ tends to
1.0. Under this condition Eq. (3.15) reduces to Eq. (3.14). Hence with only
small error, the stress intensity for the two cases of elliptical notches illustrated
in Figs. 3.11 and 3.12 can be calculated using Eq. (3.14) and the design chart in

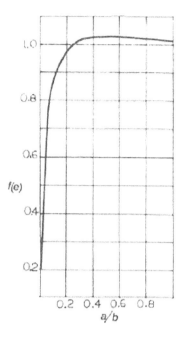

FIGURE 3.13 Approximate design curve for elliptical notches.

Fig. 3.13. Strictly speaking this interpretation is intended for the ratio of the minor to major axis of an ellipse equal to 0.5. However, Barsom and Rolfe[6] point out that the design curves for the correction factor, such as $f(e)$ (for other ratios of minor to major axes), vary only slightly.[7] This is consistent with the nature of Eq. (3.14), which represents the stress intensity at the tip of a sharp crack emanating from the surface of a notch. It is essentially different from a conventional stress–concentration effect of an elliptical opening or a notch at the edge of a plate.

Design Problem 3.7

A steel panel in a machine is subjected to 100 ksi tensile stress. The panel had a 1 in. diameter hole in the center that had a pin pressed into it. The interference of the pin resulted in two small cracks emanating from the 1 in. diameter hole, as shown in Fig. 3.8. The material is 18 Ni (250) maraging steel having plane-strain fracture toughness of 85 ksi (in.)$^{1/2}$. During service the pin was removed and the 1 in. hole was left open. Estimate the maximum crack length that can be allowed in the panel.

Solution

Rearrange Eq. (3.5)

$$a = \frac{1}{\pi} \times \left[\frac{K_{Ic}}{f(a/r)\sigma} \right]^2$$

$$= \frac{1}{\pi} \times \frac{1}{f(a/r)^2} \left(\frac{85}{100} \right)^2$$

$$= \frac{0.23}{f(a/r)^2}$$

Take

$$(a/r) = 1$$
$$f(a/r) = 1.5 \ [21]$$
$$a = 0.23/1.5^2 = 0.109$$

then

$$(a/r) = 0.109/0.5 = 0.218$$

Take

$$(a/r) = 0.2$$
$$f(a/r) = 2.4 \ [6]$$
$$a = 0.23/2.4^2 = 0.04$$

then

$$(a/r) = 0.04/0.5 = 0.08$$
$$f(a/r) = 2.8 \ [6]$$
$$a = 0.23/2.8^2 = 0.028$$

Check using Eq. (3.5)

$$K_{Ic} = 2.8 \times 100(0.028\,\pi)^{1/2}$$
$$= 83 \, \text{ksi (in.)}^{1/2}$$
$$= 91.2 \, \text{MPa (m)}^{1/2}$$

Hence

$$a = 0.028 \, \text{in.} \ (0.7 \, \text{mm})$$

Note: This result falls outside the sizable body of inspection data generated by many inspectors. In addition to the physics of crack detection, the study involved statistical methods for the development of probability-of-detection curves.[5]

Design Problem 3.8

A panel made from T-1 steel (A-517 Grade F) has been tested to have plane-strain fracture toughness of 170 ksi (in.)$^{1/2}$. The panel has a round hole of 1.2 in. diameter and two cracks 0.3 in. deep, perpendicular to the stress field, as shown in Fig. 3.8. Estimate the maximum amount of reference stress that the panel can sustain.

Solution

Rearrange Eq. (3.5)

$$\sigma = \frac{K_{Ic}}{f(a/r)(\pi a)^{1/2}}$$

$$a/r = 2 \times 0.3/1.2$$
$$= 0.5$$

From Fig. 3.9

$$f(a/r) \cong 1.8$$

Hence

$$\sigma = \frac{170}{1.8\sqrt{(0.3\pi)}}$$
$$= 97.3\,\text{ksi}$$
$$= 671\,\text{MPa}$$

The estimated stress level appears to be below the expected yield strength of the panel material although it does not allow for a customary factor of safety of, say 2 or 3, on yield.

Design Problem 3.9

Assuming the material and geometry of a panel to be the same as those given in Design Problem 3.8, estimate the minimum fracture toughness required on the premise that the reference stress has a factor of safety of 2 on the yield strength of the material, which is equal to about 110 ksi.

Solution

The reference stress is

$$\sigma = 0.5 \times 110$$
$$= 55\,\text{ksi}$$

Hence, using Eq. (3.5), we have

$$K_{Ic} = 1.8 \times 55(0.3\pi)^{1/2}$$
$$= 96\,\text{ksi (in.)}^{1/2}$$
$$= 106\,\text{MPa (m)}^{1/2}$$

This is certainly an acceptable level of fracture toughness for the steel designated as A-517 Grade F.

Design Problem 3.10

An aluminum plate which had a 0.8 in. diameter hole in the middle, failed when it was subjected to a 40 ksi tensile stress. Post mortem of the failure indicated brittle fracture due to presence of a sharp crack, which was located in the perpendicular direction to the reference stress. The plate edges were far away from the opening. Estimate the size of the crack if the plane-strain fracture toughness of aluminum was 27 ksi (in.)$^{1/2}$.

Solution

Rearranging Eq. (3.5)

$$a = \frac{1}{\pi}\left[\frac{K_{Ic}}{\sigma f(a/r)}\right]^2$$
$$= \frac{1}{\pi}\left(\frac{27}{40}\right)^2 \times \left(\frac{1}{f(a/r)}\right)^2$$
$$= \left(\frac{0.15}{f(a/r)^2}\right)$$

Take $a/r = 1$. From Fig. 3.10, the approximate function is $f(a/r) = 1.13$, and $a = 0.15/1.13^2 = 0.12$. Then

$$a/r = 0.12/0.4$$
$$= 0.3$$

From Fig. 3.10 (again taking the lower curve for a single crack), we have

$$f(a/r) = 2.1$$

and

$$a = 0.15/2.1^2$$
$$= 0.034$$

Then

$$a/r = 0.034/0.4$$
$$= 0.085$$

From Fig. 3.10

$$f(a/r) = 3.2$$
$$a = 0.15/3.2^2$$
$$= 0.015$$

Check using Eq. (3.5)

$$K_{Ic} = 3.2 \times 40(0.015\pi)^{1/2}$$
$$= 27.8 \, \text{ksi (in.)}^{1/2} = 30.5 \, \text{MPa (m)}^{1/2}$$

Hence the maximum crack size that can be tolerated is $a = 0.015$ in. (0.38 mm). This crack size is too small for reliable detection, similarly to the result noted in Design Problem 3.7.

EDGE-NOTCHED BEAMS

The general design philosophy in the case of structural beams is similar to that for panels and plates. It is only necessary to make certain that the applied stress intensity does not exceed the fracture toughness of the material. The three basic components are the reference stress, crack size, and the appropriate stress intensity factor. Hence by analogy to other expressions used in this chapter, we have

$$K_I = \sigma(\pi a)^{1/2} \cdot f(m) \tag{3.16}$$

A portion of the beam shown in Fig. 3.14 is subjected to pure bending. The classical bending stress for this case can be stated in two ways:

$$\sigma = \frac{6M}{Bw^2} \tag{3.17}$$

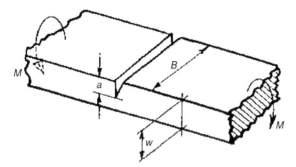

FIGURE 3.14 Bending of a beam with an edge crack.

or

$$\sigma = \frac{6M}{B(w-a)^2} \tag{3.18}$$

In the case of Eq. (3.17) we assume that the length of the crack is rather small in comparison with the depth of the beam. In practice this is probably the more likely occurrence. The use of Eq. (3.18) indicates a more conservative approach because the applied stress intensity is directly proportional to the reference stress. The degree of conservatism would be toned down a little if instead of the square one could justify, say, the 1.5 power.

For the case of a relatively small crack and unit width of a beam equal to B, the formula for the stress intensity is found by combining Eqs. (3.16) and (3.17):

$$\sigma = \frac{6Mf(m)(\pi a)^{1/2}}{Bw^2} \tag{3.19}$$

The parameter $f(m)$ is plotted in Fig. 3.15.[1,8] When the structural problem can be modeled with the aid of a three-point bending of a simply supported beam (Fig. 3.16), the bending moment term in Eq. (3.19) can be replaced by

$$M = \frac{PL}{2} \tag{3.20}$$

Hence, substituting for M in Eq. (3.19) and making $L = 2w$, we obtain

$$K_I = \frac{6P(\pi a)^{1/2} f(m)}{Bw} \tag{3.21}$$

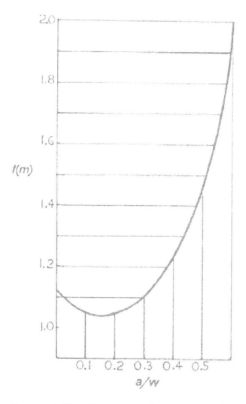

FIGURE 3.15 Stress intensity correction for moment.

When additional accuracy is required, Parker [9,10] suggests

$$K_I = \frac{3PL}{Bw^{3/2}}\left[1.93\left(\frac{a}{w}\right)^{1/2} - 3.07\left(\frac{a}{w}\right)^{3/2} + 14.53\left(\frac{a}{w}\right)^{5/2}\right.$$
$$\left. - 25.11\left(\frac{a}{w}\right)^{7/2} + 25.80\left(\frac{a}{w}\right)^{9/2}\right] \tag{3.22}$$

Here P denotes the external load and L is the effective half-length of the beam. The width of the beam is denoted by B, as before, oriented normally to the plane of the paper. The stress intensity correction factor for this case also depends on the ratio L/w.[1] To illustrate the general shape of the design curve, the $f(m)$ parameter is plotted in Fig. 3.17 for $L/w = 4$. Other cases are available.[1]

In summing up briefly the approach to estimation of the stress intensity caused by bending, it is well to note some flexibility in the choice of the reference stress and the limits of the a/w ratio. For practical reasons, however, the ratio of

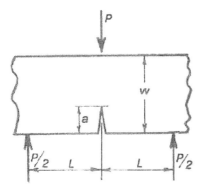

FIGURE 3.16 Three-point loading.

about 0.6 for this case may be limited by the size of the net section (ligament) as the crack length increases. The size of the ligament, as stated before, is tied to the concept of residual strength. Finally, the method of dealing with the case of pure bending (Fig. 3.14) serves as a good example of potential confusion in definition of the nominal stress because of the open choice between the net and full-section criteria.

Design Problem 3.11

Two structural beams, 1 in. thick and 3 in. long, were welded together along the 3 in. length. If crude welding leads to a single-edge crack, estimate the maximum depth of a single-edge crack that can be tolerated. The welded beams are subjected to a bending moment of 54,000 lb-in. along the weld line. The design stress

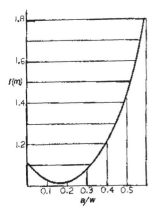

FIGURE 3.17

is 36,000 psi and the plane-strain fracture toughness of the weld section is 45 ksi (in.)$^{1/2}$. No welding stresses are present.

Solution

Substituting the foregoing numerical values in Eq. (3.19) gives

$$45 = 6 \times 54 \times 1.7725(a)^{1/2}f(m)/9$$
$$0.705 = f(m)(a)^{1/2}$$

For $a = 1$, $1/3 = 0.33$. For $a/w = 0.33$, Fig. 3.15 gives $f(m) = 1.15$. Hence

$$0.705 = 1.15(a)^{1/2}$$

or

$$a = \left(\frac{0.705}{1.15}\right)^2$$
$$= 0.376$$

For $a = 0.376$

$$(a/w) = 0.376/3$$
$$= 0.125$$

For $a/w = 0.125$, Fig. 3.15 gives $f(m) = 1.04$, and

$$a = \left(\frac{0.705}{1.042}\right)^2$$
$$= 0.46$$

The next try will show that indeed the answer is

$$a = 0.46 \, \text{in.} \, (11.7 \, \text{mm})$$

Check using Eq. (3.19)

$$K_{\text{Ic}} = K_{\text{I}} = \frac{(6 \times 541.04(0.458\pi)^{1/2}}{3 \times 3}$$
$$= 44.99 \, \text{ksi (in.)}^{1/2}$$

or

$$K_{\text{Ic}} = 45 \, \text{ksi (in.)}^{1/2}$$
$$= 49.5 \, \text{MPa (m)}^{1/2}$$

It may be noted that in our cut-and-try steps, it is also interesting to start with $a = 0$ and $f(m) = 1.125$, leading to $a = 0.393$, $a/w = 0.13$, and $f(m) \cong 1.05$ to get very close to the final result. This gets us on the flat portion of the curve between the a/w values of 0.1 and 0.2. Also, the bending moment of 54,000 lb-in. is consistent with the design stress because

$$Z = 1 \times 3^{2}/6$$
$$= 1.5 \, \text{in.}^{3}$$

and

$$M = Z\sigma$$
$$= 1.5 \times 36,000$$
$$= 54,000 \, \text{lb-in.}$$

Design Problem 3.12

A portion of a long beam is designed to a maximum stress of 32,000 psi and it is loaded in the manner shown in Fig. 3.16. The L/w ratio is 4, with B and w dimensions specified as 0.5 and 2.5 in., respectively. Estimate the maximum allowable depth of the edge crack not to exceed the stress intensity of 38.5 MPa $(m)^{1/2}$.

Solution

When the width of the beam is B and $L = 4w$, Eq. (3.21) should read

$$K_{\mathrm{I}} = \frac{12Pf(m)(\pi a)^{1/2}}{Bw}$$

Combining this equation and $\sigma = 3PL/Bw^{2}$ gives

$$K_{\mathrm{I}} = \sigma f(m)(\pi a)^{1/2}$$

Substitution of the appropriate values in English units leads to a simplified expression for further calculational steps.

$$\frac{38.5}{1.099} = 32 \times 1.772 f(m)(a)^{1/2}$$
$$0.617 = f(m)(a)^{1/2}$$

For $a = 0$, $a/w = 0$, Fig. 3.17 gives $f(m) = 1.125$, and

$$a = \left(\frac{0.617}{1.125}\right)^{2} = 0.3$$
$$0.3/2.5 = 0.12$$

and Fig. 3.17 gives $f(m) = 1.01$. Also,

$$a = \left(\frac{0.617}{1.101}\right)^2$$
$$= 0.373$$

Further iteration does not change $f(m)$ in Fig. 3.17, so that we can accept the result as

$$a = 0.37 \text{ in. } (9.4 \text{ mm})$$

and

$$1.772 \times 32 \times 1.01(0.373)^{1/2} = 35 \text{ ksi (in.)}^{1/2} = 38.5 \text{ MPa (m)}^{1/2}$$

FLAWS IN TENSION

As we are trudging through the various concepts, geometries, and applications of fracture mechanics wrapped up in modern theories and ideas, we must not lose sight of the compromises and design trade-offs in complying with fracture requirements. For instance, high-yield-strength material has relatively low fracture toughness in many instances, although one can find a material with high-strength and toughness if cost is not the ultimate criterion. The same goes for the choice of design methodology and the degree of sophistication that the responsible designer has to face. As we progress toward the more complex cases of stress intensity, more emphasis should be placed on selecting simplified solutions and approximations for practical reasons. It is much better to have a "ball park" estimate than no solution at all.

In the real world, machining marks, arc burns, and weld defects, to mention a few, can be modeled with the aid of a "thumbnail crack," shown in Fig. 3.18. The thumbnail crack is often referred to as a "surface flaw" and it is approximated by an elliptical shape with a and c being the minor and major semiaxes, respectively. The analysis of the stress intensity factors, using elliptic integrals, was first accomplished by Irwin.[11] The formula for calculating the stress intensity for a part-through thumbnail crack in a uniform tensile field can be written as

$$K_I = 2\sigma M_K \left(\frac{a}{Q}\right)^{1/2} \tag{3.23}$$

The symbol M_K denotes the front free-surface correction, which is shown in Fig. 3.19 as a function of the a/B ratio. The constant 2 is the result of the product of 1.12 (back free-surface correction) and $(\pi)^{1/2}$.

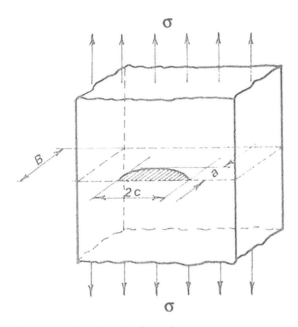

FIGURE 3.18 Thumbnail crack.

When a buried flaw of elliptical shape in an infinite body is subjected to uniform tensile stress, as shown in Fig. 3.20, the convenient formula for the stress intensity is

$$K_1 = \sigma \left(\frac{\pi a}{Q} \right)^{1/2} \tag{3.24}$$

Equations (3.23) and (3.24) are essentially identical in form and purpose. The only difference lies in the absence of surface correction factors of 1.12 and M_K in Eq. (3.24). Both formulas contain parameter Q, which has been established as a flaw shape factor as a function of the crack aspect ratio $a/2c$ and the stress ratio σ/S_y. The yield strength of the material is denoted here by S_y. Parameter Q, which applies to buried elliptical and thumbnail cracks, is given in Fig. 3.21.

The purpose of making the parameter Q a function of the stress ratio σ/S_y is to account for the effects of plastic deformation at the crack tip.[6] Although the analysis based on Fig. 3.21 is often criticized as outdated, cumbersome, and inadequate, the technique is still used in books and computer codes.

Since the elliptical shape of a flaw is the most likely model for the surface and buried cracks found in practice, Eqs. (3.23) and (3.24), the critical flaw size can be defined in rather simple terms provided $a < B/2$ and the reference

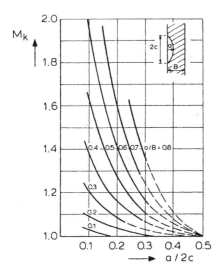

FIGURE 3.19 Front free-surface correction (from Ref. 12).

(nominal) stress is well below the yield strength of the material. Here, for the case of a surface flaw, we have

$$a_{CR} = 0.25 \left(\frac{K_{Ic}}{\sigma}\right)^2 \tag{3.25}$$

and for the buried flaw the formula is

$$a_{CR} = 0.32 \left(\frac{K_{Ic}}{\sigma}\right)^2 \tag{3.26}$$

When the reference stress σ reaches the critical value, the surface thumbnail crack will grow through the thickness of the plate to become a through-thickness flaw. Under this condition the length of the thumbnail crack is likely to reach a value of $2b$. The arrest of such a crack will depend on the crack length and the plate thickness. For thinner plates the constraint is closer to the plane-stress case, having a better chance to restrict crack growth consistent with the K_c criterion. This condition was illustrated in Fig. 2.10. However, in thicker plates we approach the condition of plane strain represented by a K_{Ic} fracture toughness that is markedly lower than K_c. In practical terms this means that once the thumbnail crack propagates through the entire thickness of the plate, there will be little chance of stopping it.

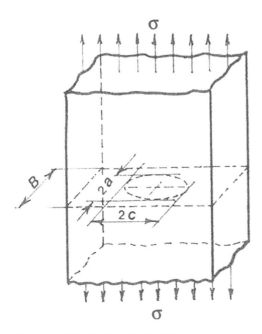

FIGURE 3.20 Buried elliptical flaw.

For a crack shape that is much closer to a circle than an ellipse, the stress intensity factor can be calculated from a very simple expression:[6]

$$K_I = 1.14\sigma(a)^{1/2} \tag{3.27}$$

This formula, for all practical purposes, can be used for surface and buried flaws alike. The error between the two applications is not expected to be more than 1%. This is certainly acceptable when the variation of material properties is of the order of 10%.

Design Problem 3.13

A rocket motor casing made of high-strength steel failed under hydrostatic test at 542 psi pressure, equal to about half of the proof value. The cause was originally expected to be an internal, longitudinal flaw 1.4 in. long and 0.1 in. deep. The casing had 260 in. diameter and 0.73 in. wall thickness. The material's yield was 240 ksi and the length-to-diameter ratio was 3.[6,12] Assuming no special influencing factors beyond the stated geometry and loading conditions, estimate plane-strain fracture toughness of the material at failure.

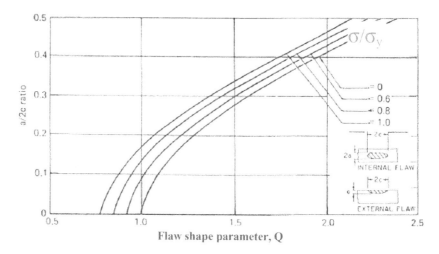

FIGURE 3.21 Flaw shape parameter.

Solution

Since in this case the conventional membrane stress theory applies, we have

$$\sigma = \frac{542 \times 260}{2 \times 0.73}$$
$$= 96.5 \, \text{ksi}$$

then

$$\frac{96.5}{240} = 0.4$$
$$= \sigma/\sigma_y$$
$$\frac{a}{2c} = \frac{0.1}{1.4}$$
$$= 0.071$$

For the above two ratios, Fig. 3.21 gives the approximate value of flaw shape factor as

$$Q \cong 1.0$$

Rearranging Eq. (3.26)

$$K_{\text{Ic}} = (\pi a_{\text{CR}})^{1/2} \sigma$$
$$= (0.05\pi)^{1/2} \times 96.5$$
$$= 38\,\text{ksi(in.)}^{1/2} = 42\,\text{MPa (m)}^{1/2}$$

This value is certainly too low for a quality, high-strength material, in spite of the expected scatter of K_{Ic} test results. However, Barsom and Rolfe[6] point out that in welding rocket casing material, K_{Ic} can be as low as 39 ksi (in.)$^{1/2}$.[14]

This experience with a large casing was highly publicized at the time and it should be kept in the records as an example of the lack of design knowledge in the process of material selection. It also points out the necessity of applying the most rudimentary rules of fracture mechanics in dealing with high-strength steels. Such rules are often given as elementary design formulas, Eqs. (3.23) through (3.27), and other examples of simplified design methodology available from the various sources in the public domain.[15–18]

Design Problem 3.14

A long pressure vessel was designed and manufactured to withstand a maximum internal pressure of 4200 psi. The vessel inner diameter is 52 in. with a wall thickness of 2.625 in. The material has a yield strength of 100 ksi. Ultrasonic inspection discovered a longitudinal surface flaw at the inner diameter. The flaw has a length of 4 in. and a depth of 0.75 in. Determine the minimum required fracture toughness so that the vessel can operate at the design internal pressure. Estimate the critical crack depth if the plane-strain fracture toughness is 90 ksi (in.)$^{1/2}$.

Solution

The membrane hoop stress is

$$\sigma = 4200 \times 26/2.625$$
$$= 41.6\,\text{ksi}$$

hence

$$\sigma/\sigma_y = 41.6/100$$
$$= 0.416$$

For the given crack dimensions

$$a/2c = 0.75/(2 \times 2)$$
$$= 0.1875$$

For the approximate ratios of 0.42 and 0.19, the design chart in Fig. 3.21 gives $Q = 1.3$. Also

$$a/B = 0.75/2.625$$
$$\cong 0.29$$

Using this ratio the surface correction M_K is found from Fig. 3.19 to be about 1.08. Therefore, applying Eq. (3.23) we obtain

$$K_I = 2 \times 41.6 \times 1.08 \times (0.75/1.3)^{1/2}$$
$$= 68.3 \, \text{ksi (in.)}^{1/2} = 75.1 \, \text{MPa (m)}^{1/2}$$

The critical crack depth follows from Eq. (3.25)

$$a_{CR} = 0.25(90/41.6)^2$$
$$= 1.17 \, \text{in. (29.7 mm)}$$

Design Problem 3.15

A large plate made of 4340 steel was designed as a tension member with an ultimate tensile strength of the material of 210 ksi. Subsequent inspection discovered a surface flaw 0.2 in. deep and 2 in. long. Estimate the tensile reference (nominal) stress at failure assuming the plate thickness to be 0.5 in. and the yield strength of the material to be equal to about 85% of the ultimate strength.

Solution

Rearranging Eq. (3.23) gives

$$\sigma = \frac{K_{Ic}(Q)^{1/2}}{2M_K(a)^{1/2}}$$

Then

$$\sigma_y = 0.85 \times 210$$
$$= 179 \, \text{ksi}$$
$$\frac{a}{2c} = \frac{0.2}{2}$$
$$= 0.1$$
$$a/B = 0.2/0.5$$
$$= 0.4$$

Hence from Fig. 2.15

$$K_{Ic} = 98 \, \text{ksi (in.)}^{1/2}$$

From Fig. 3.21, trying $\sigma/\sigma_y = 0.8$ and $a/2c = 0.1$, we have

$$Q = 0.97$$

and from Fig. 3.19 (for $a/B = 0.4$), we get $M_K \cong 1.4$, Next, substituting the foregoing data

$$\sigma = \frac{98 \times (0.97)^{1/2}}{2 \times 1.4 \times (0.2)^{1/2}}$$

$$= 77\,\text{ksi}$$

then

$$\sigma/\sigma_y = 124/179$$

$$= 0.43$$

From Fig. 3.21, $Q \cong 1.0$. Then

$$\sigma = 77 \times (0.97)\, -^{1/2} \quad \text{(next iteration not required)}$$

$$= 78.2\,\text{ksi}\ (539\,\text{MPa})$$

As long as the reference stress is lower than this value, the crack is not expected to grow, provided other variables in Eq. (3.23) remain unchanged. It may also be of interest to note that in spite of a rather simple appearance of Eq. (3.23), we deal here with eight variable quantities, some of which, of course, can be controlled by the very nature of the design problem at hand.

While the design formulas given in this subsection are intended for elliptical and circular configurations, experience shows that the majority of flaws can be modeled with the aid of elliptical geometry. When the flaw is better represented by a circle, Eq. (3.27), given for this purpose, is extremely simple in use and appearance.

SUPERPOSITION OF STRESS INTENSITY FACTORS

Since we are dealing with linear elastic problems in fracture mechanics, the individual components of stress can be added to solve a complex loading problem, which can be decomposed into several simpler loading situations. This implies that in LEFM stress intensity factors can be added as long as the mode of loading is the same. Thus,

$$K_1(F) = K_1(A) + K_1(B) + K_1(C) + \cdots$$

where the final loading F is the sum of individual loadings $A, B, C \ldots$

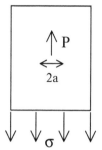

FIGURE 3.22   Crack in a riveted plate.

This principle of superposition allows us to solve complex design problems by using known simple expressions for stress intensity factors.

Design Problem 3.16

A riveted plate (12 in. wide and 1 in. thick) in a large structure has developed a crack as shown in Fig. 3.22. If the plate is fabricated from maraging steel with a plane strain fracture toughness of 100 ksi $(in.)^{1/2}$, what is the maximum stress at failure?

Solution

Apply superposition as in Fig. 3.23. Since the loadings (a) and (d) of Fig. 3.23 are identical:

$$2K_{I(a)} = K_{I(b)} + K_{I(c)}$$

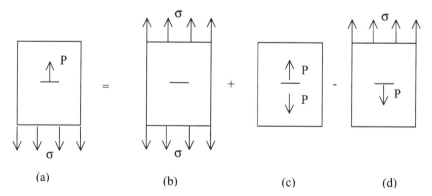

FIGURE 3.23   An example of superposition.

At failure $K_{I(a)} = 100$ ksi $\sqrt{\text{in.}}$ Also, $K_{I(b)} = \sigma\sqrt{\pi a}$ and $K_{I(c)} = P/(\sqrt{\pi a})$. From equilibrium of forces, $P = \sigma\,(12 \times 1) = 12\,\sigma$. Therefore,

$$2 \times 100 = \sigma\sqrt{0.5\pi} + \frac{12}{\sqrt{0.5\pi}}$$

$$\sigma = 18.5\,\text{ksi}\ (127\,\text{MPa})$$

Design Problem 3.17

The idea of superposition can also be successfully used in solving fracture problems that involve residual stresses in materials. The welding process in a plate in a bridge structure has developed residual stresses as shown in Fig. 3.24. Subsequent inspection detects a through crack of length 0.5 in. The plate is fabricated from tempered 4130 steel with a plane strain fracture toughness of 100 ksi$\sqrt{\text{in.}}$ and a yield strength of 160 ksi. What is the maximum applied tensile stress the plate can withstand?

Solution

The crack in Fig. 3.24 is subjected to a combination of residual tensile stress as well as remote tensile stress. Thus using the principle of superposition,

$$K_I(\text{final}) = K_I(\text{residual}) + K_I(\text{applied})$$

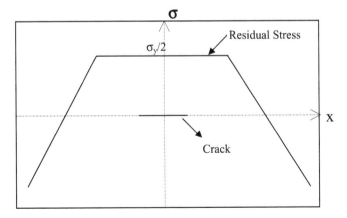

FIGURE 3.24 Description of residual stresses and crack.

At failure, $K_I(\text{final}) = K_{Ic}$. Also, $K_I(\text{applied}) = \sigma\sqrt{\pi a}$. The residual stress applies a constant crack-line pressure which produces a stress intensity factor of

$$K_I(\text{residual}) = \frac{\sigma_y}{2}\sqrt{\pi a}$$

$$K_{Ic} = \frac{\sigma_y}{2}\sqrt{\pi a} + \sigma\sqrt{\pi a}$$

$$100 = 80\sqrt{0.25\,\pi} + \sigma\sqrt{0.25\,\pi}$$

$$\sigma = 32.9\,\text{ksi} \ (227\,\text{MPa})$$

Therefore, maximum applied stress before failure is 32.9 ksi. If the plate was stress relieved before the application of remote load, the failure stress will improve to

$$K_{Ic} = \sigma\sqrt{\pi a}$$

$$100 = \sigma\sqrt{0.25\,\pi}$$

$$\sigma = 112.9\,\text{ksi} \ (778\,\text{MPa})$$

SPECIAL CASES OF STRESS INTENSITY

Before embarking on any journey through countless pages of papers and books dealing with fracture mechanics, one has to accept the state of things as they are in this field of science and technology. The task is not made easier when the majority of publications do not agree to employ the same symbols, units, notations, or even definitions to record and to convey the results of analytical and experimental studies to the practitioners whose job it is to solve "real-world" problems. Considerable progress has been made during the past three decades in the various selected areas of fracture theory and design methodology, but it is still difficult to find formulas, techniques, and case studies for the problems at hand.

It seems that beyond more elementary geometries and material characteristics, the horizon is still murky. And this is, perhaps, not very surprising because structural systems become more complex and cracks continue to occur in spite of careful design. Structures can hardly be conceived, for instance, without access or fastener holes, and there is some evidence[18] in the aircraft field that one crack out of three originates from structural holes. This is alarming news and one can only hope that rational methodology evolving from the theory of fracture mechanics can give us more design confidence.

One of the special cases of predicting the stress intensity factor involves the response of a "Griffith-type" crack (Fig. 2.5) to the line force P shown in Fig. 3.25. Barsom and Rolfe[6] quote the following expressions for the stress intensity caused by line forces applied to the crack surfaces.

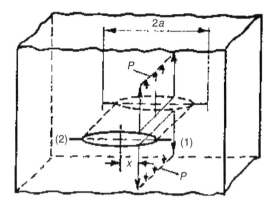

FIGURE 3.25 Crack under eccentric line loading.

Location (1) (Fig. 3.25)

$$K_1 = \frac{P}{(\pi a)^{1/2}} \left(\frac{a+x}{a-x}\right)^{1/2} \tag{3.28}$$

Location (2) (Fig. 3.25)

$$K_1 = \frac{P}{(\pi a)^{1/2}} \left(\frac{a-x}{a+x}\right)^{1/2} \tag{3.29}$$

When the line forces P are centrally placed, then $x = 0$, and the stress intensity factors at both locations become equal. This yields

$$K_1 = \frac{P}{(\pi a)^{1/2}} \tag{3.30}$$

Comparing Eq. (3.30) with Eq. (3.28) or Eq. (3.29) indicates that the stress intensity has now decreased. It appears, at the same time, that moving the wedge opening load P to the center location should increase the total length of the crack. Assuming that such a mechanism is realistic, the rate of crack propagation should decrease, creating the conditions of crack arrest. For the solution of several crack problems discussed so far, the formulas for the stress intensity factors involve the concept of reference stress, so that the term defining, say, uniform stress field σ is featured directly in the particular equation. It has been, then, always easy to see that the dimensions of the stress intensity factor or fracture toughness are, for instance, ksi (in.)$^{1/2}$ or MPa (m)$^{1/2}$, as the case may be. It has been relatively obvious which stress one should insert in the expression for K_I, K_c, K_{Ic}, or other stress intensity parameters of interest.

It was shown by Broek[5] in the case of dealing with compact specimens for measuring fracture toughness K_{Ic} that the stress intensity had to be expressed in terms of load rather than stress. Although the use of the load parameter did not prevent the scientist from obtaining the correct fracture toughness data, conventional structural analysis and computer codes tailored to crack growth analysis were based on stress.

Indeed, a first glance at Eqs. (3.28) through (3.30) gives the impression that these formulas are fundamentally different from other forms of equations describing the concept of stress intensity and fracture toughness. The problem with the compact ASTM (American Society for Testing and Materials) specimen was solved by introducing a hypothetical term defining stress. Since the foregoing equations, pertinent to Fig. 3.25, are still acceptable, there is no need for additional terms that have little physical meaning. However, a simple dimensional interpretation of Eq. (3.30) may be in order, such as

$$\frac{P}{(\pi a)^{1/2}} \times \frac{(\pi a)^{1/2}}{(\pi a)^{1/2}} = \frac{P}{a} \times \frac{(a)^{1/2}}{(\pi)^{1/2}}$$

and since the line force is measured in pounds per inch, we obtain the dimension of stress

$$\frac{\text{lb}}{\text{in.}} \times \frac{1}{\text{in.}} \times (\text{in.})^{1/2} = \text{psi (in.)}^{1/2}$$

The stress intensity in the wedge-opening mode of loading is important in dealing with ASTM standards with special regard to compact configurations of test specimens.[20] Although more detailed comments on fracture toughness test specimens belong to Chapter 4, the basic form of the stress intensity equations is discussed here because of the special feature of these equations involving the applied load rather than the applied stress. The idealized model and all the pertinent symbols are illustrated in Fig. 3.26. The two basic specimens are characterized by the ratios $H/w = 0.972$ and $H/w = 1.2$.[6] The general form of the wedge opening mode equation involves a special polynomial function of the so-called dimensionless crack length a/w. These functions pertain to specimen geometries known in the business of fracture testing as T-type and compact tension (CT).[6] For the T-type specimen the function is[20,21]

$$f\left(\frac{a}{w}\right)_{\text{T}} = 30.96\left(\frac{a}{w}\right) - 195.8\left(\frac{a}{w}\right)^2 + 730.6\left(\frac{a}{w}\right)^3$$
$$- 1186.3\left(\frac{a}{w}\right)^4 + 754.6\left(\frac{a}{w}\right)^5 \tag{3.31}$$

which is accurate within 0.5% of the experimental compliance for the range $0.25 < a/w < 0.75$. For the compact-tension case[5,6,9] the polynomial function

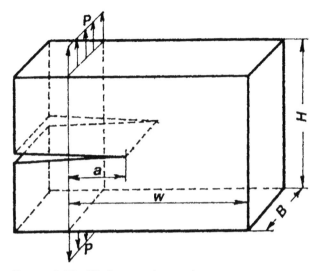

FIGURE 3.26   Wedge-opening mode.

is accurate within 0.25% for a/w ratios between 0.3 and 0.7. This function is

$$
f\left(\frac{a}{w}\right)_C = 29.6\left(\frac{a}{w}\right)^{0.5} - 185.5\left(\frac{a}{w}\right)^{1.5} + 655.7\left(\frac{a}{w}\right)^{2.5}
$$
$$
- 1017\left(\frac{a}{w}\right)^{3.5} + 639\left(\frac{a}{w}\right)^{4.5} \tag{3.32}
$$

The geometrical and dimensional details of the two specimens are given by Barsom and Rolfe.[6]

The polynomials defined by Eqs. (3.31) and (3.32) are used with the general formulas for the stress intensity factors as shown here.

T-type specimen

$$
K_I = \frac{P}{B(a)^{1/2}} \cdot f\left(\frac{a}{w}\right)_T \tag{3.33}
$$

Compact-tension (CT) specimen

$$
K_I = \frac{P}{B(w)^{1/2}} \cdot f\left(\frac{a}{w}\right)_C \tag{3.34}
$$

The more practical values of the power series $f\left(\frac{a}{w}\right)_T$ and $f\left(\frac{a}{w}\right)_C$ are plotted in Figs. 3.27 and 3.28.

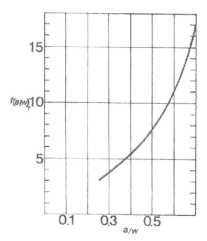

FIGURE 3.27   Plot of Eq. (3.31).

The dimensional character of Eqs. (3.33) or (3.34) is consistent with that of the stress intensity factor if we multiply numerator and denominator by $a^{1/2}$ so that

$$\frac{P(a)^{1/2}}{B(a)^{1/2}} \times \frac{1}{(a)^{1/2}} = \frac{P(a)^{1/2}}{Ba}$$

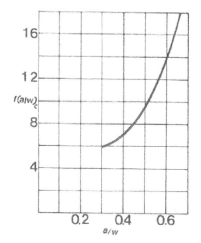

FIGURE 3.28   Plot of Eq. (3.32).

which has the dimensions of stress and K_{Ic}, or K_I, because

$$\frac{lb}{in.} \times \frac{(in.)^{1/2}}{in.} = psi \ (in.)^{1/2}$$

When the crack is subjected to internal pressure (Fig. 3.29) the stress intensity factor is expressed in simple terms as

$$K_1 = p(\pi a)^{1/2} \tag{3.35}$$

Note that p in this formula and in Fig. 3.29 denotes pressure (psi) and not the line load P, in pounds.

Another useful and practical case of dealing with the stress intensity factor is concerned with a circumferentially cracked round bar subjected to axial tension. Stress analysis of this type of a notched bar was conducted by Bueckner[22] and it was consistent with the results obtained by Irwin, and by Paris and Sih.[23] The geometry, notation, and loading are illustrated in Fig. 3.30. The basic formula for stress intensity for this notched bar is

$$K_1 = \frac{P}{D^{3/2}} \left(\frac{1.72D - 1.27d}{d} \right) \tag{3.36}$$

Again, by rewriting the original expression we can see that the stress intensity factor is, indeed, directly proportional to the reference stress.

$$\frac{P}{D(D)^{1/2}} \left(\frac{1.72D - 1.27d}{d} \right) = \frac{P(1.72D - 1.27d)}{D(D)^{1/2}d} \times \frac{(D)^{1/2}}{(D)^{1/2}}$$

$$= \frac{P(D)^{1/2}(1.72D - 1.27d)}{D^2 d}$$

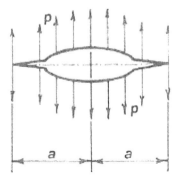

FIGURE 3.29 Crack under internal pressure.

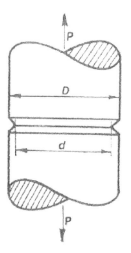

FIGURE 3.30 Round bar with circumferential crack.

or in terms of dimensional units

$$\frac{\text{lb (in.)}^{1/2} \times \text{(in.)}}{\text{(in.)}^2 \times \text{(in.)}} = \text{psi (in.)}^{1/2}$$

Introducing $n_o = d/D$, the original formula, Eq. (3.36), becomes

$$K_{\mathrm{I}} = \frac{P}{D^{3/2}} \left(\frac{1.72}{n_o} - 1.27 \right) \tag{3.37}$$

The limits for the use of Eq. (3.37) have been established in terms of the diameter ratio d/D between the values of 0.4 and 0.8,[15] as shown in Fig. 3.31. For this plot, Eq. (3.37) is transformed into a simpler form as in Eq. (3.38). The diameter ratio is $n_o = d/D$.

$$K_{\mathrm{I}} = \frac{Pf(n)}{D^{3/2}} \tag{3.38}$$

where

$$f(n) = \left(\frac{1.72}{n_o} - 1.27 \right) \tag{3.39}$$

In all practical cases of stress intensity considered in this chapter so far, the plane of the crack was assumed to be normal to the direction of the applied (reference) stress. It was also assumed that the degree to which a material can carry a load without brittle failure can be measured by the plane-strain fracture toughness

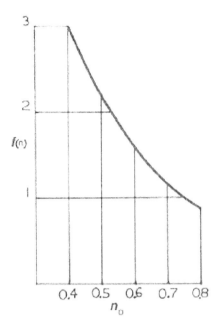

$f(n)$

0.4 0.5 0.6 0.7 0.8

n_D

FIGURE 3.31 Parameter in Eq. (3.38).

denoted by K_{Ic}. This is, of course, a conservative criterion, and as long as the term $\sigma(a)^{1/2}$ can be kept below the K_{Ic} value, the crack should be harmless. The crack can only begin to grow when the applied stress reaches the level of $K_{\mathrm{Ic}} = \sigma(a)^{1/2}$. All this, of course, happens when we operate in the so-called first mode, defined as crack opening in a tensile stress field. No shearing or tearing effects are present. However, the question arises when the plane of a through-thickness crack is tilted at an angle θ with the plane of loading which is also parallel to the surface of the plate, as shown in Fig. 3.32. The method of solution for this type of a problem was reported by Sih,[24] which resulted in a formula for the critical applied stress (σ_c). This formula can be written as

$$\sigma_c = \frac{2(E_s U_c)^{1/2}}{f(\theta)(a)^{1/2}} \tag{3.40}$$

where

$$f(\theta) = \sin \theta (1 - 2\nu \sin^2 \theta)^{1/2} \tag{3.41}$$

and E_s = shear modulus of elasticity (lb/in.^{2}), U_C = critical strain energy density factor (lb/in.), and ν = Poisson's ratio.

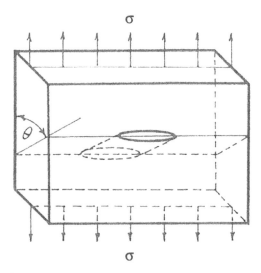

FIGURE 3.32 Slanted-crack mode.

The parameter U_C can be obtained by tests using a technique similar to that for measuring K_{Ic}. The use of this parameter is intended for those crack systems subjected to the mixed mode of loading, and the parameter U_C can be correlated with the existing K_{Ic} data with the help of another expression given by Sih:[24]

$$U_C = \frac{(1+v)(1-2v)}{2\pi E} K_{Ic}^2 \tag{3.42}$$

The reader may recall that since our encounter with the effect of Poisson's ratio on the critical crack length under plane-stress and plane-strain conditions, there was no need to be directly involved with v in calculating various stress intensity factors. However, for the first time in this chapter we are beginning to see a subtle change as evidenced by Eqs. (3.40) through (3.42). In the case of a more conventional engineering material (steel, titanium, aluminum) the extreme values v can vary, say, between 0.25 and 0.35. Other materials, of course, can have values outside this range, as shown in Table 1.2. The critical applied stress σ_C is affected by a combined influence of crack angle θ and v, represented by Eq. (3.41) and Fig. 3.33. For a given elastic modulus E and plane-strain fracture toughness K_{Ic}, Eq. (3.42) can be rearranged to emphasize the effect of v.

$$\frac{2\pi E U_c}{K_{Ic}^2} = (1+v)(1-2v) \tag{3.43}$$

This effect is illustrated in Fig. 3.34.

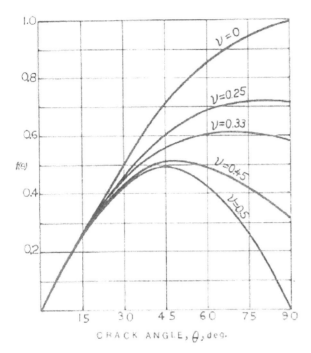

FIGURE 3.33 Crack angle parameter, $f(\theta)$.

The case of a slanted-crack problem shows clearly that as we deviate from the classical Mode I loading, we enter the area of a combined (mixed) mode effect. Under these conditions, as Sih[24] points out, Mode I or K_{Ic} does not always result in the lowest critical applied stress. Also, as the crack plane tilts further away from Mode I, the plate member becomes unable to support the external load.

Design Problem 3.18

Determine the ratio of the stress intensity factors caused by eccentric loading (Fig. 3.25) as a function of the distance of the line loading from the crack center. Calculate the larger of the two stress intensity factors for $x = 0.6a$ and a 6000 lb load acting on a 1.5 in. long through-thickness crack.

Solution

Divide Eq. (3.28) by Eq. (3.29) to give the desired ratio

$$\frac{(a + x)}{(a - x)}$$

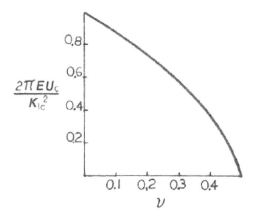

FIGURE 3.34 Effect of Poisson's ratio.

Plot this ratio as a function of x/a to obtain the curve shown in Fig. 3.35. The larger stress intensity factor is given by Eq. (3.28).

$$a = 0.5 \times 1.5$$
$$= 0.75 \text{ in.}$$
$$P = 6000/1.5$$
$$= 4000 \text{ lb/in.}$$

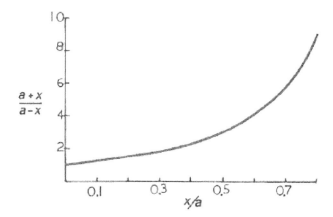

FIGURE 3.35 Plot for Design Problem 3.18.

Hence

$$K_1 = \frac{4000(0.75 + 0.6 \times 0.75)^{1/2}}{(0.75\pi)^{1/2}(0.75 - 0.6 \times 0.75)^{1/2}}$$
$$= 5.2 \, \text{ksi (in.)}^{1/2} = 5.7 \, \text{MPa (m)}^{1/2}$$

Design Problem 3.19

A test was performed on a compact-tension (CT) specimen. Calculate the approximate stress intensity factor if the maximum test load was 3150 lb and the test specimen dimensions were as follows:

Thickness of specimen $= 1.37$ in.
Width of specimen $= 2.00$ in.
Crack length $= 1.08$ in.

Solution

$$a/w = 1.08/2$$
$$= 0.54$$

From Fig. 3.25

$$f\left(\frac{a}{w}\right)_C \cong 11$$

Then, using Eq. (3.34) for the CT specimen, we have

$$K_1 = \frac{3150 \times 11}{1.37 \times (2)^{1/2}}$$
$$= 17.9 \, \text{ksi (in.)}^{1/2}$$
$$= 19.7 \, \text{MPa (m)}^{1/2}$$

Design Problem 3.20

A round bar with a circumferential crack is used in a calibration procedure and is expected to experience 18,000 lb tensile load. If the outer diameter of the bar is 1.2 in. and the depth of the radial notch is 0.22 in., estimate the stress intensity factor for the bar material.

Solution

Inner diameter is

$$d = D - 2 \times 0.22$$
$$= 1.2 - 0.44$$
$$= 0.76$$

The ratio of diameters is

$$n_o = 0.76/1.2$$
$$= 0.63$$

Using n_o and Fig. 3.31, we obtain

$$f(n) \cong 1.4$$

Hence, from Eq. (3.38)

$$K_I = \frac{18,000 \times 1.4}{1.2^{1.5}}$$

$$= 19.17 \, \text{ksi (in.)}^{1/2}$$

$$= 21 \, \text{MPa (m)}^{1/2}$$

Design Problem 3.21

A titanium alloy plate member is subjected to uniform tensile stress of 20 ksi and it contains a 2 in. crack inclined $60°$ to the surface of the plate, as shown in Fig. 3.32. Calculate the minimum required fracture toughness of the material to tolerate the existing flaw.

Solution

Substitute Eqs. (1.11) and (3.42) into Eq. (3.40)

$$\sigma_c = \frac{2\left[\dfrac{E(1 + v)(1 - 2v)K_{\text{Ic}}^2}{2(1 + v) \times 2\pi E}\right]^{1/2}}{f(\theta)(a)^{1/2}}$$

$$\sigma_c = \frac{(1 - 2v)^{1/2}K_{\text{Ic}}}{(\pi a)^{1/2}f(\theta)}$$

From Table 1.2, the Poisson's ratio for titanium alloy is $v = 0.33$. Substituting this value and solving the foregoing expression for K_{Ic}, we have

$$K_{Ic} = \frac{(\pi a)^{1/2} f(\theta) \sigma_c}{(1 - 2 \times 0.33)^{1/2}}$$
$$= 3.04(a)^{1/2} f(0) \sigma_c$$

For $\theta = 60°$ and $v = 0.33$, Fig. 3.33 yields

$$f(\theta) = 0.6$$

Hence, the required fracture toughness is

$$K_{Ic} = 3.04(1)^{1/2} \times 0.6 \times 20$$
$$= 36.5 \, \text{ksi (in.)}^{1/2}$$
$$= 40.1 \, \text{MPa (m)}^{1/2}$$

CLOSING REMARKS

The primary purpose of this chapter is to briefly review fundamental concepts of fracture mechanics and to characterize fracture behavior in terms familiar to design engineers such as stresses, crack dimensions, and special material properties affecting technical decisions. Also, the information includes several numerical examples, definitions, and results expressed in English and SI units. The technology concerned is restricted to linear elastic fracture mechanics (LEFM) and the stress intensity factors at the crack tip in the first mode of loading on the premise that the tensile stress field is normal to the plane of the crack. The conventional symbol σ is used for stress, and the symbol K with the subscript Roman numeral I denoting the opening tensile mode in which the stress intensity is considered. It is important to keep in mind that these two essential parameters carry different dimensions.

It has been shown in this chapter, under various loading and geometric conditions, that all crack tip stresses are proportional to the applied elastic stress, often referred to as nominal, gross, or reference. Fracture is normally expected to occur when the crack tip stresses become too high. The analysis includes the geometric effects and loading conditions by utilizing the appropriate dimensionless factors, represented in the open literature by different symbols and, sometimes, by lack of uniformity in definitions.

In simpler cases the value of the stress intensity factor under the general symbol K is directly related to the magnitude of the applied (nominal) stress and the square root of the crack length, which is almost universally denoted by a. However, one should note that, at times, a is defined as one-half the crack length. Usually, the formulas in the literature spell out the definition of a for a particular design case.

In more complex situations (such as finite width of the plate), reference can be made to a "tangent correction," "secant correction," or in the case of an elliptical crack, the correction can be designated as "shape factor" or "flaw shape parameter." In dealing with wedge-opening loading the correction can be made with the help of a function expressed in a polynomial form, which can also be defined as a "function of the dimensionless crack length," and so on. Hence in scrutinizing the formulas for stress intensity factors, the definitions of reference stresses and correction parameters should be critically examined.

This chapter has been limited to LEFM problems and the first mode of loading for practical reasons, some of which are noted. In the case of small crack tip displacements and negligible plastic deformation, the stress and strain fields near the crack tip must be elastic, justifying the use of plane-strain fracture toughness K_{Ic}. This is certainly consistent with the LEFM philosophy, and the cases discussed in this book, so far, are intended to be examples of elastic behavior. The parameter K_{Ic} signifies sufficient toughness of the material to prevent progressive crack extension. On the other hand, the parameter K_I defines the stress intensity ahead of a sharp crack. Again we should recall that for design purposes, K_I should be kept below K_{Ic} to avoid fracture in a manner similar to the result of holding a conventional stress σ below yield strength σ_y to prevent yielding.

Another reason for selecting the K_{Ic} parameter in the various sample problems was the desire to assure the maximum level of design conservatism. It is now necessary to put the plane-stress fracture toughness K_c into the correct frame of reference in relation to LEFM design methodology. Let us start with the definitions.

Plane-Stress Fracture Toughness (K_c): Depends on specimen thickness, geometry, and crack size. Applies to static loading and plane stress under variable constraint.

Plane-Strain Fracture Toughness (K_{Ic}): Applies to static loading and plane-strain conditions of maximum constraint. It represents a minimum value of toughness for thick plates.

The K_c and K_{Ic} parameters depend on temperature, especially for those materials prone to a transition from brittle to ductile behavior. The plane-stress toughness K_c can be used in the LEFM formulas and it is subject to a general relationship between the applied nominal stress σ and crack size a, similar to that for the K_{Ic} parameter. For instance, using the case of a through-thickness crack in a wide plate, Eq. (2.14), this relationship is as illustrated in Fig. 3.36. The area under the particular K_c curve, which represents many combinations of stress σ and crack size $2a$, shows that the fracture can only occur if K_I reaches

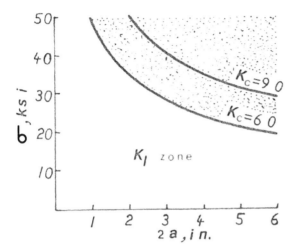

FIGURE 3.36   Stress–crack size relationship.

or exceeds the K_c level. Hence there are also numerous combinations of stress and crack size that will cause failure (fracture zone). Figure 3.36 indicates further that for a tougher material, by moving, say, from 60 to 90 ksi (in.)$^{1/2}$, the permissible crack lengths at all stress levels can be increased substantially. This is quite important because with the increased toughness, brittle fracture is less likely to occur and the material's failure in tension can only develop through a general plastic yielding similar to that in a conventional tensile test. In quality structural steels we expect to have sufficient resistance to crack propagation.

SYMBOLS

a	Length (or depth) of crack (flaw), in. (mm)
a_{CR}	Critical flaw size, in. (mm)
a_e	Effective crack length, in. (mm)
B	Thickness of plate, in. (mm)
b	Minor semiaxis (ellipse), in. (mm)
c	Dimension of thumbnail or buried crack, in. (mm)
D	Outer diameter, in. (mm)
d	Inner diameter, in. (mm)
E	Young's modulus, psi (N/mm^2)
E_s	Shearing modulus of elasticity, psi (N/mm^2)
$f(b)$	Correction parameter for plates
$f(e)$	Corre ction factor for elliptical notches
$f(m)$	Stress intensity correction for beams

$f(n)$	Circumferential crack parameter
$f(w)$	Correction factor for edge cracks
$f(\theta)$	Crack angle parameter
$f(a/r)$	Correction factor for round holes
$f(a/w)_C$	Polynomial for compact specimen
$f(a/r)_T$	Polynomial for T-type specimen
H	Depth of test specimen, in. (mm)
h	Major semiaxis (ellipse), in. (mm)
K	Stress intensity factor, ksi (in.)$^{1/2}$ [MPa (m)$^{1/2}$]
K_I	Tensile mode stress intensity, ksi (in.)$^{1/2}$ [MPa (m)$^{1/2}$]
K_{Ic}	Plane-strain fracture toughness, ksi (in.)$^{1/2}$ [MPa (m)$^{1/2}$]
K_c	Plane-stress fracture toughness, ksi (in.)$^{1/2}$ [MPa (m)$^{1/2}$]
k	Stress concentration factor
L	Half-length of beam, in. (mm)
M	Bending moment, lb-in. (N-mm)
M_K	Back free-surface correction
m	Ratio of crack length to width of plate
n	Total width of ligament, in. (mm)
n_o	Diameter ratio
P	Concentrated load, lb (N)
p	Internal pressure, psi (N/mm^2)
Q	Flaw shape factor
r	Hole radius, in. (mm)
S_y	Yield strength, ksi (MPa)
U_c	Critical strain energy density factor, lb/in. (N/mm)
w	Half (or full) width of plate, in. (mm)
Z	Section modulus, in.^{3} (mm^3)
θ	Tilt angle of crack, degrees
v	Poisson's ratio
ρ	Root radius, in. (mm)
σ	Applied (reference) stress, ksi (MPa)
σ_c	Critical applied stress, ksi (MPa)
σ_{col}	Collapse stress, ksi (MPa)
σ_{max}	Maximum elastic stress, ksi (MPa)

REFERENCES

1. Rooke, D.P.; Cartwright, D.J. *Compendium of Stress Intensity Factors*; Her Majesty's Stationery Office: London, 1976.
2. Tada, H.; Paris, P.C.; Irwin, G.R. *The Stress Analysis of Cracks Handbook*; University of St. Louis: St. Louis, 1973.

3. Sih, G.C. *Handbook of Stress Intensity Factors*; Institute of Fracture and Solid Mechanics: Lehigh University, Bethlehem, PA, 1973.

4. Paris, C.P.; Sih, G.S. Stress analysis of cracks. In *Fracture Toughness Testing and Its Applications*, ASTM STP 381; American Society for Testing and Materials: Philadelphia, 1965.

5. Broek, D. *The Practical Use of Fracture Mechanics*; Kluwer Academic Publishers: Dordrecht, 1988.

6. Barsom, J.M.; Rolfe, S.T. *Fracture and Fatigue Control in Structures*, 2nd Ed.; Prentice-Hall: Englewood Cliffs, NJ, 1987.

7. Novak, S.R.; Barsom, J.M. *Brittle Fracture (K_{Ic}) Behavior of Cracks Emanating from Notches, Cracks and Fracture*, ASTM STP 601; American Society for Testing and Materials: Philadelphia, 1976.

8. Blake, A., Ed. *Handbook of Mechanics, Materials and Structures*; John Wiley: New York, 1985.

9. Parker, A.P. *The Mechanics of Fracture and Fatigue*; E. & F.N. Spon: London, 1981.

10. British Standards Institute. *Methods of Test for Plane Strain Fracture Toughness of Metallic Materials*; BS 5447; BSI: London, 1977.

11. Irwin, G.R. The crack extension force for a part through crack in a plate. Trans. ASME J. Appl. Mech. **1962**, *29*(4).

12. Kobayashi, A.S.; Zu, M.; Hall, L.R. *Approximate Stress Intensity Factor for an Embedded Elliptical Carck Near to Parallel Free Surfaces*. Int. J. Fracture Mech. **1965**, *1*, 81–95.

13. Streit, R.D. *Design Guidance for Fracture-Critical Components at Lawrence Livermore National Laboratory*; Lawrence Livermore Laboratory, UCRL-53254, March 3, 1982.

14. Gerberich, W.W. *Fracture Mechanics Approach to Design – Application: Short Course on Offshore Structures*; University of California: California, 1967.

15. Brown, W.F.; Srawley, J.E. *Plane Strain Crack Toughness Testing of High Strength Metallic Materials*, ASTM Special Technical Publication No. 410; American Society for Testing and Materials: Philadelphia, 1967.

16. Pellini, W.S. *Principles of Structural Integrity Technology*; Department of the Navy, Office of Naval Research: Arlington, VA, 1976.

17. Holman, W.R.; Langland, R.T. Recommendations for Protecting Against Failure by Brittle Fracture in Ferritic Steel Shipping Containers up to Four Inches Thick. Lawrence Livermore Laboratory, NUREG/CR-1815 UCRL-53013, August 1981.

18. Gran, R.J.; Orario, F.D.; Paris, P.C.; Irwin, G.R.; Hertzberg, R.W. *Investigation and Analysis: Early Development of Aircraft Structural Failures*, AFFDL-TR-70-149, 1971

19. Wessel, E.T.; Server, W.L.; Kennedy, E.L. *Primer: Fracture Mechanics in the Nuclear Power Industry*; Electric Power Research Institute: Palo Alto, CA, 1991.

20. Wilson, W.K. Stress intensity factors for compact specimens used to determine fracture mechanics parameters, Research Report 73-1E7-FMP WR-R1; Westinghouse Research Laboratories: Pittsburgh, July 27, 1973.

21. Wilson, W.K. Analytical determination of stress intensity factors for manjoine brittle fracture specimen, Research Report AEC, WERL0029–3; Westinghouse Research Laboratories: Pittsburgh, August 26, 1965.

22. Bueckner, H.F. *Coefficients for Computation of the Stress Intensity Factor K_I for a Notched Round Bar, Fracture Toughness Testing and Its Applications*, ASTM STP 381; Am. Soc. Testing Mats., 1965.
23. Paris, C.P.; Sih, G.S. Stress analysis of cracks. In *Fracture Toughness Testing and Its Applications*, ASTM STP 381; American Society for Testing and Materials: Philadelphia, 1965.
24. Sih, G.C. The role of fracture mechanics in design technology, Paper No. 76-DET-58. ASME J. Eng. Ind. **1976**.

4

Test and Analysis of Fracture Toughness

OVERVIEW OF THEORY AND PRACTICE

Review of various aspects of basic fracture mechanics in Chapters 2 and 3 was supported by a number of numerical examples in order to indicate briefly how the formulas featuring the particular stress intensity values and externally applied nominal stresses can be used in design. It became clear, as expected, that linear elastic fracture mechanics (LEFM) shows limited concern with forces and stresses because, as experience has shown in a number of well-documented cases, conventional stress analysis and design failed either to predict or explain catastrophic failures. This is the historical basis on which modern fracture mechanics has grown and prospered. All structures under load contain varied amounts of strain energy available for propagating the ever present cracks and flaws. It is simply a self-destructive process on a "micro" and "macro" level if we say that "microcracks" are those that are beyond our capability to detect.

What happens when a progressively higher tensile load is applied to the structure in the manner, say, of that in a familiar tensile test designed to verify the tensile strength of the material? It is certain that the stress is directly proportional to the applied load and that, according to Eq. (2.5), the strain energy stored within the material varies as the square of the applied stress. This is, in itself, a very significant observation because additional stress can rapidly increase the total amount of energy driving the crack. Of course, the reverse is also true and important. When the crack enters the region of a lower applied stress, it is likely to be arrested.

When the increased applied load breaks the structure, it is rather easy to conclude that simple action of a tensile stress is responsible for the entire process of fracture, including the overcoming of the chemical bonds between the atoms in the material. However, in light of the modern theory of materials science, the maximum tensile stress necessary to pull the atoms apart is far higher than the conventional tensile strength of the material. The main question to be resolved before any reasonable conclusion can be reached is whether we have a reliable mechanism for the conversion of the available strain energy for the purpose of propagating the existing crack or creating a new one. The conversion of strain energy into fracture energy is here a key theoretical and practical consideration because the quantity of energy required to break the material defines its toughness.

The reader may recall that Griffith,[1] who was first to put forth a rational theory of fracture mechanics (Chapter 2) regarded Inglis's stress concentration[2] as a mechanism for converting strain energy into fracture energy. This mechanism is triggered by the size and geometry of the worst discontinuities in the structure such as sharp cracks, holes, and other defects. And the amount of energy (which is toughness) to break a given cross-section of a material is different for various solids depending on the degree of brittleness or ductility involved. Brittle materials should not be used in tensile applications, if at all possible, because the energy needed for their fracture is indeed very small. For instance, the energy required to fracture mild steel may be close to a million times as high as that needed to break the equivalent cross-section in glass.[3] What is certainly misleading is that, while the energy requirement is, in this particular case, vastly different, the static tensile strengths of these materials are not too far apart. The designer must be cautioned not to rely on the tables of tensile strengths alone in making the choice of a material for a particular application.

There is little doubt about the enormous complexity of the process by which the energy is absorbed within tough materials and how the chemical bonds are broken in most structural solids. However, some general principles may be stated in relation to brittle and ductile behavior.

From the materials science point of view the only amount of energy required to fracture a brittle material is that sufficient to break all the chemical bonds on a given cross-section. In other words, a brittle material is not judged by a low tensile strength but by a low energy level to break the chemical bond. From an engineering point of view we look at a brittle fracture as that which occurs without a prior plastic deformation and at very high speeds. The fracture appears as a flat surface, referred to by metallurgists as "cleavage," which shows a distinct glitter due to reflection of light. The fractures almost always start and continue at cracks and notches, with virtually no overall deformation at the failed cross-section. However, as far as the fracture mechanism is concerned, the views of engineers, materials scientists, and fractographers may vary.

In a tough material having fine structure and good ductility, the work of fracture can be very high, so that the material is able to absorb energy and resist fracture. This is quite different from brittle behavior. Necking and other forms of ductile fracture can develop in a ductile material because many layers of atoms in the metallic crystals can slide and absorb energy due to the mechanism of dislocation originally postulated 60 years ago.[4] This is a complicated theory showing that the mechanics of deformation of a piece of metal is as clever and intricate as in many living tissues. The dislocation process allows the metals to be tough, forged, and hardened, the very significant characteristics so much in demand.

The role of material parameters in the application of fracture mechanics to design cannot be overemphasized, particularly in the area of LEFM. If a given material shows certain characteristics of brittle behavior, then the experiments should be conducted in the laboratory to establish the minimum credible value of fracture toughness. The experimental part of such a program is a vital phase of the fracture mechanics technology. The American Society for Testing and Materials (ASTM) Committee E24 on Fracture Testing of Materials spearheaded the development of standard test methods.

All stress intensity factors corresponding to the opening K_I, edge-sliding K_{II}, and tearing K_{III} modes are obtained by calculation. The factors depend on the load, crack size, and geometry, and are the same for any structural material. The critical stress intensity factors such as K_c, or K_{Ic}, at which unstable crack growth begins, depend on temperature, specimen thickness, and constraint under static conditions. For the case of dynamic (or impact) loading, the critical stress intensity factor is denoted by K_{Id}. Because of the special importance of the plane-strain fracture toughness K_{Ic} in many problems of design, certification, and failure analysis of structures, the ASTM Committee has made a special effort to develop test methods for K_{Ic} restricted to linear-elastic response.

CONSTRAINT AND THICKNESS

Many years ago, dating back to the time of Hooke when experimental facilities did not amount to much and new concepts of solid mechanics were hard to accept, there was a good deal of confusion about whether to work with the structure as a whole or any given portion of the structural material at hand. Today we accept a "test piece" of the material (or the structure) in order to arrive at the correct stress–strain diagram or other key property of the material.

In the case of experimental studies of the critical stress intensity factors K_c and K_{Ic}, corresponding to plane-stress and plane-strain conditions, the effect of thickness (and/or constraint) comes to the forefront. It appears that it is impossible to analyze the level of a constraint without considering the effects of plate thickness on fracture toughness and through-thickness stresses, which are particularly significant in very thick plates. The limiting thickness has been defined in the ASTM standard test method for K_{Ic}.[5] The particular equation is

$$B \geq 2.5(K_{Ic}/\sigma_y)^2 \tag{4.1}$$

In this expression B is the plate thickness (in.) and σ_y denotes the yield strength of the plate material (ksi). With these dimensions, the fracture toughness is given as ksi (in.)$^{1/2}$. Then for a given set of K_{Ic} and σ_y values, the limiting thickness B comes directly from Eq. (4.1), which corresponds to the maximum constraint under plane-strain conditions. The thickness at the onset of plane strain, Eq. (4.1), is also indicated by the vertical line (m−m) in Fig. 4.1.

The lateral constraint increases with the increase of plate thickness, and a triaxial tensile state of stress develops ahead of the crack. The constraint ahead of the crack is also increased, which has the net effect of reducing notch toughness. What is particularly interesting is that notch toughness can be influenced by the

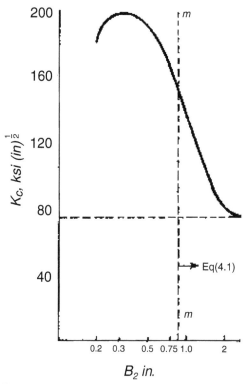

FIGURE 4.1   Effect of thickness on K_c in maraging steel (from Ref. 6).

thickness of the plate while the metallurgical properties of the material still remain the same.[6] The effect of plate thickness on toughness parameters was illustrated in Fig. 2.10. The sketch suggests that for a larger plate thickness the minimum toughness K_{Ic} is attained, consistent with the condition of a maximum constraint. However, as the plate thickness intended for a particular problem is made smaller, the notch toughness should increase to reflect the nature of plane-stress behavior. Hence in the calculation of a critical crack in this case we should use K_c in place of K_{Ic}. A design based solely on the K_{Ic} parameter in this situation would be too conservative. It will also be noticed that the structural design, based on the thinner plate concept, should be more resistant to brittle fracture. The experiments show that fractured thinner plates have a higher percentage of shear lips in relation to the total fractured surface. This feature of a cracked surface of a specimen can be regarded as a qualitative measure of a better notch toughness of the material.

Physical significance of the constraint was described by Pellini[7] with reference to a circular notch in a tensile bar as illustrated in Fig. 4.2. The extension of a smooth tensile bar (B) is essentially free and with a minimum of opposition to lateral contraction due to the Poisson ratio effect. In the case of a notched bar (C) plastic deformation at the notch can develop while other portions (unnotched) can still be elastic. Since the elastic contraction is relatively small, the plastic contraction of the notched section (which is significantly larger) must be opposed by the reaction stresses in a triaxial system. In other words, stresses (σ_x) and (σ_z) (A) tend to inhibit the extension in the direction of (σ_y), and the stress (σ_y) must always be greater than (σ_x) or (σ_z). Pellini describes the mechanism of constraint as "inhibition of plastic flow due to triaxial stress".[8]

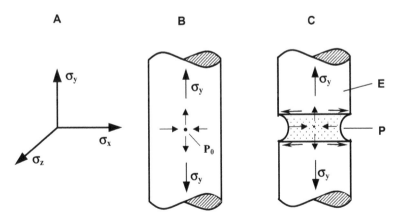

FIGURE 4.2 Model of constraint (P_0, Poisson's effect unimpeded; E, elastic region; P, plastic zone).

In general, the constraint ahead of a sharp crack can be increased by increasing the plate thickness. At the same time the critical stress intensity K_{Ic} decreases with the increase of specimen thickness. However, there is a natural limit that represents the maximum constraint capacity of a through-thickness crack. This limit is reached when the crack dimension exceeds twice the plate thickness. This approximate rule should be regarded as conservative.[8] Also, it is important to note that the fracture resistance of the metal under the maximum conditions of constraint is independent of further increases of crack size, and the parameter K_{Ic} reaches its minimum level.

It is generally accepted that the presence of sharp notches and cracks generates the conditions of a triaxial stress system involving elastic and plastic effects. Detailed evaluations of this problem are based on mathematical relationships and are available in numerous publications that still reflect traditional communication difficulties among the metallurgist, the fracture mechanics specialist, and the design engineer. Even with the development of linear elastic fracture mechanics and test procedures, we still have to deal with the intricacies and limitations inherent in applying the mathematics of the stress and crack size relationships to real-world structures. Similarly, in the area of definition of constraint capacity, we have to limit our discussion to simpler concepts and calculation tools that will help to establish a measure of confidence in mitigating the effects of crack size and geometry.

The stress concentration governed by crack geometry elevates the stress of the crack tip to yield strength of the material, and a plastic zone is developed. The size of this zone is important because the larger the crack tip ductility zone, the greater should be the fracture resistance. However, the actual fracture process takes place in an extremely small region just ahead of the tip of a crack, which is often referred to as the "fracture process zone." In addition, the area outside the plastic zone (further from the crack tip) can be denoted as the K zone[9] where the stress–strain distribution may be calculated from the stress intensity factor equations, several of which are given in Chapter 3. A schematic presentation of the remaining three zones is given in Fig. 4.3. The smallest region, defined as the process zone, is subject to very high strains and deformations resulting in microcracking and void growth. Owing to these phenomena the conventional principles of continuum mechanics do not apply.

The relatively small plastic zone, compared with the K zone, does not affect the stress field away from the crack front. The rules of LEFM should apply as long as the plastic zone is small compared to the K zone. If the situation is reversed, showing that the plastic zone is large enough to change the elastic stress distribution in the K zone, then LEFM does not apply and the K formulas cannot be used for the calculation of the stress intensity at the crack tip.

In practice, we cannot assume that there is a unique size of plastic zone or a fixed value of fracture toughness for a given material. This situation is caused by

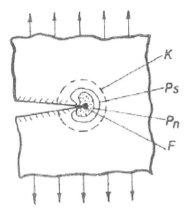

FIGURE 4.3 Schematic of stress zones. [K, stress intensity zone; P_s, plastic zone (plane stress); P_n, plastic zone (plane strain); F, fracture process zone].

changing conditions in structural design and crack geometries. However, it is known that a minimum plastic zone size corresponds to the maximum mechanical constraint under plane-strain conditions. It should be added here that in a sufficiently thick specimen, defined by Eq. (4.1), plane strain occurs in the middle portion of the thickness. At the same time, plane stress prevails near the faces of the specimen. The minimum size of the plastic zone (sometimes referred to as the plastic zone radius) can be obtained from

$$r_y = \frac{1}{6\pi}\left(\frac{K_{Ic}}{\sigma_y}\right)^2 \tag{4.2}$$

Although we deal here with an empirical definition,[10] the formula helps to check on fracture instability and validity of K_{Ic}. The three general types of fracture instability relate to residual strength analysis and other concepts such as subcritical crack growth that have been the objectives of recent research.[6,7,10,11] Under plane-stress conditions the size of the plastic zone (or plastic zone radius) at the point of fracture instability becomes

$$r_y = \frac{1}{2\pi}\left(\frac{K_c}{\sigma_y}\right)^2 \tag{4.3}$$

Utilizing Eqs. (4.1) and (4.2), we find a special requirement for the specimen thickness in the K_{Ic} test, related directly to the minimum size of the plastic zone.

$$\frac{\text{Specimen thickness}}{\text{Plastic zone size}} = \frac{B}{r_y} = \frac{2.5(K_{Ic}/\sigma_y)^2}{(1/6\pi)(K_{Ic}/\sigma_y)^2} = 2.5 \times 6\pi = 47.1$$

The specimen thickness should be large compared to the size of the plastic zone, so that any influence of the plastic zone size on the calculation of the stress intensity factors can be ignored. The generally accepted rule is that the specimen thickness should be about 50 times the minimum size of the plastic zone. This rule is also used in judging the limits of applicability of plane-strain LEFM to the structural design and in the process of establishing valid material parameters. When the material thickness b and crack size a comply with the limits of Eq. (4.1), and when the size of the plastic zone is about $1/50$ of the material thickness b, then the validity of applying LEFM techniques to design can be fully demonstrated. This technique is then applicable to a great variety of conditions and materials of special interest to the designers.

Design Problem 4.1

A fracture analysis is required on a 40 mm thick steel plate member of a machine. Check the validity of plane-strain LEFM (linear elastic fracture mechanics), if this steel plate has a yield strength of 140 ksi and a fracture toughness of 65 MPa $(m)^{1/2}$.

Solution

Convert to consistent dimensions

$$40/25.4 = 1.57 \text{ in.}$$

$$65/1.099 = 59.1 \text{ ksi (in.)}^{1/2}$$

The minimum size of the plastic zone follows from Eq. (4.2)

$$r_y = \frac{1}{6\pi}\left(\frac{59.1}{140}\right)^2$$

$$= 0.0095 \text{ in.}$$

To assure validity, use the requirement of the form $50 \, r_y$. Hence

$$1.57 \text{ in.} > 50 \times 0.0095 \text{ in.}$$

$$1.57 \text{ in.} > 0.475 \text{ in.}$$

The use of LEFM is therefore appropriate.

Design Problem 4.2

A reactor grade A533B steel is to be tested for plane-strain fracture toughness. If the yield strength of the steel is 50 ksi and the expected fracture toughness is about 180 ksi $(in.)^{1/2}$, calculate the specimen thickness for valid K_{Ic} testing.

Solution

The specimen thickness for a valid K_{Ic} test is given by Eq. (4.1)

$$B \geq 2.5 \, (180/50)^2$$
$$B \geq 32.4 \, \text{in.}$$

This requirement of thickness is highly unrealistic. Thus, very tough materials generally fail by yielding or if their fracture toughness is to be determined then the thickness of the material in actual practice is used.

Design Problem 4.3

It is proposed to use a piece of 2.5 in. thick plate for making a compact tension specimen for K_{Ic} tests. If the yield strength of the plate material is 180 ksi and the plane-strain fracture toughness is expected to be on the order of 120 ksi (in.)$^{1/2}$, verify that the specimen thickness satisfies the ASTM E-399 Standard for K_{Ic}. Show the variation of the specimen thickness with fracture toughness.

Solution

Since the required thickness of the specimen for a valid K_{Ic} test is given by Eq. (4.1), we obtain

$$B \geq 2.5 \, (120/180)^2$$
$$2.5 \geq 1.11 \, \text{in.}$$

Hence the ASTM validity limit is observed. Also, from Eq. (4.1) we have

$$B = 2.5(K_{Ic}/180)^2$$
$$B = 77 \times 10^{-6}(K_{Ic})^2$$

The variation of specimen thickness with fracture toughness is shown in Fig. 4.4.

The ratio K_{Ic}/σ_y given by Eq. (4.1), representing the state of plane strain, is an important parameter because it defines ductility of the material and constraint capacity for a particular section size. For instance, taking the ratio value of 1.0, Fig. 4.5 shows that the section thickness should be at least equal to 2.5 in. Should the actual thickness be less than this, the constraint capacity would become inadequate.

It should be added that use of Eq. (4.1), like any other analytical process, involves certain limitations. For instance, before a K_{Ic} test specimen is made,

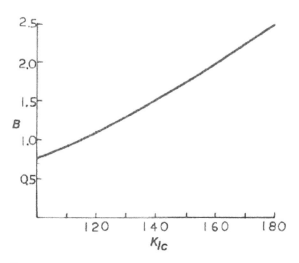

FIGURE 4.4   Variation of thickness with toughness (B is in inches and K_{Ic} in ksi (in.)$^{1/2}$).

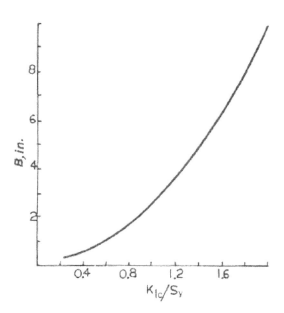

FIGURE 4.5   Section size vs. ratio limit (from Ref. 7).

the K_{Ic} value to be inserted in Eq. (4.1) must first be estimated. Hence, for practical reasons, the following initial steps may be of help:

- Use experience with similar materials to estimate the value of K_{Ic}.
- Employ empirical correlations discussed later in this chapter.
- Assume the specimen thickness to be equal to that of the plate intended for service.

Unfortunately, in many design applications involving larger structures, the section sizes may be of insufficient thickness to assure plane-strain constraint capacity under normal operating conditions.

The constraint limits under plane-strain conditions are defined by Eq. (4.1). The corresponding ratio limits can now be used to show the dependence of the section size on such ratio limits, as illustrated in Fig. 4.5.

Design thickness B, Eq. (4.1), provides an assurance that a through-thickness flaw will not propagate unless the reference (nominal) stress exceeds the yield strength of the material. In general, high K_{Ic}/σ_y ratios require large section thickness and large flaws (cracks) for plane-strain fracture initiation, as implied, in part, by Fig. 4.5. On the other hand, low K_{Ic}/σ_y ratios suggest small section size and small cracks compatible with plane-strain fracture. As shown in Fig. 4.5, at least theoretically, the K_{Ic}/σ_y ratios can vary over a wide range. A ratio above the limit of 2.0, however, is seldom attainable irrespective of the section or flaw size, because metal ductility becomes too high to support the state of plane strain.

In thicker sections, the level of the mechanical constraint is reflected by the through-thickness contraction close to the fractured surface. The mechanical constraint in thinner sections is likely to be rather small. These aspects of the constraint have been researched and reported.[8]

A special case of dealing with plane stress by means of the plane-strain technique can be defined as characterizing "constraint-relaxation" conditions in terms of "unattainable" plane-strain ratios. According to Pellini,[7] the constraint relaxation necessary to cause nominal stresses to exceed yield should develop when the following criterion is observed:

$$B \geq 1.0(K_{Ic}/\sigma_y)^2 \tag{4.4}$$

This approach should be used because K_c testing would require the application of elastic–plastic fracture mechanics (EPFM) instead of LEFM, and because the tests would otherwise have to be run on very large specimens. Hence the only practical solution is to have a conservative estimate of the yield criterion with a section size less than 0.4 times the minimum original size for the plane-strain condition.

TESTING OF FRACTURE TOUGHNESS

Traditional design involves simplified models based on the theory of elasticity and the conventional mechanical properties of the materials. Fracture mechanics, as a recent newcomer in the engineering field, offers new approaches to fail-safe design with the objective of reducing the level of conservatism by providing a theoretical stress analysis of the cracked body. This process introduces the flaw size and geometry as additional design parameters, and it requires a good deal of knowledge of relatively new material properties such as fracture toughness applicable to various modes of cracking and environmental conditions. This section outlines briefly some of the practical aspects of experimental methodology that may be of interest to design engineers and other practitioners in their search for fracture-resistant materials. One of the overriding interests in this area is concerned with the metallic materials, including carbon and low-alloy steels, high-strength steels, and nonferrous products. "Fracture toughness" and "fracture resistivity" terminology is sometimes used in the literature dealing with the mechanical and metallurgical aspects of microfracture modes.[10] It is good to keep in mind various definitions and symbols used during the earlier formative years when considering newer concepts and future developments in experimental methodology.

Fracture testing methods have developed during the past 50 years out of necessity. There were very real concerns about structural failures in some of the Liberty ships, followed by the needs of the U.S. military programs in the field of missiles and rockets. Particularly annoying and costly were brittle fractures of high-strength materials in rocket motor cases, where conventional design methodology was inadequate to cope with the technical setbacks. At about the same time (mid-1950s), the electrical power industry became aware of the potential benefits of using LEFM techniques in design because of a rash of brittle failures in turbines and generators.[12–16]

Probably the first and most widely accepted technique for a qualitative assessment of fracture toughness was the Charpy V-notch specimen test (generally known as CVN). This test was introduced in 1905 and it is still popular today in the various branches of industry for determining the effect of temperature on fracture toughness of steel.[7] Many years later (late 1940s) CVN results were correlated with direct service experience showing that crack initiation in steels occurred at 10 ft-lb of CVN energy, but that the cracks did not propagate at a CVN value greater than 20 ft-lb. Service experience with approximately 100 fractured ship structures included large welded plates and analyzed initiation, propagation, and arrest of fractures. The study resulted in a very comprehensive collection of data that provided a sound basis for establishing the 15 ft-lb notch-toughness criterion still used in the design of many steel structures. While this criterion has several merits, there are some developments and limitations of this practice that are referred to in other sections of this book. For example, Pellini[7] and other members of the Naval Research Laboratory (NRL) developed the concepts of "fracture-safe" design and extended the use of the CVN type of experiments by providing the dynamic tear test (DT) idea applicable to inter-mediate and higher strength steels. According to this methodology, the toughness information can be presented in the form of a permissible crack size for a given temperature, yield strength, DT energy, or K_{Ic}.

In general, the entire process of toughness evaluation and testing can be placed in two categories:

- Transition temperature approach;
- Fracture mechanics approach.

In the first category we look for a material with high "shelf energy" at an accep-table low temperature. In the second category the structure is designed for the tol-erable size of the crack. In other words, the crack size becomes a design parameter. However, this approach requires knowledge of the stress field and the appropriate level of toughness.

The transition temperature category disregards the presence of small flaws and allows the use of conventional stress analysis for determining the load-carrying capacity of the structure. However, the designer may be required to establish the operating temperature and to select the appropriate energy diagram from CVN or DT tests, as the more modern design practice demands. A sche-matic representation of such a diagram is presented in Fig. 4.6. This diagram is not to scale and it is only intended to show the general shape of the transition temperature curves and their respective locations. The curves have upper and lower shelf values. The vertical dashed line defines, in this particular case, the nil-ductility transition temperature, commonly denoted as NDT. The intersection

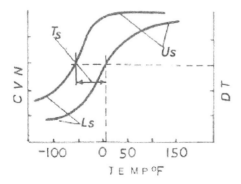

FIGURE 4.6 Energy diagram (not to scale) (T_s, temperature shift; U_s, upper shelf; L_s, lower shelf).

of NDT and DT lines defines a particular level of fracture toughness expressed in energy units.

Although the more common cases of the transition temperature involve CVN and DT techniques, similar transition characteristics can also be obtained using the drop-weight tear test (DWT) and the NRL explosion tear test, both methods developed at NRL.[7]

Fracture toughness depends on both the temperature and the rate of loading, in addition to the thickness of the material. Figure 4.7 illustrates the character of the K_{Ic} and dynamic fracture toughness K_{Id} transition curves.

Temperature strongly affects the fracture toughness across a narrow range between lower shelf (brittle plane strain) and upper shelf (elastic–plastic) regions for typical structural steels. Dynamic loading may be defined as that which has a time to fracture of 0.01 s or less. Fracture times of 1 s or more are regarded as "static phenomena." Figures 4.6 and 4.7 suggest that in testing of fracture toughness, either of the two temperature shifts may be of importance. The effect of these temperature phenomena can be significant. Experience shows that increasing the rate of loading decreases the fracture toughness of structural steels. However, if the loading rates are very high and loading occurs in microseconds the dynamic fracture toughness can be higher than the static fracture toughness. Very high rates of loading are discussed in Chapter 8. The magnitude of the temperature shift between the so-called static K_{Ic} and dynamic K_{Id} conditions in steels of various strengths (Fig. 4.7) can be estimated from a simple formula:[17–20]

$$T_s = 215 - 1.5\,\sigma_y \tag{4.5}$$

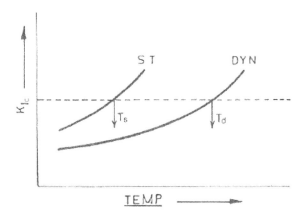

FIGURE 4.7 Effect of the rate of loading and temperature on toughness (ST, K_{Ic} static load; DYN, K_{Ic} dynamic load; T_s and T_d, minimum service temperatures for static and dynamic conditions).

The foregoing expression is applicable to steels in the range of 36 ksi and 140 ksi (or 248–965 MPa). This rule also implies that for a yield strength higher than 143 ksi, the temperature shift becomes equal to zero.[20] In Eq. (4.5) the temperature shift T_s is given in °F, and σ_y denotes room temperature yield strength in ksi. Similar temperature shift conditions exist in structural steel when we compare CVN and DT results; this is shown in Fig. 4.6, which clearly indicates that fracture toughness increases with temperature and that the transition curves exhibit well-defined trends in going from brittle (low shelf) to ductile (high shelf) behavior.

It should be stated, in contrast to the basic features implied in Figs. 4.6 and 4.7, that some structural materials such as aluminum, titanium, and many high-strength steels (yield strength above 150 ksi) do not show any evidence of transition temperature behavior.[6]

Out of the considerable body of theoretical and practical data developed so far, the interest of the designer is often centered around the K_{Ic} and K_{Id} parameters, as well as CVN information. Experience suggests that the CVN data are widely respected in the design and certification circles of industry. The CVN test specimen shown in Fig. 4.8 attracted more attention because of the specimen preparation simplicity and the straightforward experimental procedure. The specimen in Fig. 4.8 has a sharp 45° notch (groove), which is broken by the impact of a hammer swinging in a vertical plane as a pendulum. The energy absorbed by the specimen is directly proportional to the height differential before and after the strike. The specimen design is consistent with the ASTM Standard E-23, "Standard Methods for Notched Bar Impact Testing of Metallic Materials."

Since a typical CVN curve for a structural steel has upper and lower shelves corresponding to ductile and brittle behavior, the fractured surfaces should be fibrous or cleavage-like, respectively. At the transition points, such as those

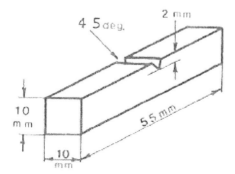

FIGURE 4.8   Charpy V-notch (CVN) test specimen.

marked in Fig. 4.6, the fractured surface is likely to be about 50% fibrous and 50% cleavage-like. However, such an even division between the fibrous and cleavage characteristics is totally unrealistic. Our choice of the transition point is often arbitrary, to say the least, and the transition temperature should not be considered as an invariant property.

Another limitation of the CVN approach can be highlighted with the help of Fig. 4.9. Our question here concerns the meaning of the measured absorbed energy. While the cross-section of the CVN specimen is relatively small, as shown in Fig. 4.8, the actual structure often has larger thickness and different cross-sectional geometry. Figure 4.9 indicates two energy curves where the test specimens and structure are made from the same material. The shapes and locations of the curves are different because the transition temperatures and fracture behavior must be different for the test sample and the structure. The result is, then, that the CVN "upper shelf" signifies ductile behavior for all points on the curve beyond the dotted line (Fig. 4.9) as temperature increases. However, the characteristic curve for the structure proper shows that the transition point is reached at a temperature higher than that for the specimen. This difference can lead to some confusion because design selection based on the CVN transition point alone suggests that the structure may respond in a brittle manner.

To overcome some of the deficiencies in CVN tests, such as the failure of the CVN technique to provide an invariant characterization of the true transition temperature for steels, the DWT and DT methods have been developed. Some of the features of this methodology, and the related developments, merit further attention, with a specific emphasis on practical aspects of this area of fracture mechanics.

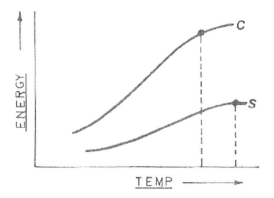

FIGURE 4.9 Potential response of CVN specimen and structure (C, CVN specimen; S, structure).

The basic principle of the DWT technique is best illustrated by the simple sketch in Fig. 4.10. The brittle weld bead has a saw cut across the bead at the center of the specimen plate. The saw cut models a thumbnail flaw with a very sharp crack tip. This model localizes the fracture of the weld, which is stressed to the yield level due to a predetermined dynamic load caused by a dropping weight. The deformation is limited to about 3° (maximum) and the tests are made at various temperatures on the premise that at higher temperatures, the specimen bends without cracking. At lower temperatures, however, crack propagation usually occurs. In order to pinpoint the NDT temperature as the highest temperature of nil-ductility break, we normally look for a flat break area. This type of a failure indicates the initiation of the fracture due to a small crack before a significant plastic deformation takes place. Between the brittle (low shelf) and ductile (high shelf) behavior of the DWT specimen, there is normally a sharp increase in fracture toughness in low-strength structural steel. An example of such a characteristic is given in Fig. 4.11.

According to Pellini,[7] the accuracy of predicting the reproducibility of experimental results of NDT is not better than ± 10°F. This is still quite good considering normal temperature variations and the shape of the transition curves. In a well-defined transition curve — that is, where the curve leaves the lower shelf (brittle region) and the line indicates a sharp rise in fracture toughness — NDT can be well defined. This is not the case, say, in Fig. 4.6 or even in Fig. 4.11. In most cases we look for a definite toe in the transition curve. An example of a temperature transition curve with a toe-type bend in the curve can be taken from impact tests of standard CVN specimens for a commonly used structural steel such as A36.[6] The approximate shape of this curve is shown in Fig. 4.12.

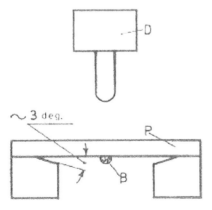

FIGURE 4.10 Drop-weight method (*D*, drop weight; *P*, plate test piece; *B*, brittle weld bead).

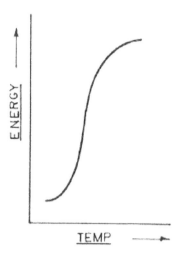

FIGURE 4.11   Typical transition temperature curve for low-strength steel.

The DWT test was established as a standard method for determining the location of the NDT for structural steels having yield strengths less than 140 ksi and a transition capability from brittle to ductile behavior. The NDT was defined as the highest temperature at which a standard specimen fails in a brittle mode of fracture. In other words, we expect the NDT to be reached at the instant of initiation of a small crack. The DWT methodology was preceded by a series of special tests at the NRL utilizing brittle weld beads deposited on

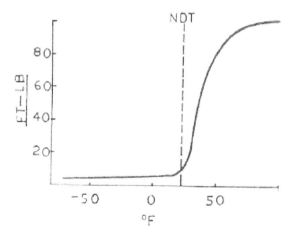

FIGURE 4.12   Approximate CVN impact curve for A36 steel.

plate specimens. These test pieces were then subjected to an explosive-type of loading. The brittle weld beads during the process of this loading provided natural sharpness of small cracks that propagated from the center to the edges of the plate specimens. The objective of the entire series of such tests[7] was to study the effects of temperature on the initiation and propagation of the fractures in steel plates during early ship failures. The test results clarified the role of a dynamically loaded small crack during the elastic fracture at a critical point on the temperature scale consistent with the definition of NDT. At much higher temperatures the brittle cracks disappeared and the plastic overload resulted in bulging phenomena, indicating a significant increase of fracture toughness. The tests have clearly shown why catastrophic ship fractures during the 1940s occurred only at winter temperatures. A simple sketch of the basic configuration of the NRL test is presented in Fig. 4.13.

The next step in the development of advanced methods for testing of fracture toughness brings us to 1964, when it became apparent that a new test of simple characteristics was needed to investigate the properties of steels that exhibit "low shelf ductility." This category includes high- and ultra-high-strength steels and those of intermediate strength that have a distinct direction of weakness (anisotropy of fracture). The new test was named the dynamic tear, or DT for short. The smallest standard test specimen for this purpose is shown in Fig. 4.14. The depth of the notch in a standard case is 0.5 in. (12 mm), introduced by an electron beam weld, by fatigue, or by slitting and sharpening a deep notch using a pressed knife edge. The DT specimens are tested in a pendulum device, utilizing a technique similar to that for CVN specimens where the upswing of the pendulum after the fracture indicates the energy absorbed.

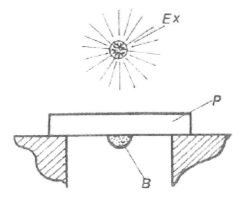

FIGURE 4.13 Explosion loading method (*Ex*, explosion force; *P*, plate specimen; *B*, brittle weld bead).

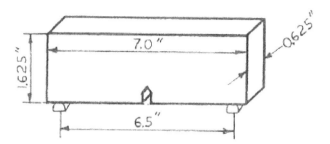

FIGURE 4.14 Standard DT specimen.

DETERMINATION OF LEFM PARAMETERS

The fracture mechanics approach is quite different from the transition temperature philosophy as described in the preceding section. In the LEFM approach the designer is given a material property to be used in establishing the maximum allowable crack size in the structure. The crack size then becomes a design parameter, as indicated in a number of numerical examples featured in Chapter 3. This rational approach may be more difficult to execute and the main objective of this section is to review the test methodology for determination of the fracture toughness LEFM parameters. It is important that the material used for K_{Ic} specimens is the same as that selected for the structural design. Also, the overall size of the specimen depends on the expected level of fracture toughness and the yield strength of the material. The test must reflect the plane-strain conditions, with respect to both the thickness b and the crack length a of the specimen, by satisfying Eq. (4.1). In other words, the dimensions b and a for the K_{Ic} test must be greater than 2.5 $(K_{Ic}/\sigma_y)^2$. When K_{Ic} is higher and the yield strength σ_y is, at the same time, lower, the required specimen thickness must be larger. The overall dimensions of the compact tension (CT) specimen for various measurement capacities are specified by the ASTM standards.[5] This specimen is of the wedge opening type of configuration referred to in Fig. 3.26. The tension loading is applied through a pin-and-yoke combination shown schematically in Fig. 4.15. All specimens in the K_{Ic} tests are notched to provide a sharp crack tip response. Specimen preparation, precracking, loading fixtures, instrumentation, and data reporting are defined by the ASTM methodology. K_{Ic} values are obtained by calculation from the records of loads and deflections.

　　Various test fixtures, recommended by the ASTM standards, have been developed to minimize friction and to assure good alignment of the interacting components. One of the key requirements is the accurate measurement of the relative displacements of two points located symmetrically on both sides of the crack plane. Since load displacement records must be reliable for obtaining

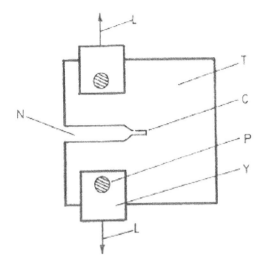

FIGURE 4.15 Loading method for CT specimen (L, load; T, test specimen; C, precracking; P, pin; Y, yoke; N, notch).

the credible value of K_{Ic}, a special gauge had to be developed for this purpose. It is known as the "double-cantilever clip-in displacement gauge." A typical continuous record is shown in Fig. 4.16 to indicate the approximate shape of the curve.[6] This shape, described as a "roundhouse" curve, represents relatively

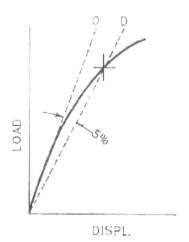

FIGURE 4.16 Load displacement curve for K_{Ic} test (O, original slope, D, displaced slope line).

tougher structural materials rather than a typical plane-strain behavior. A more likely shape for the plane-strain condition is given in Fig. 4.17.

It is not the first time in the field of fracture mechanics that we face an interesting paradox, particularly in the case of selecting the test specimen for finding the plane-strain fracture toughness K_{Ic}. The story here goes as follows. In order to determine K_{Ic}, it is necessary to specify the critical specimen dimensions consistent with a known value of K_{Ic}, Eq. (4.1). The trick is to guess a conservative value of K_{Ic} on the basis of experience with similar materials, and to check the validity of the original assumptions after the test. The second time around, we can then design the specimen more economically.

In concentrating on finding the lower limiting value of the plane-strain fracture toughness, the following requirements should be kept in mind:

- The test must be valid.
- The test specimen should adhere to the industry-wide standard.
- The test specimen should be easy to make and be economical.
- The testing method should be direct and proven to yield reproducible results.

The length of the crack a consists of two elements. One is the notch (with a taper section) and the other is a sharp length of precracking created by cyclic loading and unloading of the specimen. This technique is acceptable provided the effect of residual stresses can be avoided, which can be accomplished if we

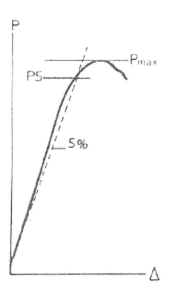

FIGURE 4.17 General type of test record.

assure that the maximum stress intensity level, during the cyclic loading, stays significantly below the anticipated level of K_{Ic}. It is generally recognized that the cyclic loading method produces a very sharp crack as a requirement for a valid K_{Ic} test. The stress intensity for a compact tension specimen can be verified with the aid of Eqs. (3.32) and (3.34), with key dimensions a, W, and b as illustrated in Fig. 3.26.

If the clip gauge is placed at the mouth of the notch, the crack opening can be measured and the corresponding load can be recorded to develop a load displacement curve with a longer straight line portion than that shown in Fig. 4.16. The usual testing procedure is to increase the tensile load (Fig. 4.15) until fast fracture is initiated. This can be noted on the load displacement chart as a gross nonlinearity. It is customary here to be concerned with the region between the two dotted lines, one of which intersects the load displacement curve at a point shown in Fig. 4.16. Over many years of research various criteria to define the load consistent with the level of K_{Ic} were considered. This is not surprising because the load displacement curves obtained from the K_{Ic} tests were not perfectly elastic and often showed various degrees of nonlinearity. After a great deal of experimentation,[5,21] a 5% offset (Fig. 4.16) was selected to establish K_{Ic} as the critical stress intensity factor at which the original crack length increased by about 2%. This decision is somewhat arbitrary and can be compared to the 0.2% offset for yield strength of materials that do not have a well-defined yield point on the stress–strain diagram. Be that as it may, the 5% criterion allows us to calculate the stress intensity factor, Eqs. (3.32) and (3.34), which may be accepted as a valid plane-strain fracture toughness, as long as the size criteria of Eq. (4.1) are observed. In other words, if we calculate the term $2.5(K_{Ic}/\sigma_y)^2$, where K_{Ic} follows from Eq. (3.34) and where σ_y denotes the yield strength of the material, then the result, in English units, can be expressed in inches. When this quantity is smaller than both the thickness and the crack length of our test specimen, then the K_I parameter is assumed to be equal to K_{Ic}. Note that in this calculation, K_I is expressed in ksi (in.)$^{1/2}$, and σ_y is given in ksi. In the event that $2.5(K_I/\sigma_y)^2$ is larger than our specimen dimensions b and a, then it is necessary to use a larger specimen to determine the valid K_{Ic}.[6]

The load defined as P_5 in Fig. 4.17 is determined at the intersection of the 5% line (slope reduction) and the test curve. The maximum load on the chart is denoted by P_{max}. For the K_{Ic} test to be valid, the ratio (P_{max}/P_5) should be lower than 1.10. For ratios higher than 1.10 the metals are too tough to exhibit plane-strain response.[6] This condition of toughness is consistent with the general yielding type of behavior, and the shape of the load displacement curve is closer to that given in Fig. 4.16.

The essential dimensions of a compact tension (CT) specimen for measuring plane-strain fracture toughness are given in Fig. 4.18. It should be noted that the active length of the crack a is taken from the line of loading passing through

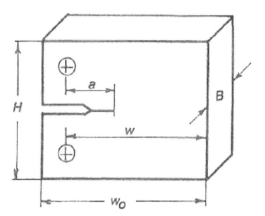

FIGURE 4.18 Compact tension specimen.

the hole centers, although the entire length of the notch extends to the edge of the plate. The dimension a resembles the distance in a bending moment term (load times the distance) in a conventional stress analysis. The lowercase symbol w is reserved in this book for the so-called width of the specimen plate, and the symbol for thickness of the plate B is almost exclusively used throughout the literature of fracture mechanics.

The second most common specimen used in the studies of K_{Ic} is known as a single-edge cracked bend specimen, subjected to three-point bending, as shown in Fig. 4.19. This specimen design became a part of the ASTM standard[5] that

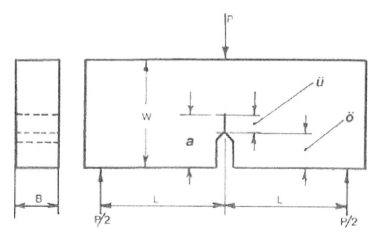

FIGURE 4.19 Bend specimen ($\ddot{u}$ precracked length; $\ddot{o}$, machined notch).

was developed after an extensive series of tests to prove that the test procedures using the compact tension and three-point bend specimens were, indeed, reproducible. To date, experience suggests that the K_{Ic} values from various sources can agree within 15%.[6]

The sizes of the essential dimensions a and B are governed by Eqs. (4.1) and (4.2) under plane-strain conditions. As far as the overall size of the test specimen is concerned, the thickness B, height h, and the overall width w_0 are well proportioned and represent quite a range of dimensions. Such dimensions are illustrated briefly in Fig. 4.18 and Table 4.1.

Design Problem 4.4

The test record for a 170 ksi yield strength steel has been obtained from the compact tension specimens indicating a P_5 load of 52,000 lb and a maximum load of 55,160 lb as P_{max}. Calculate the plane-strain fracture toughness K_{Ic} assuming the following specimen dimensions:

> Width of the specimen: $w = 4$ in.
> Thickness: $B = 1.8$ in.
> Crack size: $a = 1.5$ in.

Check all the requirements for the valid K_{Ic}. The appropriate configuration is shown in Fig. 4.18.

Solution

The dimensionless crack parameter is

$$a/w = 1.5/4$$
$$= 0.375$$

TABLE 4.1 Overall Dimensions of Test Specimens for CT Configuration.

Type	Thickness, B (in.)	Height, H (in.)	Width, w_0 (in.)
1T-CT	1	2.4	2.5
2T-CT	2	4.8	5.0
3T-CT	3	7.2	7.5
4T-CT	4	9.6	10.0
6T-CT	6	14.4	15.0
8T-CT	8	19.2	20.0
10T-CT	10	24.0	25.0
12T-CT	12	28.8	30.0

From Eq. (3.32), related to compact tension specimens:

$$f\left(\frac{a}{w}\right)_c = 29.6(0.375)^{0.5} - 185.5(0.375)^{1.5} + 655.7(0.375)^{2.5}$$
$$- 1017(0.375)^{3.5} + 639(0.375)^{4.5}$$
$$= 18.13 - 42.60 + 56.47 - 32.84 + 7.74$$
$$= 6.9$$

Note that for a ratio of 0.375, the chart in Fig. 3.28 indicates a reasonably close value of $f(a/w)_c$. The charts are helpful in verifying the order of magnitude of the result calculated from the polynomial functions.

The next step involves Eq. (3.34):

$$K_I = \frac{52,000 \times 6.9}{1.8 \times (4)^{1/2}}$$
$$= 100 \, \text{ksi (in.)}^{1/2}$$

The validity conditions can now be set up as follows:

$$2.5\left(\frac{100}{170}\right)^2 = 0.87 \qquad \text{from Eq. (4.1)}$$

$1.8 > 0.87$ and $1.5 > 0.87$

$$\frac{1}{6\pi}\left(\frac{100}{170}\right)^2 = 0.018 \qquad \text{from Eq. (4.2)}$$

$1.8/0.018 = 100$. Required ratio is 50.

$$P_{max}/P_5 = 55,160/52,000 = 1.06$$

Here $1.06 < 1.10$, and therefore all four conditions of K_{Ic} validity are satisfied.

$$K_I = K_{Ic} = 100 \, \text{ksi (in.)}^{1/2}$$
$$= 109.9 \, \text{MPa (m)}^{1/2}$$

Design Problem 4.5

The tests to be run on aluminum CT specimens have the following initial specifications:

$\sigma_y = 55 \, \text{ksi}$
$K_{Ic} = 20 \, \text{ksi (in.)}^{1/2}$ (preliminary estimate)
$B = 1.2 \, \text{in.}$
$w = 2.0 \, \text{in.}$

Assuming the test load P_5 (see Fig. 4.17) to be of the order of 3000 lb, select the approximate crack length a to be consistent with the initial specifications. Check the validity of the plane-strain conditions for this experiment.

Solution

From Eq. (3.34), using $P = P_5$, and $K_I = K_{Ic}$

$$f\left(\frac{a}{w}\right)_c = \frac{K_{Ic}B(w)^{1/2}}{P_5}$$

$$= \frac{20 \times 1.2 \times (2)^{1/2}}{3}$$

$$= 11.3$$

From the chart in Fig. 3.28, the rough estimate gives

$$a/w = 0.55$$

From this

$$a = 0.55w$$
$$= 0.55 \times 2.0$$
$$= 1.1 \text{ in.}$$

From Eq. (4.1):

$$2.5(20/55)^2 = 0.33$$

$1.2 > 0.33$ and $1.1 > 0.33$
     From Eq. (4.2):

$$\frac{1}{6\pi}\left(\frac{20}{55}\right)^2 = 0.007$$

$1.2/0.007 = 171$ and $171 > 50$
Hence the plane-strain conditions are satisfied. An additional check can be obtained from Eq. (3.32).

$$f\left(\frac{a}{w}\right)_c = 21.95 - 75.66 + 147.10 - 125.48 + 43.36 = 11.27$$

This number shows a rather fortuitous comparison between the exact calculation and a chart readout. However, the case only strengthens the importance of using charts as a quick check on a more complicated numerical exercise.

Design Problem 4.6

The three-point bend method (Fig. 4.19) is proposed for the determination of the valid K_{Ic} parameter, using the following specimen dimensions:

$B = 1$ in.
$w = 2$ in.
$a = 1$ in.
$L = 4.5$ in.

Assuming a test load P of 6000 lb, and an A517 structural steel with yield strength of 110 ksi, calculate the valid fracture toughness to resist crack propagation through the test specimen. Compare the LEFM result with the conventional stress calculation using a minimum stress concentration factor k of 3.

Solution

The dimensionless crack size is

$a/w = 1/2$
$\quad\quad = 0.5$

From Eq. (3.22)

$$K_I = \frac{3 \times 6000 \times 4.5}{1 \times (2)^{1.5}} [1.93(0.5)^{0.5} - 3.07(0.5)^{1.5} + 14.53(0.5)^{2.5}$$
$$\quad - 25.11(0.5)^{3.5} + 25.80(0.5)^{4.5}]$$
$$\quad = 28{,}642(1.365 - 1.085 + 2.569 - 2.219 + 1.140)$$
$$\quad = 50.7 \text{ ksi (in.)}^{1/2}$$
$$\quad = 55.7 \text{ MPa (m)}^{1/2}$$

The validity check is from Eq. (4.1):

$$2.5 \left(\frac{50.7}{110} \right)^2 = 0.53$$

$1 > 0.53$ (for b and a terms) and from Eq. (4.2):

$$\frac{1}{6\pi} \left(\frac{50.7}{110} \right)^2 = 0.011$$

$1/0.011 = 91$ ($91 > 50$ required). The stress in the net section (ligament) follows from

$$\sigma = \frac{M}{Z}$$

where

$$M = PL/2$$

and

$$Z = B(w - a)^2/6$$

Combining gives

$$\sigma = \frac{3kPL}{B(w - a)^2}$$

where $k = 3$ (minimum stress concentration factor). Hence, substituting gives

$$\sigma = \frac{3 \times 3 \times 6000 \times 4.5}{1 \times 1}$$

$$= 243 \, \text{ksi}$$

The bending stress, even with a low value of k at the tip of the crack, may be unacceptable, pointing out in this case that conventional stress analysis is not a realistic tool for the evaluation of the effect of sharp notches and cracks. This comment is also appropriate for the Inglis formula [2] and its limitations in view of debates of the "rounding-off" mechanism for local plastic phenomena.

It was already mentioned in connection with Fig. 4.7 that fracture toughness in general is sensitive to temperature and the rate of loading, while the K_{Id} parameer has a special role. K_{Id} is a critical stress intensity factor for impact loading under plane-strain conditions of maximum constraint. Some test results of K_{Id} for structural steel are available.[6,7] The effect of the rate of loading on the plane-strain fracture toughness is shown approximately in Fig. 4.20. Structural materials such as aluminum, titanium, or a steel having yield strength higher than 150 ksi do not show loading rate effects.

The test specimen currently used for K_{Id} measurements is illustrated in Fig. 4.21. It is a three-point bend specimen loaded by a free-falling weight. The methodology and specimen configuration resemble the K_{Ic} slow-bend arrangement. The dynamic test procedure for the K_{Id} setup in line with Fig. 4.21 was developed first by Shoemaker and Rolfe.[22] The working spans (long and short) were employed during the development work, including different cushion materials and drop heights. Currently the test method is probably restricted to $K_{Id}/\sigma_{yd} = 0.7$. The basic design equation is

$$K_{Id} = \frac{6M}{Bw^2}(a)^{1/2}f\left(\frac{a}{w}\right) \tag{4.6}$$

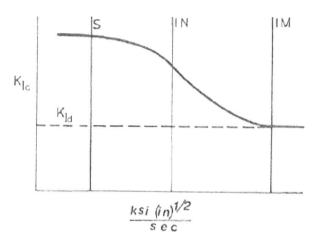

FIGURE 4.20 Effect of loading rate (S, slow; IN, intermediate; IM, impact).

where M = applied bending moment (lb-in), a = length of crack (in.), and B = specimen thickness (in.), and w = specimen width (in.).

The correction factors $f(a/w)$, featured in Eq. (4.6), are given in graphical form for the long and short spans in Fig. 4.22. The bending moment is normally obtained from the elementary strength of materials by calculating the equivalent static load at the time of crack initiation. The strain gauges are used here to detect crack initiation as well as to measure the nominal elastic stress. The strain gauges were also used in developing another test method[23] and comparing the new approach with the existing ASTM procedures for testing K_{Ic}. Other standard procedures, available from ASTM, relate to a more specialized area of fracture

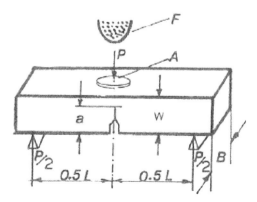

FIGURE 4.21 Dynamic bend test (F, free-falling tup; A, aluminum cushion).

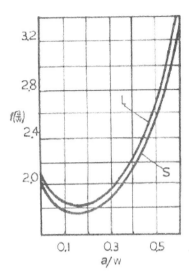

FIGURE 4.22 Correction factors for dynamic test (L, long span; S, short span).

mechanics involving elastic–plastic behavior. It should also be noted that some of the theoretical investigations, test methods, and specimen configurations are similar to those found in the ASTM standards.

In dealing with the effects of temperature on the K_{Ic} and K_{Id} parameters (Fig. 4.7), certain materials did not show any evidence of a transition-type behavior. It is again of interest to note that the same materials (aluminum, titanium, and very high-strength steel) are not affected by the rate of loading. The term "very high-strength" in this case signifies yield strength above 150 ksi for steel.

As stated previously, the maximum constraint is only possible when the plate thickness is sufficiently large in line with Eq. (4.1). Essentially, the same rule applies to dynamic fracture toughness, so that

$$B \geq 2.5 \left(\frac{K_{Id}}{\sigma_{yd}} \right)^2 \tag{4.7}$$

where σ_{yd} = dynamic yield strength in tension. This mechanical property is obtained in "rapid" tensile tests, experiencing high rates of loading. Based on rather difficult experiments on structural steel, the following practical formula may be recommended:[6]

$$\sigma_{yd} = \sigma_y + 30 \tag{4.8}$$

Equation (4.8) is expressed in ksi. It can be argued that the numerical term in this equation should vary gradually from 30 to 0 when the upper limit of σ_{yd} is

140 ksi.[24] Because of the relative difficulties in measuring dynamic parameters, several investigators[25,26] attempted to define a relationship between K_{Id} and σ_{yd}. This work resulted in a simple and practical rule for preliminary design estimates:

$$K_{Id} = 0.6\sigma_{yd} \tag{4.9}$$

For lower strength steels, Eq. (4.9) can be restated as

$$K_{Id} = 0.6\sigma_y + 18 \tag{4.10}$$

In Eq. (4.10) K_{Id} is expressed in ksi (in.)$^{1/2}$ and σ_y has the dimension of ksi. In SI units, we have MPa (m)$^{1/2}$ and MPa, respectively. In structural steels, K_{Id} is generally lower than K_{Ic} at a given temperature.

Design Problem 4.7

A dynamic test rig using a short-span test piece had the following specimen dimensions:

$a = 1.2$ in.
$B = 1.0$ in.
$W = 3.0$ in.

The maximum elastic stress in bending was recorded as 58 ksi. Estimate valid dynamic fracture toughness if the static yield strength of the material is 80 ksi.

Solution

From conventional stress analysis of the net cross-section (Fig. 4.21) $\sigma = M/Z$

$$Z = \frac{B(w - a)^2}{6}$$

then

$$M = \sigma Z$$

$$= \frac{\sigma B(w - a)^2}{6}$$

Substituting in Eq. (4.6)

$$K_{Id} = \frac{6\sigma B(w - a)^2}{6Bw^2}(a)^{1/2} f\left(\frac{a}{w}\right)$$

$$K_{Id} = \frac{\sigma(w - a)^2}{w^2}(a)^{1/2} f\left(\frac{a}{w}\right)$$

Since $(a/w) = 1.2/3 = 0.4$ from Fig. 4.22, using the short-span curve, we obtain

$$f\left(\frac{a}{w}\right) = 2.08$$

Hence, substituting gives

$$K_{Id} = \frac{58(3 - 1.2)^2}{3^2} \times (1.2)^{1/2} \times 2.08$$

$$K_{Id} = 47.6\,\text{ksi (in.)}^{1/2}$$

From Eq. (4.8)

$$\sigma_{yd} = 80 + 30 = 110\,\text{ksi}$$

Hence, from Eq. (4.7)

$$2.5\left(\frac{47.6}{110}\right)^2 = 0.47$$

Both a and b are larger than 0.47. Also from Eq. (4.2)

$$\frac{1}{6\pi}\left(\frac{47.6}{110}\right)^2 = 0.01$$

so that

$$1/0.01 = 100 \qquad \text{and} \qquad 100 > 50$$

All requirements for the validity of K_{Id} are satisfied, and

$$K_{Id} = 47.6\,\text{ksi (in.)}^{1/2}$$

$$= 52.3\ \text{MPa (m)}^{1/2}$$

Although it is clear from a number of design problems that for many years the procedures for design of metal structures were inadequate for assuring reliability in the presence of cracks, it is also clear that the best approach to problem solution is the proper blend of conventional stress analysis and specialized fields of fracture mechanics. However, to achieve the proper blend is not a simple matter because the science of fracture mechanics is still evolving and because this topic is a mixture of scientific fundamentals, design criteria, analytical procedures, and certification requirements. Although in some instances the theoretical effort has evolved into a mature engineering technology in spite of the economic constraints, the process still requires more effort from engineering and regulatory circles for development and compliance with the standard methodologies. The educational part of this process must be relegated to the field of practice and generalists. As we dare to cross deeper into the territory of specialists, the numerical illustrations of the design-related matters will, it is hoped, be stripped of some murky details.

CORRELATION TECHNIQUES

The material on practice of fracture mechanics presented so far has several references to quantitative features of stress intensity and fracture toughness, transition temperature characteristics of structural materials, and some of the correlations involving fracture behavior under test conditions. It appears that even today, after general acceptance of the LEFM tools, the adoption of ASTM plane-strain methodology, with the rigor of a "valid" fracture toughness, is somewhat limited because of the unfavorable economics of laboratory experiments. When we add to this problem the difficulty of applying the mathematics of the stress crack relationship to real structures, the correlations between the stress intensity and fracture energy become absolutely essential in engineering design. It is fortunate for designers that some of the conventional structural steels can be analyzed more easily because of the sharp and well-defined temperature transitions. As mentioned before, the titanium and aluminum alloys show no temperature transition. The ultimate goal is a complete integration of the mechanical and metallurgical elements into a single analytical model.[10] In the meantime we have to use practical engineering test methods and quantitative evaluation of experiments to have a valid number representing a given fracture toughness parameter.

Various methods of measuring fracture toughness have been reviewed in a previous section, and the correlation of dynamic toughness with the dynamic yield is given by Eqs. (4.9) and (4.10). The transition in performance is tied to the NDT temperature. Below the NDT, as experiments indicate, the fracture toughness is very low and even a small crack can initiate a failure. Here rapid loading or static loading plus small "pop-ins" caused, say, by arc strikes in the brittle weld region, can cause a fracture. The term "pop-in" represents a sudden crack growth that is sometimes used in establishing the K_{Ic} range of values for a structural steel.[6] The "pop-in" condition at the crack tip on a "micro" scale is supposed to be caused by a sudden separation of a few metal grains smaller than 0.01 in. The more likely mechanism, however, would be that of a sudden separation of a stiffener or a gusset to affect the local stress distribution on a "macro" rather than a "micro" scale. The test specimen in NDT evaluation should have a minimum section of 5/8 in. to assure sufficient constraint for a plane-strain fracture. For sections smaller than 5/8 in., we can expect an elastic–plastic condition of fracture. Specimen sizes of 5/8 in. and 1 in. are used in the dynamic tear (DT) test method.

The DT test method has also been extended to very thick sections, such as in experiments with A533 steel for the construction of reactor pressure vessels, in order to establish the validity of temperature transition for materials up to 12 in. thick. This work represented the ultimate link between LEFM and the transition temperature approach. It was established that the maximum shift of the midpoint in the temperature transition was between 60 and 80°F toward the higher temperature. The dynamic plane-strain limit for A533 steel was obtained by running 5/8 in. DT tests and adding the temperature increment of 60–80°F. The reader may recall a section on testing in this chapter concerned with the two basic categories of toughness evaluation, emphasizing the temperature transition and LEFM.

The illustration in Fig. 4.23 highlights the patterns of initiation, propagation, and arrest of cracks based on the experience with the ship plate steels of World War II. The energy here is expressed in CVN (Charpy V-notch) test values, indicating 10–20 ft-lb over a narrow temperature range, representing a sharp rise in fracture toughness. Although this illustration is based on rather early data, it shows a clear correlation between the energy and crack behavior. However, Lange[10] cautions against extending the CVN data (Fig. 4.23) to steels with different metallurgical characteristics such as those found in quenched and tempered (Q&T) steels.[27] The reason for this is that Q&T steels can shift the CVN–temperature correlation toward the lower temperature by as much as 120°F.

Useful correlation between the CVN and DT energy levels is given in Fig. 4.24. It represents 1 in. thick plates starting with a medium-yield steel such as HY80 and progressing through 4330 Vanadium Mod., 4340, 4140, D6AC, and HP150 to maraging and other steels of similar strength. The energy values given here correspond to the upper shelf response.[28] In general, it is

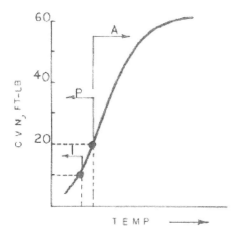

FIGURE 4.23 CVN energy based on ship plate cracks (I, initiation; P, propagation; A, arrest).

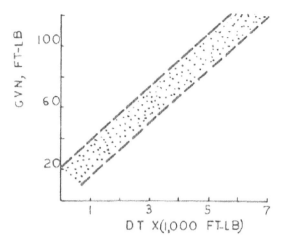

FIGURE 4.24 Correlation of CVN and DT shelf energy.

easier to conduct CVN tests because of good availability of the test equipment and better familiarity with the test procedure. However, when the structural materials do not show a significant dependence on temperature within a given fracture toughness corresponding to K_{Id}/σ_{yd} ratios of about 0.3–0.6, the performance criteria should better be based on dynamic tear energy (DT) values obtained with 5/8 in. or 1.0 in. DT test specimens.

Practical design problems involving fracture mechanics are seldom well defined. Crack size and crack geometry are often assumed rather than verified dimensionally, and the calculation of nominal stresses may or may not be easy to make. Moreover, the third basic parameter such as K_{Ic} may be open to question, not only because of the nature of the material, but also due to the actual degree of constraint present. One way, of course, out of this dilemma is to adopt a conservative design philosophy, such as to select a priori the numerical value of plane-strain fracture toughness. It is hoped that we can find the appropriate level of K_{Ic} from a materials handbook or a research report in order not to have to resort to a K_{Ic} test, the equipment and skills for which may not be readily available. However, there is an alternative in the form of utilizing a correlation of this LEFM parameter and the basic energy tests such as CVN or DT.

In the first place, it is rather fortunate that both DT and K_{Ic} tests well reflect the process of resisting, in one form or another, the propagation of a fracture. This similarity alone suggests that there must be a close relationship between the two parameters, such as, for instance, that shown in Fig. 4.25. This particular correlation is intended for high-strength steels with 1.0 in. thick specimens used in the K_{Ic} experiments, and it represents a fair degree of coupling between an

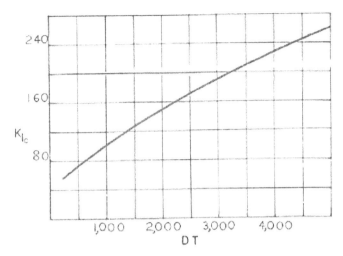

FIGURE 4.25 Correlation of K_{Ic} [ksi (in.)$^{1/2}$] and DT (ft-lb).

engineering test and the LEFM parameter.[10] This development has a highly practical meaning to designers because the DT energy data can be translated into K_{Ic} values under plane-strain conditions. Furthermore, the ratio K_{Ic}/σ_y normalizes the fracture toughness to yield strength of all structural metals and improves our LEFM analytical capability by establishing the design relationships featuring crack sizes, reference stresses, and K_{Ic}/σ_y ratios. With the appropriate corrections this can be a design tool because the ratio represents a level of fracture toughness related to a specific mechanical performance.[10]

In dealing with the various correlations, it is customary to lump together high-strength steels that are designed metallurgically by applying quench and temper (Q&T) or quench and age (Q&A) techniques. The two extremes of microfracture are recognized as cleavage (brittle mode) and ductile (dull appearance). However, there is a transition region where complex alloying and rapid cooling can cause a variety of microstructures with at least two major modes involving boundary separation and quasicleavage, as shown in Fig. 4.26. Although these modes still appear as brittle, they correspond to a fracture toughness of pearlitic steels above that associated with pure cleavage. Figure 4.26 provides a quick assessment of microstructure between a well-known, fracture-tough steel of medium range, such as HY-80, and very high-yield steels. Finer points of this comparison, including nonferrous metals, are described in practical terms by Lange.[10] All such information is backed by empirical studies of extreme importance to engineers and designers.

Following a number of investigations in fracture mechanics from the 1940s through the 1960s, Barsom and Rolfe[29] proposed establishment of empirical

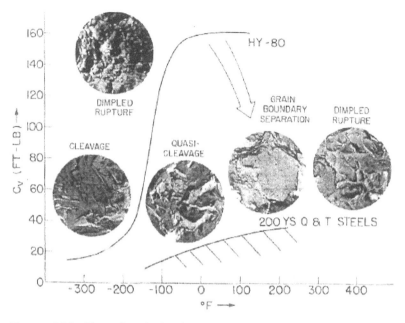

FIGURE 4.26 Examples of microfracture of high-strength steels, related to Charpy V-notch and temperature (from Ref. 10) (4000×).

correlations between K_{Ic} and CVN test results, because the effects of temperature and rate of loading on CVN and K_{Ic} test values were essentially the same. It also seems remarkable that, with the use of the J-integral approach, there was a possible theoretical basis for justifying an empirical correlation.[30,31] Hence the concept of the J-integral represents an extension of LEFM into plane-stress fracture mechanics. It is formally defined as a path-independent integral that is an average measure of the elastic–plastic stress–strain field ahead of a given crack.

The development of the K_{Ic}–CVN empirical correlation involved 11 steels having a range of yield strength from 110 to 246 ksi. The plane-strain fracture toughness for these steels varied between 87 and 246 ksi (in.)$^{1/2}$. The relevant CVN values covered the range 16–89 ft-lb. The result known today as the Rolfe-Novak-Barsom correlation is

$$\left(\frac{K_{Ic}}{\sigma_y}\right)^2 = 5\left(\frac{CVN}{\sigma_y} - 0.05\right) \tag{4.11}$$

This formula applies to upper shelf values on the premise that the effects of loading rate and notch sharpness are not critical and should not invalidate the process of correlation. The variables in Eq. (4.11) are defined as

K_{Ic} = plane-strain fracture toughness at slow loading rates, ksi (in.)$^{1/2}$
CVN = standard Charpy V-notch test values at upper shelf, ft-lb
σ_y = 0.2% offset yield strength at upper shelf temperature, ksi

Normally, Eq. (4.11) is limited to steels having yield strengths higher than 100 ksi. However, Barsom and Rolfe[6] suggest that Eq. (4.11) may also be applicable to yield strength lower than 100 ksi provided the dynamic yield strength is used instead. This gives

$$\left(\frac{K_{Ic}}{\sigma_{yd}}\right)^2 = 5\left(\frac{CVN}{\sigma_{yd}} - 0.05\right) \tag{4.12}$$

Here σ_{yd} denotes the yield strength (in ksi) that can be estimated from Eq. (4.8). Both formulas, Eqs. (4.11) and (4.12), are workable in spite of the fact that K_{Ic} is a static test and CVN represents impact environment.

Design Problem 4.8

A low-alloy quenched and tempered (Q&T) steel, which has a yield strength of 110 ksi, is being used as a tension member in constructing a ship. Estimate the plane-strain fracture toughness K_{Ic} if the CVN energy is 30 ft-lb at the upper shelf of the Charpy V-notch test curve. Show graphically the variation of K_{Ic} with the CVN energy between 10 and 50 ft-lb.

Solution

Rearrange Eq. (4.12) to give

$$\begin{aligned}
K_{Ic} &= (5\,\text{CVN}\,\sigma_y - 0.25\sigma_y^2)^{1/2} \\
&= (5 \times 30 \times 110 - 0.25 \times 110^2)^{1/2} \\
&= 116\,\text{ksi (in.)}^{1/2} \\
&= 127\,\text{MPa (m)}^{1/2}
\end{aligned}$$

Calculate variation of K_{Ic} with CVN using the following expression

$$\begin{aligned}
K_{Ic} &= (5 \times 110\,\text{CVN} - 0.25 \times 110 \times 110)^{1/2} \\
&= (550\,\text{CVN} - 3025)^{1/2}
\end{aligned}$$

The result is shown in Fig. 4.27.

Over many years of development and use of the DT and CVN test techniques for characterizing transition temperatures of Navy and Department of Energy structural materials,[32,33] there was a definite need for correlating the test data. In spite of the superiority of the DT approach compared to the process of developing and analyzing the transition temperature curves, the

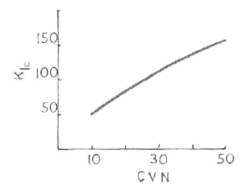

FIGURE 4.27 Variation of K_{Ic} [ksi (in.)$^{1/2}$] with CVN (ft-lb), Design Problem 4.7.

CVN technique is still popular in industry. Experience has shown so far that the following approximate correlation formula can be used in dealing with structural steels.

$$CVN = 0.12DT + 15 \qquad (4.13)$$

Design Problem 4.9

High metallurgical quality steel used in design and fabrication of nuclear submarine components has a nominal yield strength of 90 ksi. The conservative level of DT energy for this material at room temperature is expected to be on the order of 700 ft-lb. Estimate the corresponding plane-strain fracture toughness.

Solution

Rearrange Eq. (4.13) to give

$$K_{Ic} = (5CVN \, \sigma_{yd} - 0.25\sigma_y^2)^{1/2}$$

From Eq. (4.8)

$$\begin{aligned} \sigma_{yd} &= \sigma_y + 30 \\ &= 90 + 30 \\ &= 120\,ksi \end{aligned}$$

From Eq. (4.13)

$$\begin{aligned} CVN &= 0.12\,DT + 15 \\ &= 0.12 \times 700 + 15 \\ &= 99\,ft\text{-}lb \end{aligned}$$

Hence

$$K_{Ic} = (5 \times 99 \times 120 - 0.25 \times 120^2)^{1/2}$$
$$= 236 \, \text{ksi (in.)}^{1/2}$$
$$= 260 \, \text{MPa (m)}^{1/2}$$

Special emphasis on the correlation techniques discussed in this section is a matter of pragmatic necessity. The key LEFM parameters, such as K_{Ic}, are seldom available to the designer because of technical and economical constraints. It is therefore of utmost importance to utilize all the correlation tools available. However, it is also important to keep in mind various limitations and subtle points of the K_{Ic} selection process in relation to the actual structure that we try to analyze and design. These topics alone can fill volumes,[7,20] and the purpose of this book is only to touch on some of the more rudimentary elements of LEFM and experiments related to the task of design.

SYMBOLS

a	Length (or depth) of crack (flaw), in. (mm)
B	Thickness of plate, in. (mm)
$f(a/w)$	Correction for dynamic bend test
$f(a/w)_C$	Polynomial for compact specimen
H	Depth of test specimen, in. (mm)
K_c	Plane-stress fracture toughness, ksi (in.)$^{1/2}$ [MPa (m)$^{1/2}$]
K_{Ic}	Plane-strain fracture toughness, ksi (in.)$^{1/2}$ [MPa (m)$^{1/2}$]
K_{Id}	Dynamic fracture toughness, ksi (in.)$^{1/2}$ [MPa (m)$^{1/2}$]
k	Stress concentration factor
L	Length, in. (mm)
M	Bending moment, lb-in. (N-mm)
P	Concentrated load, lb (N)
P_5	Special load, lb (N)
P_{max}	Maximum test load, lb (N)
r_y	Plastic zone radius, in. (mm)
T_s	Temperature shift, °F
w	Half- (or full) width of plate, in. (mm)
w_o	Total width of specimen, in. (mm)
Z	Section modulus, in.^{3} (mm^3)
σ	Applied stress, ksi (MPa)
σ_y	Yield strength, ksi (MPa)
σ_{yd}	Dynamic yield strength, ksi (MPa)
$\sigma_x, \sigma_y, \sigma_z$	Three-dimensional stresses, ksi (MPa)

REFERENCES

1. Griffith, A.A. The phenomena of rupture and flow in solids. Phil. Trans. Royal Society **1920**, *221*, 163–198.
2. Inglis, C.E. Stresses in a plate due to the presence of cracks and sharp corners. Trans. Instit. Naval Architects (*London*) **1913**, *55*, 219–241.
3. Gordon, J.E. *Structures, or Why Things Don't Fall Down*; Plenum Press: New York, 1978.
4. Cottrell, A. *Mechanical Properties of Matter*; John Wiley: New York, 1964.
5. American Society for Testing and Materials. Standard method of test for plane-strain fracture toughness of metallic materials. In *ASTM Annual Standards*, ASTM Designation E-399-83; ASTM: West Conshohocken, PA 1987; Vol. 03.01.
6. Barsom, J.M.; Rolfe, S.T. *Fracture and Fatigue Control in Structures*, 2nd Ed.; Prentice-Hall: Englewood Cliffs, NJ, 1987.
7. Pellini, W.S. *Principles of Structural Integrity Technology*; Department of the Navy, Office of Naval Research: Arlington, VA, 1976.
8. Pellini, W.S. Principles of fracture safe design. Weld. J. (Suppl.) **1971**, March (Part I), 91s–109s; April (Part II), 147s–162s.
9. Irwin, G.R. Analysis of stresses and strains near the end of a crack traversing a plate. J. Appl. Mech. **1957**, *24*, 361–364.
10. Lange, E.A. Fracture toughness of structural metals, NRL Report 7046; Department of the Navy, Naval Research Laboratory: Washington, DC, 1970.
11. Broek, D. *The Practical Use of Fracture Mechanics*; Kluwer Academic Publishers: Dordrecht, 1988.
12. Winne, D.H.; Wundt, B.M. Application of the Griffith–Irwin theory of crack propagation to bursting behavior of discs. ASME Trans. **1958**, *80*.
13. Wessel, E.T. The influence of pre-existing sharp cracks on brittle fracture of a nickel-molybdenum-vanadium forging. ASME Trans. **1960**, *52* (277).
14. Thum, E.E. Recent accidents with large forgings. Met. Progr. ASM **1956**, *69*.
15. Paris, P.C.; Erdogan, F. A critical analysis of crack propagation laws. ASME Trans. J. Basic Eng., **1963**, *85*, 528–534.
16. Johnson, H.H.; Willner, A.M. Moisture and stable crack growth in a high strength steel. Appl. Mater. Res. **1965**, *4*.
17. Barsom, J.M. Relationship between plane-strain ductility and K_{Ic}. ASME Trans. J. Eng. Ind. Ser. B, **1971**, *93*(4).
18. Barsom, J.M.; Sovak, J.F.; Novak, S.R. AISI Project 168—Toughness criteria for structural steels: fracture toughness of A36 steel. Research Laboratory Report 97.021-001(1); Transportation Research Board, National Research Council: Washington DC, 1972.
19. Barsom, J.M.; Sovak, J.F.; Novak, S.R. AISI Project 168—Toughness criteria for structural steels: fracture toughness of A572 steels. Research Laboratory Report 97.021-001(2); Transportation Research Board, National Research Council: Washington DC, 1972.
20. Barsom, J.M. Development of the AASHTO fracture toughness requirements for bridge steels. In *Engineering Fracture Mechanics*; Pergamon Press: London, 1975; Vol. 7.

21. British Standards Institute. Methods of Test for Plane Strain Fracture Toughness of Metallic Materials, BS 5447; BSI: London, 1977.

22. Shoemaker, A.K.; Rolfe, S.T. Static and dynamic low-temperature K_{Ic} behavior of steels. ASME Trans. J. Basic Eng. **1969**, *91*, 512–518.

23. Madison, R.B.; Irwin, G.R. Dynamic K_{Ic} testing of structural steel. J. Structural Div. **1974**, ASCE 100, No. ST7, Proc. Paper 10653.

24. Lange, E.A. Fracture toughness measurements and analysis for steel castings. AFS Trans. **1978**.

25. Irwin, G.R.; Krafft, J.M.; Paris, P.C.; Wells, A.A. Basic aspects of crack growth and fracture. NRL Report 6598: Washington, DC, 1967.

26. Shoemaker, A.K.; Rolfe, S.T. The static and dynamic low-temperature crack toughness performance of seven structural steels. Eng. Fract. Mech. **1971**, *2*, 319–339.

27. Lange, E.A. Significance of Charpy-V Test Parameters as Criteria for Quenched and Tempered Steels, NRL Report 7483, October 1972.

28. American Society for Testing and Materials. *Impact Testing of Metals*, ASTM STP 466; ASTM: Philadelphia, 1970.

29. Barsom, J.M.; Rolfe, S.T. Correlations between K_{Ic} and Charpy V-notch test results in the transition-temperature range. In *Impact Testing of Metals*, ASTM STP 466; American Society for Testing and Materials: Philadelphia, 1970.

30. Begley, J.A.; Landes, J.D. The *J* integral as a fracture criterion, in Stress Analysis and Growth of Cracks, ASTM STP 514; American Society for Testing and Materials: Philadelphia, 1972.

31. Landes, J.D.; Begley, J.A. The effect of specimen geometry on K_{Ic}. In *Stress Analysis and Growth of Cracks*, ASTM STP 514; American Society for Testing and Materials; Philadelphia, 1972.

32. Lange, E.A.; Cooley, L.A. *Fracture Control Plans for Critical Structural Materials Used in Deep-Hole Experiments*, NRL Memorandum Report 2497; Naval Research Laboratory, 1972.

33. Blake, A. *Fracture Control and Materials Technology for Downhole Emplacement*, UCRL-53398; Lawrence Livermore National Laboratory: Livermore, CA, 1983.

5

Crack Mechanics

GENERAL COMMENT

It is hoped that by now the reader can agree that the primary objective of a practical book such as this should emphasize the design and materials aspects of fracture mechanics, on the premise that simple as well as complex structures can suffer from either man-made or nature-induced discontinuities. The majority of cracks and flaws encountered are obviously nature controlled and our selection of the term "crack mechanics" is meant here only to indicate our interest in both static and dynamic response of cracks. By appropriate design, of course, we would hope to mitigate the incidence of brittle fracturing and perhaps to minimize the effect of stress concentrations. And if by now we firmly believe in the Griffith theory of fracture, it is easy to accept the idea that stress concentration in general is simply a mechanism for converting strain energy into fracture energy. And the stress concentration can be triggered by the various discontinuities such as holes, grooves, and, of course, sharp cracks. At the same time, the probability of brittle fracture is known to increase with an increase in the complexity of larger structures such as bridges, ships, aircraft, pressure vessels, and large field cranes, to mention a few.

  This chapter is mainly concerned with the mechanical aspects of crack behavior. However, the stress corrosion and the corrosion–fatigue behavior topics are also included under the general category of special effects.

SURFACE AND CORNER CRACKS

Surface and corner cracks have been brought together in this section in order to indicate the extent of theoretical and experimental complexities as well as differences involved. When we try to model weld defects, arc burns, or machining marks, the idea of a "thumbnail crack" comes to mind, as already shown in Fig. 3.18. The current practice is to put this type of a flaw in the category of "surface cracks" in general. The assumption is that plane strain prevails, and that the task of calculating the stress intensity factor, say, for a "thumbnail crack" is based on the elliptical shape. The resulting formula, Eq. (3.23), involves the use of a flaw shape factor Q and the front free-surface correction denoted by M_K.[1] There is no doubt that Eq. (3.23) is only an approximation, as elegantly shown by Broek [2] and other investigators by resorting to three-dimensional analysis. In brief, the analysis indicates that there must be different stress intensities at different points of the crack surface, with certain consequences for crack behavior. In other words, we have a problem because we are dealing with part-through cracks and because, in the real world, surface (as well as corner) cracks may have irregular shapes while our approximate correction factors have been originally derived for elliptical geometry.

A schematic view of several crack configurations, including surface flaws, is given in Fig. 5.1. As long as a plane-strain condition exists, the K_{Ic} parameter should be used in the analysis, without special attention to thickness. It should be noted that in practically all situations the K_{Ic} values have been derived from a compact tension (CT) test specimen.

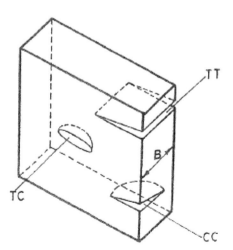

FIGURE 5.1   Part-through cracks (TT, through-thickness crack; CC, corner crack; TC, thumbnail crack).

Although considerable progress has been made in the analysis of all kinds of part-through cracks during the past 30 years, crack behavior is best discussed in terms of a uniform tension field and the elementary linear elastic fracture mechanics (LEFM) solutions. This is particularly true in the case of surface and corner cracks, which probably represent the majority of cases where special analytical procedures apply. Interested readers are directed to the reviews of such problems as interactions between the existing or postulated cracks and circular openings.[3] Although surface flaws may have partly irregular shapes, practical analysis dictates that all part-through cracks are best assumed to be elliptical. However, in the case of a highly irregular geometry one should expect considerable analytical discrepancies. Other uncertainties may include anisotropy, fracture modes, and fatigue criteria.

Many questions and problems arise when cracks have to interact with holes. The case of a through-thickness crack emanating from a hole is somewhat simpler to analyze, as shown in Chapter 3. However, when the crack starts out as a corner (or surface) flaw from the hole wall, various complications set in due to such matters as the presence of a fastener, load transfer across a structural boundary, interference fit, or residual stresses. It is amazing how much work has already been done in this area, and how many of the basic issues have not yet been settled. Here the usual complaints are insufficient data and the fact that too many test parameters are still on the back burner. The net result is that we often deal with a problem in a speculative manner.

Over the years of LEFM development many fatigue tests have been run on through-thickness cracks emanating from holes, grooves, and notches. The results indicate that crack growth from holes is rather similar to the growth of a central crack in a panel without a hole. Any differences are expected to be of the order of magnitude of the normal scatter in crack growth. Hence the practice of using the concept of equivalent crack length, Eq. (3.8), is justified. The only thing to know is that cracks emanating from holes may grow slightly faster. Also, the difference between the asymmetric and symmetric cases of crack growth and stress intensity is relatively small, as may have been noticed in Fig. 3.10. However, there are some aspects of the comparison that are still difficult to resolve. For instance, drilling a hole at one tip of a central crack was reported to have reduced the total growth rate by a factor of 2.[3] It is also fascinating to note how often in our comparative studies we refer to the work of Bowie,[4] including the cases of residual strength of through-cracks near the holes, in the regime of plane stress. Bowie's work dates back more than 40 years.

For many years now a corner crack at a hole has been considered an important element in fracture control although rigorous solutions have been limited by the three-dimensional stress intensity. This situation, however, has not stopped experimentalists and theoreticians in their efforts to propose various solutions for estimating the stress intensities. Many of these solutions gave good correlations with the specific sets of fracture data. In the case of a quarter-circular flaw geometry, the correlations were remarkably consistent.[3] Beyond this point, however, where other crack shapes are involved, further analysis and testing are still required in spite of continuing progress. This is essential before we can ever attempt multiparameter solutions of problems such as cracks emanating from fastener holes, interference, and cold work at filled fastener holes as well as crack propagation in reinforced or built-up structures. Such areas of investigation are heavy on analysis and multivariable experiments, which are certainly beyond the scope of practical fracture mechanics intended for designers.

One of the basic questions of fracture propagation is: What happens when a crack runs into a hole? In looking back on past developments in fracture mechanics, it is easy to come across a statement that such a crack should be arrested for a significant length of time. Holes were considered to be crack mitigators and crack stoppers, at least until about the early 1970s.[5] It was reasoned that as the moving crack tip came very close to the hole surface, the stress intensity had to become very high, or theoretically, the stress should tend to infinity. It was also reasonable to assume that the crack must have run into the hole at a very high velocity, and that the effective crack length was suddenly increased by the hole diameter. These two effects appeared to be additive so that the resulting crack propagation was practically independent of the size and spacing of the holes.[6]

In general it is difficult to predict whether a crack will run into a fastener hole or pass between holes, unless we can define the effect of a reinforcement such as a stringer. Unfortunately, each particular geometry of a reinforcement requires special analysis, and in real situations cracks cannot be forced into a specific mode of behavior. The role of holes in the crack arrest process is complicated because of a combined effect of dynamic stress intensity, elastic energy release rate, and kinetic energy.

In dealing with the arrest capabilities of holes, the following three actions should be recognized:

- Reduction of stress intensity at the crack tip;
- Introduction of residual compressive stresses in the fastener hole through the provision of a tight fit;
- Reduction of a conventional stress concentration.

In the case of surface and corner flaws in thicker wall structures, the critical crack sizes are relatively small and difficult to detect. To enhance the safety of such a structure we have to look for an increase in fracture toughness, better resistance to cyclic crack growth, and methods for crack arrest.

Once the crack is discovered through the appropriate inspection techniques, corrective action may be required. Very small cracks can be removed, say, by reaming or drilling, although there is no guarantee of complete crack removal.

With a larger and longer crack, repair may consist of drilling so-called stop holes to interfere with crack propagation. Such holes may not be very effective unless they are expanded by cold deformation to induce a layer of residual stresses.

ELEMENTS OF FATIGUE TECHNOLOGY

It is well known that a structural component may fail after a given number of load applications even at a maximum nominal stress much lower than the yield strength of the material. The nominal stress σ may be as illustrated in Fig. 5.2 for the simplest cases. Two-way arrows merely indicate the nature of a cyclically changing loading. The variation of stress with time between some maximum and minimum levels is shown in Fig. 5.3. When a maximum stress is plotted against the number of cycles to failure, we can obtain a curve similar to that shown in Fig. 5.4. There is a rather well-defined change in the shape of the curve at about 10^6 to 10^7 cycles, beyond which a constant stress S_e is approached, known as the endurance limit of the material. According to the original ASTM rules,[7] the $S-N$ diagram defines the fatigue life prior to which 50% of the specimens failed at the given maximum stress. A significant amount of scatter cannot be avoided in the majority of fatigue experiments. For tension–compression or push–pull tests, an endurance limit of approximately 75% of that obtained in bending may be assumed for correlation purposes. For design purposes, the

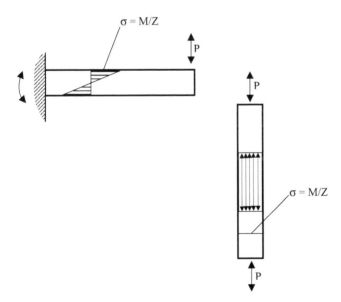

$\sigma = M/Z$

P

P

$\sigma = M/Z$

P

FIGURE 5.2 Components under varying loading.

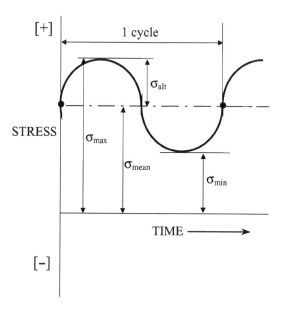

FIGURE 5.3 Typical cyclic loading.

fatigue strength ratios S_e/S_u for a number of structural materials can be taken from Table 5.1. The symbol S_u denotes the ultimate strength of the material.

The prediction of a fatigue life is not a simple matter because the stress spectrum, stress history, mechanical variables, metallurgical factors, and the

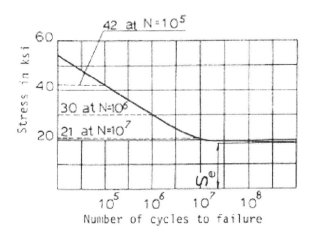

FIGURE 5.4 Example of *S–N* diagram.

TABLE 5.1 Recommended Fatigue Strength Ratios (S_e/S_u) for Preliminary Design.

Cast aluminum, 220-T4	0.17
Cast aluminum, 108	0.52
Cast aluminum, F132,-T5	0.38
Cast aluminum, 360-T6	0.40
Wrought aluminum, 2014-T6	0.29
Wrought aluminum, 6061-T6	0.45
Beryllium copper, HT	0.21
Beryllium copper, H	0.34
Beryllium copper, A	0.47
Naval brass	0.35
Phosphor bronze	0.32
Gray cast iron (No. 40)	0.48
Malleable cast iron	0.56
Magnesium, AZ80A-T5	0.29
Titanium alloy, 5A1, 2.5 Sn	0.60
Steel, A7-61T	0.50
Steel, A242-63T	0.50
Spring steel, SAE 1095	0.36
Steel, SAE 52100	0.44
Steel SAE 4140	0.42
Steel SAE 4340	0.43
Stainless steel, Type 301	0.30
Tool steel, H.11	0.43
Maraging steel, 18 Ni	0.45

Note: T, heat-treated; H, hard; HT, hardened; A, annealed.

environment can influence the result. A useful summary of practical S–N diagrams for structural steels is given by Tall.[8] Since any stress history of a structural component is likely to be based on several amplitudes and periods of operation, a useful approximation can be obtained from the cumulative damage criterion. When N_0/N fractions add up to less than unity no failure is expected.

$$\sum \frac{N_0}{N} = 1 \tag{5.1}$$

In this formula N_0 denotes the number of applied cycles at a particular working stress while N is the number of cycles to failure. This value can be obtained from a fatigue diagram such as that given in Fig. 5.4. In this particular case $N = 10^7$, corresponding to the endurance stress limit of S_e. It should be noted that Fig. 5.4 represents a semilogarithmic plot of data. The vertical axis has a regular stress scale while the abscissa is a logarithmic scale. This type of plot is probably the most popular and it does not distort the overall picture of structural response. The concept of the endurance limit is of special importance in design of rotating machinery and in other applications where cyclic loading is encountered. For a working stress below the level of S_e the number of cycles to failure is said to be infinite. It has also been established that structural materials indicating a sharp knee in a conventional stress–strain diagram tend to have an abrupt change in the $S-N$ curve when N is plotted as log N.

Although in a ductile material the stress concentration can be mitigated by inelastic behavior, in fatigue we tend to use the full value of a theoretical elastic stress concentration factor in order to be on the safe side. However, numerous fatigue experiments conducted with notched bars and sharp corner radii led to the concept of the Neuber effect.[9] It is of interest to observe that although we are still talking about a conventional process of fatigue, the Neuber theory suggests that there is a small limiting value of the notch radius below which no additional stress increase in fatigue should be expected. The ratio between the apparent stress increase and that predicted by elastic theory has been defined by Peterson[10] as the notch sensitivity factor, q:

$$q = \frac{k_f - 1}{k - 1} \tag{5.2}$$

In Eq. (5.2), k_f is the stress concentration factor derived from fatigue tests, while k is the classical elastic stress concentration factor. At $q = 0$, we have zero notch effect while at $q = 1$, the maximum theoretical notch sensitivity is attained. The intrinsic nature of the Neuber–Peterson contribution appears to fall between the classical stress concentration and the stress intensity factor in fracture mechanics.

Before leaving the area of a more conventional fatigue analysis it is well to mention some of the physical and mechanical effects on metal endurance.

In addition to the applied stress, geometrical discontinuities, surface conditions, and the environment must influence the fatigue life. Highly polished specimens experience at least 10% improvement and the surface sensitivity to a life-increase tends to be enhanced with higher strength steels. However, residual tensile stresses in these materials can be induced by milling and grinding. With poor practice a residual tensile stress in the surface layer of a high-strength steel was found to be much higher than 100,000 psi.[11] Fortunately such effects can be mitigated by stress relieving, abrasive tumbling, or shot-peening. The surface can be made stronger by carburizing, flame-hardening, or nitriding processes. The application of shot peening can be very beneficial because this process creates compressive residual stresses in the skin of the metal, so that the tensile fatigue stress is ultimately reduced.[12]

The machining process can be detrimental to fatigue life when the surface is subjected to grinding burns, stress-corrosion, cracks, distortion, and harmful residual stresses. For instance, abusive grinding can reduce the endurance limit by more than 10%, which in the case of nickel-base and titanium alloys can go up to 30%.

The effect of temperature and creep on the endurance limit of conventional structural materials is not significant up to about 650°F. At higher temperatures, a simplified approach for judging a combination of endurance limit and creep is best illustrated by Fig. 5.5.[13] Although the experimental results fall closer to an elliptical rather than a straight line between points A and B in Fig. 5.5, the proposed method is simple and conservative.

The combination of fatigue and corrosion is a rather serious problem because the corrosion products can act as a wedge, opening the crack. In the case of a carbon steel the reduction of the endurance limit is about 20%. In salt water and other corrosive media the reduction can be even higher. To combat this effect, alloying elements and protective coatings are used.

In usual design practice we deal with a fatigue life involving millions of cycles. The other extreme, of course, is represented by a conventional tensile test. The middle range, if such a term can be used, is characterized by a logarithmic scale. The only reasonable definition is that of a "low-cycle fatigue" below 10,000 cycles, where a structure is loaded beyond the elastic limit. To predict the low-cycle fatigue strength, an empirical approach appears to be fully justified,[14] because of its simplicity.

$$\sigma_a = S_e + \frac{CE}{2N^{1/2}} \tag{5.3}$$

In Eq. (5.3) E denotes the modulus of elasticity of the material and N is the number of cycles to failure at a given stress amplitude σ_a. The other stress term is the

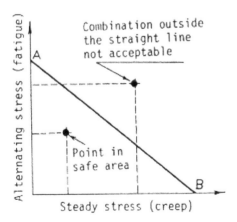

FIGURE 5.5 Simplified approach to judging a combination of creep and fatigue.

endurance limit S_e. The factor C depends on the property called the "reduction of area" ($R.A.$) in a conventional tensile test:

$$C = 0.51 \ln\left(\frac{100 - \%R.A.}{100}\right) \tag{5.4}$$

The basic formula given by Eq. (5.3) agrees with the experimental data on common carbon steel, copper, aluminum, nickel, stainless steel, and titanium.

The foregoing analysis has made no distinction between fatigue crack initiation and crack growth to failure. The S–N diagrams are not unique to the material representing the structure but are rather dependent on the various factors, ranging from surface roughness to environment. To develop an S–N curve reflecting the actual conditions, it would be necessary to employ special modifying factors derived from extensive test data. With standard practices covering high- and low-cycle fatigue testing[15,16] we have learned the essential steps in the analysis of metal fatigue, which can be described as the classical and traditional.

The second and more modern approach to the analysis of metal fatigue starts with the crack growth from an assumed or preexisting crack size. The idea here is to calculate the cyclic stress intensity at the tip of the crack and then to establish the number of cycles necessary to propagate the crack a given amount. Generally failure is expected when the crack grows to such a length that the uncracked ligament on its own can no longer support the design load.

Although there is no clear distinction between the various stages of metal fatigue, it is customary in fracture mechanics to divide the fatigue process into crack initiation, growth, and arrest (or failure). The formation of a macrocrack from a crack-free material can be defined as initiation. The size of a macrocrack is expected to be rather large compared to many grain diameters. Also, the crack initiation phase can extend over 50–75% (or more) of the life of the component. Hence the presence of a manufacture- or service-induced initial defect can have a serious repercussion on the fatigue life of the component.

It is difficult to determine which of the theories of crack initiation is universally accepted. However, it is safe to assume that all mechanisms, outside the corrosive environment, involve some plastic deformation at points of high stress concentration, such as holes or fillets. These may also be inclusions and grain boundaries of a localized nature. Since the surface is the prime location for crack growth, the appropriate surface treatment can help to protect the structural integrity of the components and systems involved.

It appears that on a microscale the basic mechanism of fatigue includes cycling shearing of material near the surface. However, as the crack initiation progresses by linking many microcracks, the mode of shear deformation changes into the tensile mode, and the crack propagation can be recognized as the "striation" markings, a familiar term in fractography. The direction of crack propagation here is normal to the direction of the maximum tension. The failure analysis may be stress or fracture controlled, that is, it may use either conventional stress analysis or fracture mechanics methodology. It should be noted that fatigue striations are not developed during all phases of the fatigue process. They do not appear during crack growth under very low loads or during the final stages of the fatigue process.

Assuming that we can have a smooth component, entirely free of imperfections and discontinuities, the cyclic elastic stress should not cause this component to fail by fatigue. The situation, however, would be quite different if this component were loaded beyond the yield strength of the material. Barsom and Rolfe[1] describe the steps by which plastic deformation of the component creates the nucleation sites for fatigue cracks and what we can learn about the stochastic character of this process.

In the majority of cases fatigue cracks start and propagate from stress raisers in the form of surface imperfections and geometrical changes. This is not the case with the embedded flaws encountered in welded components, which are reviewed later in this book.

It is known that a fatigue crack propagates normal to the tensile stress field while the crack front maintains constant stress intensity. A penny-shaped crack is obtained from an embedded source while a semicircular part-through crack develops from a source on the surface. The rate of crack propagation is said to depend on the stress gradient. For the case of fatigue cracks developing in complex stress fields, and when this process depends on certain metallurgical parameters related to crack propagation in steels, the interested reader is referred to other publications.[17,18]

The objective of laboratory tests is to establish information on the fatigue behavior of small test specimens that can simulate the behavior of actual structures if tests on full-scale components are not practical. The information should be reliable for the purpose of selecting the material and design. This can only happen if the conditions of material, stress history, and environment are closely reflected by the test.

It is customary to employ small test specimens that have simple geometries. This geometrical simplicity makes it easier to calculate the relevant stress intensity factors, and this is why center-cracked panels and compact tension specimens are used so often in laboratory work. However, there is one general problem related to the modeling of the actual structural behavior. In some cases of fatigue life studies no differentiation is made between the initiation and propagation phases of the fatigue process. Barsom and Rolfe[1] recommend running the initiation and propagation tests separately; these, when required, can be combined to predict the total life.

While the mechanism of crack initiation can be investigated, for instance, by testing a series of polished specimens in the so-called rotating beam fatigue experiment, most fatigue crack propagation tests require precracked specimens. One of the popular specimen designs for crack propagation tests, the compact tension type,

works in a wedge-opening mode, as shown in Fig. 3.26. The change in the crack
length can be monitored visually, ultrasonically, or by means of an electrical potential at any specific number of the elapsed load cycles during the test. The resulting
plot of crack length a vs. the number of load cycles N is illustrated schematically in
Fig. 5.6. The fatigue life of a test specimen is known to decrease with the increase
of the cyclic load. The approximate shape of the curve in Fig. 5.6 results from a
series of constant-amplitude stress cycles and indicates that the crack length
does not change very much during the useful life range of the specimen.

Design Problem 5.1

A short-life bearing component in a rotating machinery undergoes cyclic
dynamic loads due to synchronous rotor vibration. The rotor was operated at
50 rpm for 33 hrs and 20 min and then the speed was increased to 100 rpm and
the rotor was run for additional 8 hrs 20 min. Assuming the methodology and
numerical data of Fig. 5.4, estimate the allowable number of operational fatigue
cycles and maximum run time at 200 rpm so that the cumulative damage factor of
0.75 is not exceeded. The maximum stresses that the component is subjected to at
the abovementioned speeds are:

$$\sigma_{max} = 21 \text{ ksi at } 50 \text{ rpm}$$
$$\sigma_{max} = 30 \text{ ksi at } 100 \text{ rpm}$$
$$\sigma_{max} = 50 \text{ ksi at } 200 \text{ rpm}$$

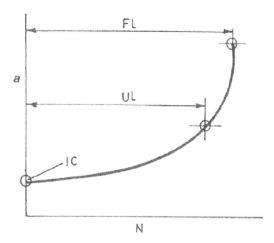

FIGURE 5.6   Illustration of constant fatigue crack growth (IC, intitial crack; UL, useful
life; FL, fatigue life) (from Ref. 1).

Solution

The cycles to failure at the above maximum stresses can be obtained from the curve in Fig. 5.4.

$$N = 10^6 \text{ at } 30 \text{ ksi}$$
$$N = 10^7 \text{ at } 21 \text{ ksi}$$

The number of cycles the components was subjected to are:

$$100 \times 500 = 50{,}000 \text{ cycles at } 30 \text{ ksi}$$

and

$$50 \times 2000 = 100{,}000 \text{ cycles at } 21 \text{ ksi}$$

From Eq. (5.1)

$$\left(\frac{N_0}{N}\right) + \frac{50{,}000}{1{,}000{,}000} + \frac{100{,}000}{10{,}000{,}000} = 0.75$$

then

$$\frac{N_0}{N} = 0.69$$

For the maximum stress of 50 ksi, the number of cycles to failure from Fig. 5.4 is 36,000. Hence the required number of operational cycles is

$$N_0 = 0.69 \times 36{,}000$$
$$= 24{,}840$$

or

$$\frac{24{,}840}{36{,}000} + \frac{50{,}000}{10^6} + \frac{100{,}000}{10^7} = 0.75$$

The maximum allowable service life at 200 rpm will be

$$\frac{24{,}820}{200} = 124.1 \text{ min} \cong 2 \text{ hrs } 4 \text{ min}$$

Since the sum of the cycle ratios is less than unity, the structure represented by Fig. 5.4 can be presumed safe if it is run for 2 hrs. In practice the cumulative damage ratio, Eq. (5.1), should at least be less than 0.8.

Design Problem 5.2

Estimate the numerical value of the reduction of area parameter for a low-cycle fatigue experiment using SAE steel 4140 specimens with 115 ksi ultimate strength and a stress amplitude of 36 ksi. The structural component made of this grade of steel is expected to survive 60,000 stress cycles. Assume the elastic modulus to be 30×10^6 psi. Show the variation of $R.A.$ with the number of stress cycles.

Solution

Solve Eq. (5.3) for parameter C, to give

$$C = \frac{2(\sigma_a - S_e)N^{1/2}}{E}$$

From Table 5.1, the ratio (S_e/S_u) for 4140 steel is 0.42. Hence the endurance limit S_e, is

$$0.42 \times 115 = 48 \text{ ksi}$$

Next substituting

$$C = \frac{2(36 - 48) \times 10^3 \times 60,000^{1/2}}{30 \times 10^6} = -0.2$$

Then, from Eq. (5.4)

$$C = -0.2$$
$$= 0.5 \ln\left(\frac{100 - R.A.}{100}\right)$$

and

$$R.A. = 33\%$$

This is probably a reasonable number, considering that the elongation for the SAE 4140 steel is about 28%. Although the empirical formulas Eqs. (5.3) and (5.4) are quite rational and simple to use, the information or "reduction of area" should be, but may not always be, easy to find in the literature. It should also be added that the maximum fatigue stress in low-cycle fatigue can be described as the "pseudo elastic limit" because the stress is expected to extend beyond the elastic limit. The maximum and minimum stresses for uniaxial fatigue loading can be estimated from Soderberg's law[19]

$$\sigma_{max} = S_e + \sigma_{mean}\left(1 - \frac{S_e}{S_y}\right) \tag{5.5}$$

$$\sigma_{min} = S_e + \sigma_{mean}\left(1 + \frac{S_e}{S_y}\right) \tag{5.6}$$

For stress definitions, see Figs. 5.3 and 5.4. Yield strength is denoted by S_y. The variation of R.A. with N can be obtained from the following expression (in percent).

$$R.A. = 100\left(1 - e^{-0.0016N^{\frac{1}{2}}}\right)$$

Derivation of this formula is given in Design Problem 5.3; see the section headed Design Procedures, at the end of this chapter. The result is plotted in Fig. 5.7 as the R.A.% as a function of N. This curve applies only to the conditions specified in Design Problem 5.2.

Although the plot in Fig. 5.7 is limited to the conditions spelled out in Design Problem 5.2, the curve shows the general trend of a decreasing number of cycles to failure with a decrease in the reduction of area. This is consistent with the mechanical concept of an increasing brittleness of the material as the R.A. parameter tends to zero. In other words, the process of low-cycle fatigue is governed essentially by the ductility of the material, while high-cycle fatigue depends heavily on the static strength of the material.

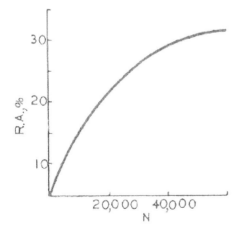

FIGURE 5.7   R.A. plot for Design Problem 5.2.

The selection of formulas for illustrating the problem of low-cycle fatigue, Eqs. (5.3) and (5.4), is convenient because of direct use of engineering stresses in presenting the empirical results.[14] Other approaches, using primarily strain–life relationships in the empirical world, are also available.[1,20,21]

INITIATION OF FRACTURE

Any structure can be perceived as a device designed specifically to delay some event involved in transformation of energy. All structures will no doubt be broken in the end, and the only difference is the time interval of safe life built into the structure by an experienced designer. The basic problem is, however, that planning safe time intervals is contingent upon probabilities, estimates, and compromises, resulting always in some finite risk of premature failure, which we have learned to accept. A rather splendid example of trading the technical, operational, and safety values comes from the history of World War II. The aircraft designers on the Allied side had to make subtle choices when the losses of bomber aircraft by enemy action were 1 out of 20 in each sortie vs. the losses from structural causes of the order of 1 aircraft in 10,000. The design decisions were often made even more difficult by the opinions of the airmen, who preferred the big risk of being shot down by the enemy to the much smaller risk of the structural failure of the plane in flight.[22] It appears that this general perception of structural failures continues to persist in all areas of technology and product development, and in all walks of life, reflecting a part of human nature.

In descending now to the level of engineering reality, we can find that a significant portion of design effort boils down to identifying the weakest link in a load-supporting system. When the stiffness of a structure governs the design, the prediction of a structural behavior is made a little simpler for us, because we can see, so to speak, what is happening. Unfortunately, the search for a weak link involves the prediction of strength of the material where the metallurgical processes, flaws, cracks, and geometrical discontinuities have something to say under static and dynamic conditions. And during the most insidious loss of strength under fatigue conditions, the initiation and propagation of cracks deserve special recognition by designers. There are too many horror stories to tell about the role of a single fatigue crack during the past 100 years. Whether we think about the dropping-off of the propeller in the Bay of Biscay, Comet disasters in flight, or fracture of railway train axles on the ground,[22] significant changes take place in a material under fluctuating stresses at the tip of a notch or a crack. These changes reduce the work of fracture (which is toughness) of the metal and make it easier for the crack to grow and accelerate, with disastrous results.

Although fracture mechanics methodology has developed and progressed along independent and unique lines, it still has features closely related to stress analysis because it involves the concepts of a stress-field not only at the crack tip but also at the nominal stress (applied to the structure) locations further away from the crack tip. These two types of stresses are tied together by the principle of linear elastic fracture mechanics (LEFM) and open the door for better understanding of a fatigue crack initiation and propagation until the crack size reaches the critical proportions. Schematically, the general stress situation can be described as shown in Fig. 5.8. The stress field corresponds to Mode I deformation, and it depends on the size, shape, and orientation of the crack line flaw, having a sharp tip, where the magnitude of the stress intensity is denoted by K_I in the immediate vicinity of the crack tip. This stress intensity is the same for all materials. The "immediate vicinity" term is further illustrated in Fig. 5.9.

The elastic stress equations describing stress intensity are given in the literature in terms of K_I, r, θ, and ρ.[23] The degree of sharpness of the tip in the theoretical equations is governed by the curvature radius ρ (Fig. 5.9). While the exact equations may not be of prime interest to the practicing designer, the general relation between the theoretical stress field and the more conventional concept of stress concentration can be of some value in analyzing crack initiation behavior. Each equation for the two-dimensional system consists of two parts. One deals with the sharp crack while the second term of the equation provides a correction for a blunt-tip radius. The material close to the tip is shown in Fig. 5.9 as a formed blunt-tip geometry with radius ρ. This location is subjected to the maximum stress σ_{max} and the corresponding stress amplitude $\Delta\sigma_{max}$, and is therefore the area of crack initiation. In mathematical terms,[23] the stress

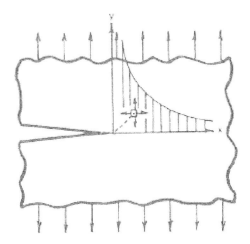

FIGURE 5.8 Elastic stress field.

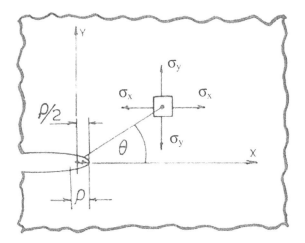

FIGURE 5.9 Immediate crack tip vicinity parameters.

parameters are given by Eqs. (5.7) through (5.10). The stress definitions are based on Fig. 5.3, where $\Delta\sigma_{max}$ is equal to σ_{alt} as stress amplitude.

$$\sigma_{max} = 1.13K_I/(\rho)^{1/2} \tag{5.7}$$
$$\sigma_{max} = k\sigma \tag{5.8}$$

and

$$\Delta\sigma_{max} = 1.13(\Delta K_I)/(\rho)^{1/2} \tag{5.9}$$
$$\Delta\sigma_{max} = k(\Delta\sigma) \tag{5.10}$$

Here k denotes the conventional stress concentration factor for the notch (or crack) geometry. However, Eqs. (5.7) and (5.9) are expected to be accurate only for very small values of the notch tip radius ρ. The correlation of fatigue crack initiation with the stress concentration factors for several types of notched specimens has, so far, involved the range of k values between 2 and 10.[24] Other techniques, aiming at a quantitative prediction of fatigue crack initiation behavior, involved elastic–plastic finite-element stress analysis.[25]

It is only fair to note that the finite-element solution for the uncracked body is worthwhile, although the stress distributions so obtained may be of doubtful accuracy. Also, the popular notion that finite-element analysis is the most rigorous stress analysis technique available is, as stated by Broek,[2] "exaggerated and naive." In many structural applications, extremely large stress gradients and coarse modeling must be a major cause of limited accuracy. Furthermore, in the field of fracture mechanics, finite-element models of cracked real structures may require many assumptions of boundary conditions, load distribution, and stress gradients in addition to one real, but at times overlooked, parameter — the cost.

Since the nominal stresses in most structural members are elastic, the zone of plastic deformation at the tip of the crack must be surrounded by an elastic stress field. The concepts and equations touched on in this section so far are consistent with LEFM and the behavior of notched specimens during the process of crack initiation. The geometry and the degree of notch sharpness are of special importance in establishing the dependence of crack initiation on stress amplitude in fatigue, as shown, for instance, in Fig. 5.10 for HY-130 steel specimens.[26] The steel specimens were subjected to "zero-to-tension" loading in fatigue as a function of the number of cycles to crack initiation. Barsom and McNicol [26] also show the plot of $K/(\rho)^{1/2}$, approximated roughly in Fig. 5.11, to indicate the trend only. The vertical axis dimension is ksi because the square root term for the stress intensity factor cancels out. The threshold of fatigue crack initiation for HY-130 steel was stated as 85 ksi. The maximum elastic stress at the root is given by Eq. (5.7).

The concept of a threshold for fatigue cracking was first noted by Frost.[27] It is marked by a significant deceleration in the growth rate of a crack at low stresses, below which fatigue cracks are not expected to propagate. The existence of a threshold was also predicted by McClintock, using an elastic–plastic analysis.[28] Later work by Paris[29] provided a basis for using the rules of LEFM in studying the fatigue crack propagation threshold and the stress intensity factor range of ΔK_I. This effort was followed by many investigators in the field, resulting in a two-volume book on the subject.[30]

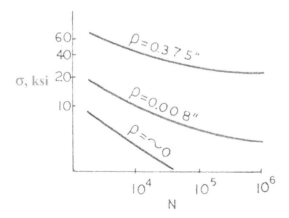

FIGURE 5.10 Approximate crack initiation in various notches of HY-130 steel.

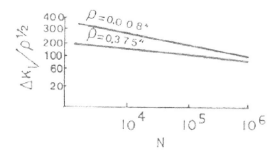

FIGURE 5.11   Stress intensity in crack initiation for HY-130 steel.

There is sufficient evidence to show that the fatigue crack initiation threshold depends to some degree on the tensile strength of the material,[1] as sketched roughly in Fig. 5.12. The materials used for the development of the correlation given in Fig. 5.12 cover a range of steels of varying chemical compositions and tensile strength between 77 and 233 ksi, as reported by Barsom and Rolfe,[1] and by Clark.[31] In this range of properties we can find such steels as A36, A537-A, HY-80, 403 stainless, A517-F, HY-130, and AISI 4340. The major practical significance of Fig. 5.12 is that for steels having tensile strength higher than 150 ksi, the fatigue crack initiation threshold remains constant. The relevant experimental results were obtained from three-point bending at a stress ratio of 0.1. This value is normally obtained by dividing σ_{min} by σ_{max} as illustrated in Fig. 5.3. The test data also suggest [1] that for the range of tensile strength of steel between 70 and 150 ksi, the threshold parameter can be estimated from the following equation:

$$\Delta K_I/(\rho)^{1/2} = 0.9 S_u \qquad\qquad (5.11)$$

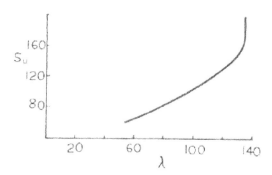

FIGURE 5.12   Threshold vs. tensile strength (S_u, tensile strength in ksi; $\lambda = \Delta K_i/(\rho)^{1/2}$, which is the threshold diameter).

In Eq. (5.11), S_u denotes the ultimate strength of the material. It is still customary to use the traditional definitions. For instance, the member under load deforms plastically if the uniaxial stress equals the yield strength of the material. Also this member is expected to fracture when the uniaxial stress equals the ultimate tensile strength S_u. This point we are likely to find on the stress-strain curve where the load on the test piece is a maximum. In practice, however, we seldom use S_u as the critical parameter in LEFM considerations, or even in the more traditional design deliberations.

Since the ratio S_y/S_u for high-strength materials such as martensitic steels varies between 0.8 and 1.0, the crack initiation threshold parameter as a function of yield strength S_y can be represented by Eq. (5.11). Barsom and Rolfe[1] also indicate that for the yield strength of steel between 40 and 140 ksi, the correlation can be stated as

$$\Delta K_I/(\rho)^{1/2} = 5(S_y)^{2/3} \tag{5.12}$$

The variables in Eq. (5.12) are expressed as ΔK_I in ksi (in.)$^{1/2}$, yield strength S_y in ksi, and notch radius ρ in in. The plot of Eq. (5.12) is given in Fig. 5.13, and it should not be used with yield strengths higher than 140 ksi.

It is generally recognized that Hooke's law describes the first portion of the stress–strain curve for a typical structural material in terms of a straight-line relationship. In the same fashion one would like to use a mathematical equation to fit the remaining portion of the curve in simple and practical terms. And it is

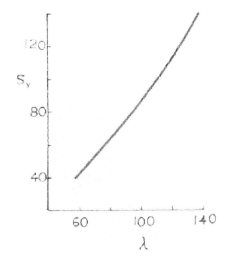

FIGURE 5.13 Yield strength vs. threshold parameter (S_y, yield strength in ksi; λ, crack initiation threshold parameter in ksi).

totally incredible that 264 years had to pass before the engineering world would see the solution of this problem in the form of the Ramberg–Osgood equation.[32] This equation involves the exponent n, which controls the shape of the stress–strain curve beyond the elastic range, consistent with the phenomenon of "strain harden- ing," known also in engineering circles as "work hardening." Hence for the purpose of a correlation technique the exponent n becomes a "strain-hardening exponent." While the use of this exponent in this chapter relates to LEFM parameters, it is only fair to add that the original form of the Ramberg–Osgood equation can well be uti- lized in the regime of elastic–plastic fracture mechanics (EPFM).[2]

The correlation of the strain-hardening exponent with yield rather than ulti- mate tensile strength was given by Clausing.[33] Also the fatigue crack initiation threshold of steels can be expressed as a function of the strain-hardening expo- nent on the condition that the yield strength of a given steel is not greater than 140 ksi. The corresponding relationship can be described by a simple formula:[1]

$$\frac{\Delta K_1}{(\rho)^{1/2}} = \frac{30}{(n)^{1/2}} \tag{5.13}$$

The plot of Eq. (5.13) is shown in Fig. 5.14. According to Barsom and Rolfe,[1] the accuracy of the chart given in Fig. 5.14 decreases when the yield strength of the steel exceeds the level of 140 ksi. The loss of accuracy is due to the increased scatter of data for stronger steels.

General interest in the effect of strain hardening on structural integrity of pressure vessels and the related technologies has led to a number of reports in the public domain, which are beyond the scope of this book. However, a brief reference can be made to the studies of the tensile properties of the selected steels

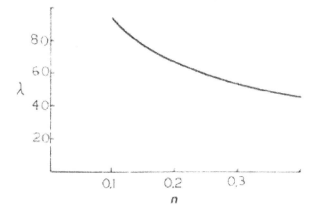

FIGURE 5.14 Crack initiation threshold vs. exponent (n, strain hardening exponent; $\lambda = \Delta K_i/(\rho)^{1/2}$, the crack initiation threshold parameter).

in relation to the strain-hardening exponent.[33,34] The tensile properties covered essentially martensitic steels as well as ferrite–pearlite steels. The corresponding relationships are illustrated in Fig. 5.15. However, Barsom and Rolfe[1] do not recommend using the information in Fig. 5.15 for austenitic stainless steels at this time.

The analysis of the fatigue crack initiation threshold discussed so far has indicated that the effect of the mechanical and metallurgical properties is negligible. For other topics such as the difference in fatigue damage during compressive and tensile cycling, as well as the nature of crack behavior in and around the plastic zone, the reader may wish to examine other sources.[35,36]

Whether we use smooth, unnotched, or notched specimens in fatigue crack initiation experiments, any surface damage and surface irregularities can reduce the initiation life drastically. This effect is expected to be even more pronounced as the volume of the plastically deformed material at the tip of the crack is increased.

CRACK PROPAGATION

Before entering the regime of crack propagation it is well to summarize some of the characteristics of the crack initiation process.

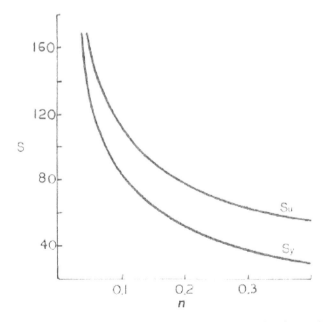

FIGURE 5.15 Tensile strength vs. strain hardening (*n*, strain hardening exponent; S_u, tensile strength; S_y, yield strength).

The initiation of cracks can only take place in the regions of plastic deformation. As long as the strains are elastic, even in the face of geometrical discontinuities and residual stresses, fatigue cracks do not start.

In the great majority of structures we find fields of nominal elastic stresses surrounding the localized, plastically deformed pockets of material. Under cyclic loading these pockets can be transformed into nuclei for fatigue crack initiation. Such pockets of plasticity are caused by strain raisers embedded in the elastic field.

The transition from a plastic region to an elastic field is strongly affected by the size of the notch (or crack) and the size of the plastic zone in relation to the size of the specimen.

Contrary to general intuition, cracks can also be initiated under compressive cycling loading. However, they can only propagate through the plastically deformed region. On the other hand, cracks initiated under tensile–cycling loading will propagate beyond the plastically deformed region to cause a fracture of the test specimen.

We normally deal with two crack realities. These are preexisting flaws and the cracks that gradually develop during the service life of the structure. The nondestructive techniques and proof-testing procedures tend to identify the upper limits of the cracks and establish the minimum fatigue life criteria for specific structural components. The prediction of a service life, in turn, depends on the degree of understanding of the rate of fatigue crack growth. The general principle of this phenomenon is briefly illustrated in Fig. 5.16. It deals with crack growth

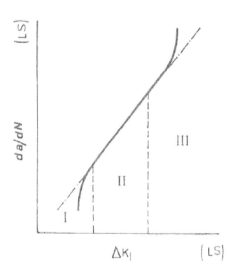

FIGURE 5.16 Example of crack growth in steel (LS, log scale; Region I, fatigue threshold; Region II, crack growth; Region III, static failure).

under constant-amplitude load cycling, which is the preferred method of fatigue experimentation. In simple terms, the increase in crack length is given as a function of the number of elapsed load cycles. When the crack growth rate per cycle of loading, da/dN, and the fluctuation of the stress intensity factor (ΔK) are brought together in one equation, we obtain

$$\frac{da}{dN} = A(\Delta K)^m \tag{5.14}$$

where a = crack length, N = number of loading cycles, ΔK = fluctuation of stress intensity factor, A = constant, and m = constant. This formula is known as the Paris equation.[37] The correct presentation of fatigue crack growth data is a log–log plot of the two major parameters, da/dN and ΔK_I, as shown in Fig. 5.16. The parameter ΔK_I defines the mechanical driving force and it does not depend on geometry. Region I corresponds to the fatigue threshold below which cracks do not propagate. This area is known as the "near threshold" or "threshold" regime. The cyclic stress intensity at the fatigue threshold is rather low so that the crack growth per cycle is almost zero. The fatigue crack threshold for a number of common metals can be approximated by a simple expression where the stress intensity factor is denoted by (ΔK_{th}).

$$\Delta K_{th} = 6.4(1 - 0.85R) \tag{5.15}$$

The load or stress ratio R in Eq. (5.15), for ferritic–perlitic, martensitic, and austenitic steels, can be taken as >0.1.[38] Predictions based on Eq. (5.15) are likely to be generally conservative. Region I can be characterized as that of noncontinuum mechanisms sensitive to microstructure, mean stress, and environment. The parameter R here is the ratio of minimum to maximum cycling loading.

Region II fatigue crack growth represents a linear relationship between $\log da/dN$ and $\log (\Delta K)$, Fig. 5.16, in line with the Paris equation, Eq. (5.14). The exponent m for most metals should be between 2 and 4. This region of crack growth is consistent with the model of a continuum growth and the mechanism of striations observed on the fatigue surface. The crack growth rates in this region vary typically between 10^{-6} and 10^{-3} mm/cycle. These rates appear to be relatively insensitive to microstructure, plane-stress, or plane-strain conditions. There is also a limited influence of load ratio R and benign environments. The models of crack growth in Region II are intended for a constant value of ΔK.[39] When a generally uniform load environment experiences occasional higher loads, a "crack retardation" may be observed. What apparently happens is that after a random overload, a period of essentially no crack growth takes place, followed by an increase in crack growth rate until the original rate is attained. For a random change in amplitude loading the cyclic stress intensity factor ΔK can be taken as the root-mean-square value of the ΔK values.

Region III is often referred to as the static failure mechanism because the actual failure occurs after a relatively small number of cycles, with the corresponding ΔK values approaching static conditions. The crack growth rates cannot be extrapolated from Region II without risking a nonconservative result: The crack growth rate is likely to be accelerated because of a combined effect of fatigue and the static mode of ductile tear. In summary, the fatigue crack growth per cycle in Region III is higher than that in Region II. The fracture surface shows fatigue striations, and ductile tear occurs when the critical strain forms at the tip of the crack.[40] The acceleration in the rate of crack growth takes place when the stress intensity factor is actually a little lower than the critical stress intensity factor, K_{Ic}.

While the entire discussion of the three regions of crack growth may not be absolutely essential to pragmatists in the area of mechanical design, the quantitative predictions of fatigue crack growth are certainly of interest. In this respect we can see the role played by the Paris equation and understand how to evaluate the threshold parameter ΔK_{th} in terms of the conventional load ratio R. It becomes rather clear that Region II is reserved for the primary mechanisms of fatigue crack growth, while Region III is restricted to essentially static failure modes. It is also evident that the purpose of fatigue testing is to characterize the two major aspects of the LEFM technique. One is the threshold, K_{th}, below which the crack is not expected to grow. The other is the regime of continuum crack growth defined by the relationship between da/dN and ΔK.

The testing of fracture behavior involves a crack propagated in the material under conditions of a known cyclic stress intensity and the recorded crack length. The crack length must be monitored continuously in order to determine the crack growth per cycle da/dN. Hence the crack length is used in establishing the growth rate and in calculating the stress intensity factor. The change in the crack length can be followed by a direct observation of the surface of the test piece, by calibrating the compliance related to crack length, or by use of the potential-drop method, as noted previously. In the last case we need an external power supply to run a current through the body of the test piece so that a difference in the electric potential can be measured between the two sides of the crack. Since the measured potential changes with the crack length, the appropriate correlation between the potential and the crack size can be established. The procedure for evaluating constant-amplitude fatigue crack growth for rates above 10^{-8} m/cycle has been regulated by the ASTM standard.[41]

Before looking at some specific formulas and plots of fatigue crack propagation for typical structural materials, it may be helpful to briefly review some of the basic parameters affecting the process of growth rates, because such data are often hard to pinpoint for the case at hand. Endless numbers of variables and circumstances will seldom reflect the actual service conditions. This situation still seems to be murky in spite of the many investigations, computer programs, and publications. For instance, as Broek remarks,[2] "no single model can explain the influence of the environment on the rate of propagation of fatigue cracks."

In the case of thin parts such as sheets, fatigue cracks start in a perpendicular direction to the sheet surface. As the crack grows, the size of the plastic zone increases and the plane-stress condition is reached where the size of the plastic zone becomes equal to or greater than the sheet thickness. Experiments also indicate that crack propagation in plane stress is slower than that in plane strain, on the premise that the stress intensity is the same for both conditions.

Many fatigue experiments run on martensitic steels with yield strength greater than 80 ksi indicate significant dependence of crack growth rate on the range of fluctuation of the stress intensity factor, as illustrated in Fig. 5.17. The two lines shown in this log–log plot define the approximate boundaries for the experimental results, which can be represented by the following equations:[1]

Curve D (Fig. 5.17)

$$\frac{da}{dN} = 0.27 \times 10^{-8}(\Delta K_{\mathrm{I}})^{2.25} \tag{5.16}$$

Curve H (Fig. 5.17)

$$\frac{da}{dN} = 0.66 \times 10^{-8}(\Delta K_{\mathrm{I}})^{2.25} \tag{5.17}$$

Both equations have the symbols $a =$ crack size (in.), $N =$ number of loading cycles, $K_{\mathrm{I}} =$ stress intensity factor, ksi (in.)$^{1/2}$. Here Eq. (5.17) is shown as the upper bound for martensitic steels (80–300 ksi yield) tested in an air environment. The expressions defining the boundaries are essentially of the Paris equation type, with the following numerical constants: $A = 0.27 \times 10^{-8}$ and 0.66×10^{-8}, $m = 2.25$.

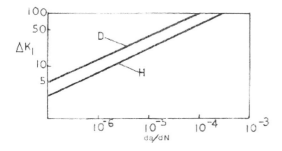

FIGURE 5.17 Fatigue crack growth in martensitic steels (D, Eq. (5.16); H, Eq. (5.17)). The term ΔK_{I} is given in ksi (in.)$^{1/2}$ and da/dN in in./cycle.

This case of fatigue crack growth is consistent with the Region II diagram in Fig. 5.16. The symbols A and m come from Eq. (5.14). The validity of Eq. (5.17) for dealing with martensitic steels has been proven.[18,42]

The boundary curve for the fatigue crack growth data based on numerous tests of ferrite–pearlite steels[1] is shown in Fig. 5.18. This group of structural steels includes A36, ABS-C, A302-B, and A537-A designations, for which the empirical formula in calculation of the fatigue crack growth is

$$\frac{da}{dN} = 3.6 \times 10^{-10}(\Delta K_{\mathrm{I}})^{3.0} \tag{5.18}$$

In this case also ΔK_{I} is given in ksi (in.)$^{1/2}$ and da/dN in in./cycle. For the unit conversion we have 1 in. = 25.4 mm and 1 ksi (in.)$^{1/2} = 1.099$ MPa (m)$^{1/2}$.

All data points obtained from the experiments on ferrite–pearlite specimens are found above the boundary curve described by Eq. (5.18).[18] The formula given by Eq. (5.18) is expected to yield reliable results. The ferrite–pearlite steels investigated cover the yield strength range between 30 and 80 ksi. Also it is noted that the rate of fatigue crack growth for a given stress intensity factor range ΔK_{I} is lower in ferrite–pearlite than in martensitic steels. The difference in the rate of crack growth between the two types of steel can be related to the composite character of the microstructure of ferrite–pearlite steel and other subtle elements of matrix behavior.[18]

Special importance is attached to the mechanical properties of pressure vessel steels, generally put into three categories such as martensitic, ferrite–pearlite, and austenitic stainless. Again, extensive experimentation with the austenitic stainless steels in the regime of fatigue crack growth rate corresponding to Region II, Fig. 5.16, resulted in a practical formula (of the Paris equation type)

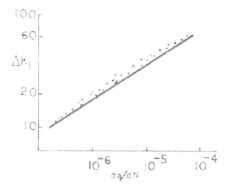

FIGURE 5.18   Fatigue crack growth in ferrite–pearlite steels.

applicable to a room temperature air environment:

$$\frac{da}{dN} = 3.0 \times 10^{-10} (\Delta K_I)^{3.25} \tag{5.19}$$

The symbols and unit conversions here are the same as those given for Eq. (5.18). The range of yield strength for these steels (type 304 and 316 stainless) is 30–50 ksi, with the strain-hardening exponent n greater than 0.30. The exponent for martensitic steels is smaller than 0.15, while the n value for ferrite–pearlite steels falls in the range 0.15–0.30.

Because of special interest in austenitic stainless steels (such as the 304 and 316 type), considerable effort was spent on developing test data for crack growth rate da/dN vs. stress intensity factor range ΔK. This information concerned cold-worked 316 stainless and solution-annealed 304 stainless.[43] The approximate envelope of these results is shown in Fig. 5.19.

The design formula given by Eq. (5.19) is expected to provide a conservative estimate for austenitic steels in an air environment at room temperature. Hence in order to cover a wide range of materials such as martensitic, ferrite–pearlitic, and austenitic stainless steels, Eqs. (5.17), (5.18), and (5.19) provided a practical design tool for estimating fatigue crack growth. These formulas are illustrated in Fig. 5.20.

The advantage of using Eq. (5.17) is that it can be used to calculate fatigue crack growth in martensitic steels for various values of stress ratio R. This feature is illustrated in Fig. 5.21, obtained from a number of samples having different

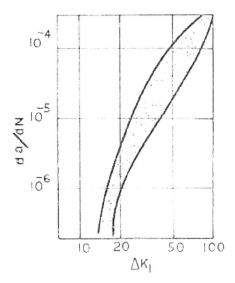

FIGURE 5.19 Fatigue crack growth for stainless steel.

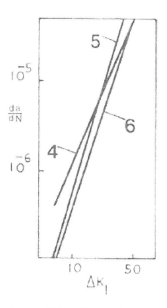

FIGURE 5.20 Comparison of crack propagation models (Curve 4, Eq. (5.17); Curve 5, Eq. (5.19); and Curve 6, Eq. (5.18)).

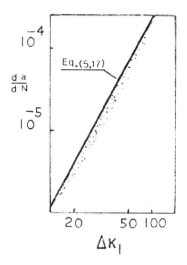

FIGURE 5.21 Upper bound scatter equation for 140 ksi martensite steel.

stress ratios ranging from 0.39 to 0.94. The upper bound in Fig. 5.21 is obtained from Eq. (5.17). However, the effect of stress ratio $R \geq 0$ can also be estimated from Eq. (5.20):

$$\frac{da}{dN} = \frac{A(\Delta K)^{2.25}}{(1 - R)^{0.5}} \tag{5.20}$$

where A is a constant.

Little is known about the effect of a conventional stress concentration, although the following relationship has been proposed:[44]

$$\frac{da}{dN} = A(\Delta K_{\text{eff}})^m \tag{5.21}$$

where

$$\Delta K_{\text{eff}} = k_{\text{t}}(a) \times \Delta\sigma(a)^{1/2} \tag{5.22}$$

and $\Delta\sigma$ = nominal stress amplitude, a = crack length, $k_{\text{t}}(a)$ = stress concentration factor as a function of crack length, with crack propagating outside the notch "shadow," A = constant, and m = constant.

Conventional steels that are not exposed to a harmful environment are largely independent of mechanical and metallurgical effects, and one would naturally ask whether other metals such as aluminum and titanium can be characterized by a similar behavior. The answer lies, of course, in the results of fatigue crack growth rate experiments with aluminum and titanium alloys, as illustrated in Figs. 5.22 and 5.23.[45] It appears that the respective scatter bands are larger than those for steels. The results forming an aluminum scatter band were derived by testing six aluminum alloys ranging in yield strength from 34 to 55 ksi. The alloy designations were 2219-T87, 5456-H321, 6061-T651, 7005-T63, 7039-T6X31, and 7106-T63. These designations signify a group of high-strength aluminum alloys. The data points for the titanium scatter band in Fig. 5.23 were obtained from experimenting with four titanium alloys ranging in yield strength from 110 to 150 ksi. The alloy designations were Ti-6Al-4V, Ti-6Al-6V-2Sn, Ti-7Al-2.5 Mo, and Ti-7Al-2Cb-1Ta. These designations are certainly consistent with the high-strength category of titanium alloys. The specifics of the fatigue rate transition from Region II to Region III for aluminum and titanium alloys are discussed by Barsom and Rolfe.[1] The general concept of crack growth in steel, in relation to the three regions of crack propagation, is outlined graphically in Fig. 5.16.

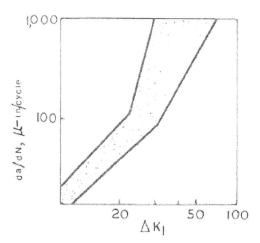

FIGURE 5.22 Approximate outline of scatter band for aluminum.

SPECIAL EFFECTS

It is shown in the preceding section that Eqs. (5.17) through (5.19) can be used for estimating fatigue crack growth rates in martensitic, ferrite–pearlite, and austenitic stainless steels. Although the establishment of these groups reflected the concern for differences in microstructure and mechanical properties, a closer

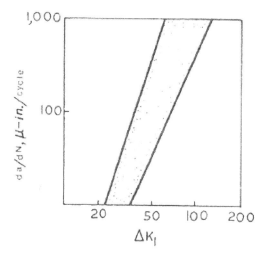

FIGURE 5.23 Approximate outline of scatter band for titanium alloys.

look at the three working equations shows that a single formula, such as Eq. (5.17), may be used for developing a conservative estimate of the rate of fatigue crack growth in various steels.[31]

Some of the special effects on fatigue crack initiation and propagation include the mean load (or stress) as well as the stress ratio, often denoted by R. In relation to Fig. 5.3 we have the mean stress $\sigma_{mean} = 0.5 \, (\sigma_{min} + \sigma_{max})$ and $R = \sigma_{min}/\sigma_{max}$. Several investigators studied the effects of these parameters in spite of the difficulties of measuring crack depth. Also, the general premise was maintained that the crack shape did not change as the crack size increased. The original studies indicated also that the stress ratio had to constitute a second-order effect on the rate of crack propagation. There was, however, a measurable effect of this ratio on the fatigue crack initiation threshold. For other details of the investigation of crack growth concepts and consequences in relation to the σ_{mean} and R parameters, the reader may wish to consult the various sources quoted in recent books.[1,2] The technical papers and books of the 1960s and 1970s in particular abound in theoretical and experimental studies in the new science of fracture. It is stated in Chapter 4, in connection with Figs. 4.7 and 4.20, that the rate of loading can significantly influence toughness. In general, increasing the rate of loading moderately causes the fracture toughness of steels to decrease. At the same time the toughness increases with a decrease in yield strength. The dynamic effects are complex at best, involving material properties, stress intensity, and kinetic energy of a fast-moving fracture.[2] The material dynamic toughness, K_{Id}, is a material's property under impact dynamic loading and conditions of a maximum constraint. This property is quite similar to K_{Ic} and it can be estimated using Eq. (4.9) or (4.10) in the case of structural steels. It relates to the dynamic tensile yield strength, which is still very difficult to measure. Indirectly, the effect of loading rate can also be analyzed in terms of the cyclic frequency because the change of loading frequency can be interpreted as a change in the rate of dynamic or impact loading. Barsom and Rolfe[1] reported that, so far, the "garden variety" A36 steel in a benign environment was insensitive to the effect of a large range of the frequency of the cyclic stress fluctuations on the rates of fatigue crack growth. Similarly, a quality maraging steel (12Ni-5Cr-3Mo), with 180 ksi yield, tested under various cyclic stress frequencies and different stress-time profiles (sinusoidal, triangular, square, etc.), has shown that cyclic frequency and waveform had virtually no effect on crack growth rates of this type of steel in a benign environment and at room temperature. The outline of the scatter bands comparing A36 and maraging steels is given in Fig. 5.24. The results of these investigations still indicate that the primary parameter governing the rate of crack growth per cycle of loading is ΔK_I.

Various preliminary comments concerning the concept of a threshold for fatigue cracking were made in connection with the potential use of Eqs. (5.11) and (5.12). It should also be stated that numerous factors may affect the fatigue crack propagation threshold, and much more work will need to be done before

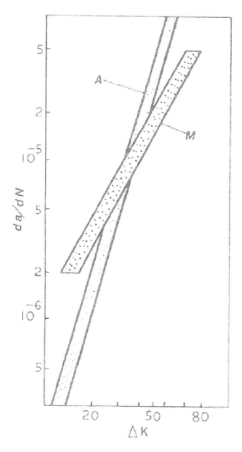

FIGURE 5.24 Comparison of scatter bands for A36 and maraging steels (*A*, A36 steel; *M*, maraging steels; ΔK is in ksi (in.)$^{1/2}$, da/dN, in./cycle).

proper use of ΔK_{th} in design is assured. This is equally important during constant and variable amplitude of cycling loading. In the meantime the test data for long fatigue cracks in "constant-amplitude cycling" support the idea that the stress ratio R is the most influential factor. It also appears that conservative predictions of ΔK_{th} can be made for martensitic, bainitic, ferrite-pearlitic, and austenitic steels, using the following criteria:[38]

$$\Delta K_{th} = 6.4(1 - 0.85R) \tag{5.23}$$

or

$$\Delta K_{th} = 7(1 - 0.85R) \tag{5.24}$$

Here, the result from Eq. (5.23) is given in ksi (in.)$^{1/2}$, and from Eq. (5.24) we have MPa (m)$^{1/2}$, for the English and SI units, respectively. The scatter in the data observed so far could be, at least in part, corrected by complying with standard testing procedures.

The basic problem of dealing with a rather complex situation such as crack growth across weldments requires a good deal of engineering pragmatism, which dictates simpler approaches. As a rule, fatigue cracks initiated in the heat-affected zone (HAZ) propagate into the adjacent weld metal. To date experience[1] suggests that the rate of fatigue crack growth in the weld and HAZ is less than that in the base metal. This conclusion is based on experiments with welded joints involving steels such as HY-140 and A645 as well as 308 and 316 weld metals.[46,47]

Whether we consider the crack growths across a welded region or weld-free areas, it is customary to assume a constant-amplitude load application for reasons of simplicity. Incremental increase of crack length can be recorded for a given number of elapsed load cycles N. Unfortunately, engineering structures in the real world are subjected to complex fluctuating loads of significant variation from cycle to cycle. This, of course, may affect the fatigue life of a structure and it would normally call for a better understanding of the entire process of fatigue. Again we are confronted with a situation beyond our control since the effects of variable-amplitude loading on fatigue life are still not well known, and whatever is known applies only to crack propagation in metals.

The simplest case, of course, deals with the constant, sinusoidal loading indicated in Fig. 5.3. Such a stress history is well defined by the stress range $\Delta\sigma$, which relates to σ_{alt} or the nominal stress amplitude.

$$\Delta\sigma = \sigma_{max} - \sigma_{min} \tag{5.25}$$

or

$$\sigma_{alt} = \frac{\sigma_{max} - \sigma_{min}}{2} \tag{5.26}$$

The corresponding mean stress follows directly from Fig. 5.3.

$$\sigma_{mean} = \frac{\sigma_{max} - \sigma_{min}}{2} \tag{5.27}$$

The case of a total random stress history may be thought of as illustrated in Fig. 5.25. It is quite distressing to note the multitude of applied stress patterns and the complexity of analytic functions that can be imagined between the two extremes of the stress histories of Figs. 5.3 and 5.25.[48]

Since important structures such as bridges, ships, large buildings, and similar engineered systems must carry variable-amplitude loading, it is necessary to fall back on some form of probability methodology to characterize the

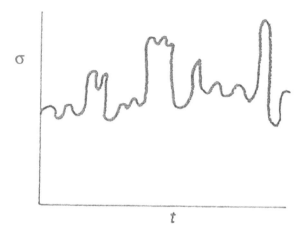

FIGURE 5.25 Random stress history (σ, applied stress; t, time).

variable-amplitude behavior. One example of this type of approach is the development of a nondimensional mathematical formula for defining probability–density curves to fit field data for bridges.[49]

The currently used technique for predicting fatigue crack growth under variable-amplitude loading involves a superposition of high tensile load fluctuations on constant-amplitude cycling, indicated roughly in Fig. 5.26. This type

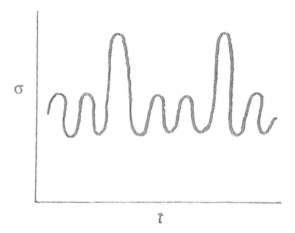

FIGURE 5.26 Superposition of local and constant-amplitude cycling (σ, applied stress; t, time).

of loading, analyzed by a number of investigators and reported in recent textbooks,[1,2] has raised the issue of the observed phenomenon of crack growth delay caused by the process of mixing the modes of loading. Much data concerning this problem has been published, attributed to such behavior as crack tip blunting, residual stresses, and crack closure. And the field of fracture mechanics has been blessed by yet another special term "delayed retardation",[50] or simply "retardation".[2] The mathematical complexity of the retardation models has certainly delayed any development of an all-encompassing formula for design purposes. This topic is definitely beyond the scope of practical fracture mechanics envisaged for this book, and indeed beyond the majority of books in the area of fracture technology. The general reader will have, most likely, no use for a detailed review of the literature devoted to various theories and computer simulations of the retardation models. However, a pragmatic view of the retardation effects is well described by Broek[2] for those who may be tempted to venture into the confines of research.

The most practical approach to dealing with variable-amplitude load fluctuations is the Barsom model,[51] which relates fatigue crack growth rate per cycle to an "effective stress intensity" factor. This approach was also designated the root-mean-square (rms) model. The increase of crack length under variable-amplitude conditions is measured at the particular number of elapsed load cycles. The magnitude of the stress intensity factor, however, has to be recorded as ΔK_I for each cycle. This procedure also requires a technique for correlating the crack length, cyclic load amplitude, and the load sequence. The objective of this correlation is to determine a single stress intensity parameter that can be used to define the crack growth rate under constant and variable loading conditions. Also the resulting curves of crack length a vs. the number of load cycles N should be reasonably continuous and smooth, in spite of the frequent overloads, such as those illustrated in Fig. 5.27. This type of curve, of course, is possible only when there is a series of frequently applied overloads. The experiments indicate that the fatigue life under a constant amplitude of load fluctuation is longer than that under a random sequence, having the same level of applied loading under both conditions.

The best correlation between the constant amplitude and variable amplitude (random sequence) so far was obtained when the root mean square represented the square of the mean of the squares of the individual load cycles. Barsom and Rolfe[1] also suggest that the average fatigue crack growth rates per cycle da/dN, under variable-amplitude and random-stress spectra conditions, can be estimated from the constant-amplitude data using the following equation.

$$\frac{da}{dN} = A(\Delta K_{rms})^m \tag{5.28}$$

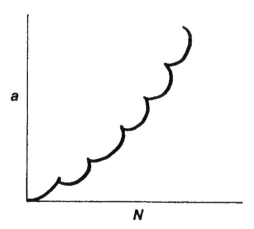

FIGURE 5.27

where A and m are constants (Paris equation) and ΔK_{rms} denotes the root-mean-square stress intensity factor. Extensive crack growth experiments under constant and variable amplitude conditions were conducted on A36, A588 Grade A, A588 Grade B, A514 Grade E, and A514 Grade F steels.[52] All these tests confirmed that the average fatigue crack growth rates per cycle, da/dN, can be evaluated with the help of Eq. (5.28).

DESIGN PROCEDURES

The subject matter of this section is intended to be a brief calculational supplement utilizing some of the formulas quoted in this chapter and illustrating a couple of numerical procedures related to constant and variable amplitudes in fatigue crack growth. Engineering pragmatism dictates that special attention be given here to calculational details in order to help the reader save time and concentrate on more demanding issues. It is often too easy for technical writers to exclude certain steps in derivation and numerical work, sometimes labeled "obvious" and thereby left to the chores of others.

Design Problem 5.3

Provide detailed steps for deriving the expression for the term $R.A.$ (reduction of area) as a function of the number of fatigue cycles N plotted in Fig. 5.7. This calculation supplements Design Problem 5.2.

Solution

From Eq. (5.3), the parameter C is

$$C = \frac{2(S_a - S_e)N^{1/2}}{E}$$

For the numerical values used in Design Problem 5.2

$$S_a = 36\,\text{ksi}$$
$$S_e = 48\,\text{ksi}$$
$$E = 30{,}000\,\text{ksi}$$

we obtain

$$C = \frac{2(36 - 48)N^{1/2}}{30{,}000}$$
$$C = -0.0008N^{1/2}$$

From Eq. (5.4)

$$C = 0.5\,\ln x$$
$$x = e^{2C}$$

where

$$x = \frac{100 - R.A.}{100}$$

and

$$R.A. = 100(1 - x)$$

since

$$2C = -0.0016N^{1/2}$$

Substituting for x and $2C$ gives the required expression as

$$R.A. = 100(1 - e^{-0.0016N^{1/2}})$$

Design Problem 5.4

Estimate the ultimate strength of a martensitic steel having a strength ratio of 0.85 if the crack initiation parameter under the threshold conditions is known to be of the order of 110 ksi. Plot the variation of the stress intensity factor ΔK_I with the notch radius between the limits of 0.1 and 0.4 in. Note that by definition the crack initiation threshold parameter is given by the ratio $\Delta K_I/\rho^{1/2}$.

Solution

From Eq. (5.12)

$$S_y^{2/3} = 110/5$$
$$= 22$$
$$S_y = 22^{1.5}$$
$$= 103.2 \, \text{ksi}$$

Since the stress ratio is

$$S_y/S_u = 0.85$$
$$S_u = S_y/0.85$$
$$= 103.2/0.85$$
$$S_u = 121.4 \, \text{ksi}$$

Rearranging Eq. (5.12)

$$\Delta K_I = 5 \times 103.2^{2/3} \times (\rho)^{1/2}$$
$$\Delta K_I = 110(\rho)^{1/2}$$

The plot of this equation is given in Fig. 5.28.

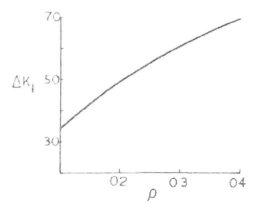

FIGURE 5.28 Stress intensity as a function of notch radius, Design Problem 5.4 (ρ, notch radius in inches; ΔK_I, variation of stress intensity factor in ksi (in.)$^{1/2}$).

Design Problem 5.5

A bearing pedestal of a rotating equipment has a martensitic steel plate with an edge notch. The plate is subjected to a cyclic load in tension due to very high rotor vibrations. Estimate the rate of crack growth on the premise that the applied maximum and minimum stresses on the plate in fatigue are 58 and 18 ksi, respectively. The approximate notch size for this case can be assumed to be 0.45 in.

Solution

The amplitude stress component follows from Fig. 5.3 as

$$\Delta\sigma = 2\sigma_{alt} = \sigma_{max} - \sigma_{min}$$

or

$$\Delta\sigma = 58 - 18$$
$$= 40\,\text{ksi}$$

From Eq. (3.4), the stress intensity factor for an edge crack in a semi-infinite plate is

$$K_I = 1.12(\pi)^{1/2}\sigma(a)^{1/2}$$

Hence the rate of change, or the "stress intensity factor fluctuation," proportional to the alternating stress in fatigue, is

$$\Delta K_I = 2(\Delta\sigma)(a)^{1/2}$$

which, on substituting the numerical values, gives

$$\Delta K_I = 2 \times 40(0.45)^{1/2}$$
$$= 53.7\,\text{ksi (in.)}^{1/2}$$

For a conservative prediction (assuming K_I to be smaller than K_{Ic}), one can use Eq. (5.17).[1]

$$\frac{da}{dN} = 0.66 \times 10^{-8}(\Delta K_I)^{2.25}$$
$$= 0.66 \times 10^{-8}(53.7)^{2.25}$$

$$\frac{da}{dN} = 0.00052\,\text{in./cycle}$$
$$= 0.0013\,\text{mm/cycle}$$

It is interesting to note that the design engineer would certainly have a less confusing task if the notation and definitions of, say, ΔK or ΔK_I could conform to a more general practice of symbols and descriptions. For instance, the knowledge of fracture mechanics is still confined to the repertoire of a few specialists and researchers in the field, with limited features of the uniformity of information needed for design purposes.

In terms of the usual references to ΔK, we find such descriptions as "stress intensity factor fluctuation," "delta K" "stress intensity amplitude," and "stress intensity range," to mention a few. And who knows how many other variations are possible considering that four recent outstanding works in the field[1,2,53,54] required over 1000 references for developing the book material.

Another point of interest relates to the use of a very useful formula, Eq. (5.17), which illustrates the inherently subtle features affecting design procedures. According to the test data,[1] the rate of fatigue crack growth (cycling zero to tension) occurs at a fixed value of crack tip displacement to give

$$K_T = 0.04(ES_y)^{1/2} \tag{5.29}$$

The term 0.04 is equal to the square root of 0.0016 in. Here K_T is the stress intensity factor range when the crack accelerates its motion in transition from Region II to Region III; see Fig. 5.16, where the vertical axis marked da/dN is the rate of crack growth. The value of K_T, Eq. (5.29), has a certain bearing on the design procedure using Eq. (5.17). When ΔK_I becomes larger than K_T, or when the K_{Ic} is numerically close to K_T, estimates of fatigue crack growth rate based on Eq. (5.17) may be in question.[1]

Although Eq. (5.29) is, under normal conditions, intended for martensitic steels working in a benign environment, its use can be extended to aluminum and titanium alloys.[45]

Since the primary mission of this book is the application of the various formulas and rules of fracture mechanics to engineering design, it is necessary, even at the risk of some repetition, to comment on the nature of basic equations defining static and fatigue behavior of cracked (or notched) structural components. Since most practical design and failure issues correspond to Mode I of crack surface displacements, only K_I relations need be discussed. The key characteristic of the stress intensity factor K_I is that it defines the local stress ahead of a sharp crack, and it relates this concept to the so-called global stress existing further away from the crack. The designer needs also to realize that the majority of brittle failures occur under Mode I conditions of loading.

The basic general form of the stress intensity factor in Mode I was given in Chapter 2 as Eq. (2.13). For the purpose of this brief discussion we assume that the crack geometry parameter is equal to unity. Hence the basic equation for the stress intensity calculation (at $f(g) = 1$) becomes

$$K_I = \sigma(\pi a)^{1/2} \tag{5.30}$$

This elementary case and all similar formulas given in this book and other publications show that the stress intensity factor depends on the applied stress σ and the term $(a)^{1/2}$, in addition to a dimensionless parameter accounting for geometry, loading, and so on, which have been the subject of many investigations and which have been quoted in numerous publications.

For the purpose of a direct calculation or a numerical integration, the designer may have to use the term ΔK_I, which follows from a stress intensity formula such as, for example, Eq. (5.30):

$$\Delta K_I = \Delta \sigma \times (a)^{1/2} \times \text{(numerical constant)}$$

which, for the case at hand, is

$$\Delta K_I = 1.77 \Delta \sigma (a)^{1/2} \tag{5.31}$$

where a can be the average crack size in the numerical integration routine and $\Delta \sigma$ is the alternating stress.

To obtain the critical crack size, one can use Eq. (5.30) as a basis for the calculations:

$$K_{Ic} = \sigma_{max}(\pi a)^{1/2} \tag{5.32}$$

or, solving for the crack size, we obtain

$$a_{CR} = \frac{1}{\pi}\left(\frac{K_{Ic}}{\sigma_{max}}\right)^2 \tag{5.33}$$

Equations (5.30) through (5.33) serve here as a brief introduction to the procedure that starts from the general form of the stress intensity relationship applicable to the stress field ahead of a sharp crack found in all kinds of structures such as construction beams, plates, aerospace parts, machine elements, or pressure vessels. However, the geometrical and loading parameters should be established first.

Design Problem 5.6

An electric furnace air-melted steel of the HY-130 type was used to fabricate a wide panel to carry a maximum tensile stress of 100 ksi and a minimum tensile stress of 40 ksi in constant-amplitude stress cycling. The panel can be modeled as a double edge notched infinite plate with the allowed initial crack size of 0.15 in. The yield strength of the material is 180 ksi and the minimum plane-strain fracture toughness is expected to be about 200 ksi $(\text{in.})^{1/2}$.

Estimate the critical crack size and the rate of crack growth based on the average length of the crack.

Solution

From Fig. 5.3 the alternating stress range is

$$\Delta\sigma = 2\sigma_{alt}$$
$$= \sigma_{max} - \sigma_{min}$$
$$= 100 - 40$$
$$\Delta\sigma = 60\,\text{ksi}\ (413.7\,\text{MPa})$$

The basic formula for the double edge notched wide plane is given by Eq. (3.4).

$$K_I = 1.98\sigma(a)^{1/2}$$

For calculating the critical crack size, we can take

$$K_I = 1.98\sigma_{max}(a_{CR})^{1/2}$$

Hence, rearranging gives

$$a_{CR} = 0.255\left(\frac{K_{Ic}}{\sigma_{max}}\right)^2$$
$$= 0.255\left(\frac{200}{100}\right)^2$$
$$a_{CR} = 1.02\,\text{in.}\ (25.9\,\text{mm})$$

Assuming the average crack size a_{ave} yields

$$a_{ave} = \frac{0.15 + 1.02}{2}$$
$$= 0.59\,\text{in.}$$

From the basic equation, we have

$$\Delta K_I = 1.98\Delta\sigma(a_{ave})^{1/2}$$
$$= 1.98 \times 60(0.59)^{1/2}$$
$$= 91.3\,\text{ksi (in.)}^{1/2}\ (100.3\,\text{MPa m}^{1/2})$$

From Eq. (5.29)

$$K_T = 0.04(ES_y)^{1/2}$$
$$= 0.04(30,000 \times 180)^{1/2}$$
$$= 93 \, \text{ksi (in.)}^{1/2} \, (102 \, \text{MPa m}^{1/2})$$

Since both K_{Ic} and ΔK_I are of the right order in relation to parameter K_T, the use of Eq. (5.17) is justified.[1] Hence

$$\frac{da}{dN} = 0.66 \times 10^{-8} \Delta K_I^{2.25}$$
$$= 0.66 \times 10^{-8}(91.3)^{2.25}$$

$$\frac{da}{dN} = 0.00017 \, \text{in./cycle}$$
$$= 0.0043 \, \text{mm/cycle}$$

Design Problem 5.7

A structural steel panel supports the spindle of a cutting machine and experiences cyclic load when machining components with an interrupted cut. The panel steel has a yield strength of 75 ksi and a plane-strain fracture toughness of 100 ksi $(\text{in.})^{1/2}$ and has two cracks emanating from a 2.5 in. hole. The observed length of each crack is 0.2 in. The panel is subjected to a cyclic tensile stress of 32 ksi maximum and 10 ksi minimum during machining operation. Estimate the critical crack length (each side of the hole) and calculate the rate of crack growth based on the average crack size. Show the variation of the rate of crack growth with the rate of the stress intensity.

Solution

From an approximate formula for the stress intensity factor, Eq. (3.6), putting $k = 3$, we have

$$K_I = 5.32\sigma(a)^{1/2}$$

The conventional stress concentration factor of 3 corresponds to a round hole. Hence

$$(\pi)^{1/2} \times 3 = 5.32$$

To obtain the critical crack length, the formula is

$$K_{Ic} = 5.32\sigma_{max}(a_{CR})^{1/2}$$

and

$$a_{CR} = 0.035\left(\frac{K_{Ic}}{\sigma_{max}}\right)^2$$

$$= 0.035\left(\frac{100}{32}\right)^2$$

$$= 0.34\,\text{in. (8.6 mm)}$$

The average crack length calculates as

$$a_{ave} = 0.5(0.34 + 0.20) = 0.27\,\text{in. (6.9 mm)}$$

and the cyclic tensile stress range is

$$\sigma = 32 - 10 = 22\,\text{ksi (151.7 MPa)}$$

Transforming the original formula, Eq. (3.6), we find

$$\Delta K_I = 5.32\Delta\sigma(a_{ave})^{0.5}$$

$$= 5.32 \times 22 \times (0.27)^{1/2}$$

$$= 60.8\,\text{ksi (in.)}^{1/2}\ (66.8\,\text{MPa m}^{1/2})$$

From Eq. (5.29) follows the numerical value of the stress intensity factor in acceleration of crack growth:

$$K_T = 0.04(ES_y)^{1/2}$$

$$= 0.04(30{,}000 \times 75)^{1/2}$$

$$= 60\,\text{ksi (in.)}^{1/2}\ (66\,\text{MPa m}^{1/2})$$

Since the K_{Ic} and ΔK_I values are of the right order of magnitude in comparison with the K_T parameter, the rate of crack growth in this problem will be estimated from Eq. (5.17). Here

$$\frac{da}{dN} = 0.66 \times 10^{-8}\Delta K_I^{2.25}$$

$$= 0.66 \times 10^{-8}(60.8)^{2.25}$$

$$= 68 \times 10^{-6}\,\text{in./cycle (0.0017 mm/cycle)}$$

To this point, the design problems in this chapter involve the estimates of the change in crack length during one stress cycle for a given rate of stress intensity. Hence the da/dN formula requires the input of ΔK_I, which, in turn, is a function of the stress range in cyclic tension $\Delta\sigma$ and the average crack length a_{ave}. However, the most important problem in design is to postulate the presence of an initial flaw and to estimate the number of fatigue stress cycles for the crack (flaw) to grow to a critical length (size). The number of cycles of loading during the crack growth must be greater than the design life of the structural member involved.

The calculation procedure for analyzing the crack growth and the number of applied load cycles requires a conservative methodology based on empirical formulas such as Eqs. (5.17) and (5.18), developed specifically for groups of steels of immediate interest to industry, government, and the design profession. The crack growth for this problem is shown in Fig. 5.29.

Design Problem 5.8

A very wide plate, 0.625 in. thick, is made from ferrite–pearlite steel and is subjected to a cyclic tensile stress alternating between 36 and 6 ksi. Routine inspection has uncovered a fabrication flaw that can be described as a thumbnail-type,

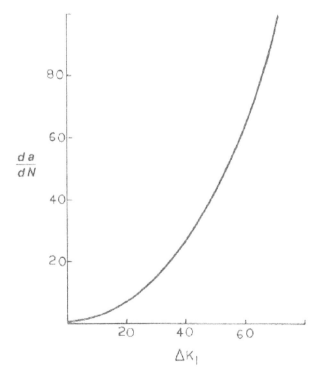

FIGURE 5.29 Crack growth as a function of stress intensity rate, Design Problem 5.7 (ΔK_I, ksi (in.)$^{1/2}$; (da/dN), (in./cycle) $\times$ 10^{-6}).

surface crack 0.1 in. deep and 0.4 in. long. Determine the rate of crack growth and the approximate number of stress cycles to reach the condition of a through-thickness crack. The yield strength is 40 ksi.

Solution

The stress intensity formula for this problem is obtained from Eq. (3.23) as follows.

$$K_I = 2\sigma M_K \left(\frac{a}{Q}\right)^{1/2}$$

From notation in Fig. 3.18

$$a/B = 0.1/0.625$$
$$= 0.16$$
$$a/2C = 0.1/0.4$$
$$= 0.25$$

In order to find the parameter Q from Fig. 3.21, calculate

$$\sigma/S_y = 36/40$$
$$= 0.9$$

Hence for $a/2C = 0.25$ and $\sigma/S_y = 0.9$, Fig. 3.21 gives the flaw shape parameter

$$Q = 1.29$$

The stress intensity factor for the surface crack also requires the use of the back free-surface correction factor M_K. For the fixed a/B ratio of 0.16 in this approximation, the factor can be taken from Fig. 3.19 as 1.0. Then the appropriate formula for the stress intensity factor is now obtained by substitution.

$$K_I = 2\sigma \times 1.0 \left(\frac{a}{1.29}\right)^{1/2}$$
$$= 1.76\sigma(a)^{1/2}$$

The average crack size is taken here as

$$a_{ave} = 0.5(0.1 + 0.625)$$
$$= 0.3625 \text{ in. } (9.2 \text{ mm})$$

Hence the stress intensity factor rate can be written as

$$\Delta K_I = 1.76(\Delta\sigma)(a_{ave})^{1/2}$$

and for

$$\Delta\sigma = 36 - 6 = 30\,\text{ksi} = 206.9\,\text{MPa}$$

$$\Delta K_I = 1.76 \times 30(0.3625)^{1/2}$$

$$\Delta K_I = 31.8\,\text{ksi (in.)}^{1/2}\ (35\,\text{MPa m}^{1/2})$$

Using Eq. (5.18)

$$\frac{da}{dN} = 3.6 \times 10^{-10}\ (\Delta K_I^3)$$

$$= 3.6 \times 10^{-10}\ (31.8)^3$$
$$= 0.0000116\,\text{in./cycle}$$
$$= 0.000295\,\text{mm/cycle}$$

Hence the approximate number of stress cycles to grow the crack through the entire plate thickness of

$$0.625 - 0.1 = 0.525\,\text{in.}$$

can be obtained as

$$N = 0.525/0.0000116 = 45,260\ \text{cycles}$$

In this calculation da/dN is essentially replaced by $\Delta a/\Delta N$, consistent with the concept of a numerical integration, assuming, however, only a single increment of crack growth. This is obviously a crude, preliminary approach. The problem should be concerned with a more realistic set of calculations based on a reasonable number of smaller increments. There is, of course, no need to compute the critical crack dimension a_{CR} because we can assume this to be equal to the plate thickness of 0.625 in., consistent with the design problem definition of a through-thickness crack. If instead of $a = 0.525$ in. we can assume the average crack length $a_{ave} = 0.3625$ in., then the crude approach to estimating the total number of cycles can bring yet another result of approximate nature:

$$N = 0.3625/0.0000116 = 31,250\ \text{cycles}$$

A more realistic approach to estimating the fatigue life in this problem is by a numerical integration, which is now given in full detail.

One of the first steps in this procedure is to assume an increment of crack growth Δa and to review the variable elements of the appropriate expression for calculating the stress intensity factor range ΔK_I consistent with the geometry and nature of the surface crack described as the thumbnail configuration. As shown by Eq. (3.23), there are essentially two corrections, the flaw shape parameter Q and the free-surface correction factor M_K, which depend on crack size and thickness of the plate. While the Q parameter varies with the crack geometry, assumed to be elliptical, it may be postulated that the $a/2c$ ratio (shown in Fig. 3.21) remains relatively constant as the crack depth a increases. Also, with only a small error, the ratio σ/S_y can be calculated at $\sigma = \sigma_{max}$ equal to 36 ksi, so that as before both ratios, $a/2c = 0.25$ and $\sigma_{max}/S_y = 0.9$, give $Q = 1.29$, from Fig. 3.21. The situation with the parameter M_K, however, is somewhat different because the M_K factor can vary. Therefore, the formulas for the stress intensity have to be stated as follows:

$$K_I = 2\sigma M_K \left(\frac{a}{1.29}\right)^{1/2}$$

and

$$\Delta K_I = 1.76\,M_K(\Delta\sigma)(a_{ave})^{1/2}$$

The corresponding expression for an increment of the fatigue life in ferrite–pearlite steel in terms of the number of stress cycles becomes

$$\Delta N = \frac{\Delta a}{3.6 \times 10^{-10}(\Delta K_I)^3}$$

Let us assume an increment of crack growth of $\Delta a = 0.1$ in. The design procedure can now be summarized. The expression for ΔK_I in this particular case is given in terms of the M_K correction factor (Fig. 3.19). The fatigue stress range at each cycle $\Delta\sigma$ and the average crack size a_{ave} are computed between the two consecutive crack increments. Hence the calculation of the stress intensity factor range ΔK_I is also required at each crack increment. Since the crack size (a_{ave}) changes with each increment, the correction factor M_K is refigured at each step in providing a set of gradually increasing values of the term ΔK_I.

In the next step, already having the appropriate formula for the crack growth rate ΔN, the increments of the number of cycles are computed and added at each step to give a total as shown in Table 5.2. The crack sizes a_i and a_f denote the initial and final values at each step, respectively.

The result of numerical integration provides adequate response to the question raised in this design problem. However, it may be of interest to note that in spite of the overall complexity of the equations involved, the crude approach based on the total crack length underestimated the total number of cycles by only 8%. Although such a close agreement in this case may well be fortuitous, the power and pragmatic utility of simplified, "ballpark" engineering solutions should never be underrated.

TABLE 5.2 Calculation of Fatigue Life, Design Problem 5.8.

a_i (in.)	a_f (in.)	a_{ave} (in.)	$\dfrac{a_{ave}}{B}$	M_K	$\dfrac{\Delta K_I}{(ksi\ in.^{1/2})}$	$\dfrac{\Delta N}{(cycles)}$	$\dfrac{\sum N}{(cycles)}$
0.1	0.2	0.15	0.24	1.04	21.26	28907	28907
0.2	0.3	0.25	0.40	1.10	29.04	11340	40247
0.3	0.4	0.35	0.56	1.21	37.80	5143	45390
0.4	0.5	0.45	0.72	1.41	49.94	2230	47620
0.5	0.6	0.55	0.88	1.70	66.57	942	48562
0.6	0.625	0.613	0.98	1.80	74.41	674	49236

Design Problem 5.9

A wide plate made of A514 steel contains an initial edge crack $a_i = 0.24$ in. The plane-strain fracture toughness of the material is 140 ksi $(in.)^{1/2}$. The plate is subjected to fluctuating tensile stresses of 20 ksi minimum and 50 ksi maximum. Calculate the number of cycles for the crack to reach the critical length.

Solution

The appropriate formula for the stress intensity factor in this case is

$$K_I = 1.12(\pi)^{1/2}\sigma(a)^{1/2}$$

or

$$K_I = 2\sigma(a)^{1/2}$$

Hence the critical crack length is

$$a_{CR} = 0.25\left(\frac{K_{Ic}}{\sigma_{max}}\right)^2$$

$$= 0.25(140/50)^2$$

$$= 1.96 \text{ in. } (50\,mm)$$

The live-load stress range, as defined in the problem, is

$$\Delta\sigma = 50 - 20$$

$$= 30\,ksi\ (207\,MPa)$$

The increment of crack growth for this calculation is taken here, for instance, as

$$\Delta a = 0.12 \text{ in. } (3\,mm)$$

The expression for computing the appropriate values of the stress intensity factor range is obtained from the foregoing formula for K_I:

$$\Delta K_I = 2(\Delta\sigma)(a_{ave})^{1/2}$$

or

$$\Delta K_I = 60(a_{ave})^{1/2}$$

The empirical formula for fatigue crack growth per cycle is given by Eq. (5.17).

$$\frac{\Delta a}{\Delta N} = 0.66 \times 10^{-8}(\Delta K_I)^{2.25}$$

and

$$\Delta N = \frac{0.12}{0.66 \times 10^{-8}(\Delta K_I)^{2.25}}$$

or

$$\Delta N = \frac{0.18 \times 10^8}{(\Delta K_I)^{2.25}}$$

With this preparation of the formulas for the numerical integration, the computational details can be compiled (see Table 5.3).

TABLE 5.3 Calculation of Fatigue Life, Design Problem 5.9.

a_i (in.)	a_f (in.)	a_{ave} (in.)	K_I (ksi in.$^{\frac{1}{2}}$)	N (cycles)	ΣN (cycles)
0.24	0.36	0.30	32.9	6944	6,944
0.36	0.48	0.42	38.9	4763	11,707
0.48	0.60	0.54	44.1	3592	15,299
0.60	0.72	0.66	48.7	2873	18,172
0.72	0.84	0.78	53.0	2375	20,547
0.84	0.96	0.90	56.9	2024	22,571
0.96	1.08	1.02	60.6	1757	24,328
1.08	1.20	1.14	64.1	1548	25,876
1.20	1.32	1.26	67.3	1388	27,264
1.32	1.44	1.38	70.5	1250	28,514
1.44	1.56	1.50	73.5	1138	29,652
1.56	1.68	1.62	76.4	1043	30,695
1.68	1.80	1.74	79.1	965	31,660
1.80	1.96	1.88	82.3	882	32,542

For a rough estimate prior to the numerical integration, the following steps can be made:

$$a_{ave} = \frac{0.24 + 1.96}{2} = 1.1$$

$$\Delta K_1 = 60(1.1)^{1/2} = 62.9$$

$$\frac{\Delta a}{\Delta N} = 0.66 \times 10^{-8}(62.9)^{2.25}$$

$$= 0.000074$$

$$1.96 - 0.24 = 1.72$$

$$1.72/0.000074 = 23{,}243 \text{ cycles}$$

Although such estimates may fall within 10–30% of the numerical integration, no general observations should be made because of the various complexities involved in the entire analytical process.

The approach to a variable-amplitude load fluctuation problem is discussed in general terms in the section dealing with the special effects in this chapter. The most practical technique known to the author is the Barsom model, which involves the following simple equations:

$$\Delta\sigma_{rms} = \left\{ \sum_i \alpha_i (\Delta\sigma_i)^3 \right\}^{1/3} \qquad (5.34)$$

$$\alpha_i = \frac{N_i}{\sum_i N_i} \qquad (5.35)$$

$$N_y = \frac{N_f}{\sum_i N_i} \qquad (5.36)$$

where $\Delta\sigma_{rms}$ = root-mean-square stress range, α_i = parameter in Eqs. (5.34) and (5.35), $\Delta\sigma_i$ = stress range for N_i cycles, and N_f = total number of fatigue cycles, and N_y = fatigue in life years.

The problem of a variable amplitude starts normally with the assumed number of cycles per year corresponding to a particular stress range. The basic elements of this methodology are illustrated in the next design example.

Design Problem 5.10

A welded thick plate is temporarily used as a structural member in a bridge. Is it safe to replace this plate after three years in service if it is designed to resist a variable-amplitude load fluctuation? Assuming that we are given the numbers of cycles per year and their corresponding stress ranges (columns 2 and 3 of Table 5.4), estimate the fatigue life of this member. The design equation for the fatigue curve (stress range vs. the number of load cycles) for this category

TABLE 5.4 Determination of Fatigue Data for Design Problem 5.10.

i	Number of cycles per year (N_i)	Stress range, ΔS_i (ksi)	$\alpha_i = \dfrac{N_i}{\sum_i N_i}$	$\dfrac{\alpha_i}{(\Delta S_i)^3}$
1	3,500,000	0.18	0.627	0.004
2	1,200,000	0.52	0.215	0.030
3	160,000	3.60	0.0287	1.339
4	720,000	5.10	0.1290	17.112
5	2,400	11.90	0.00043	0.725
	$\sum_i N_i = 5{,}582{,}400$		$\sum \alpha_i (\Delta S_i)^3 = 19.21$	

of a welded structure is given by Barsom and Rolfe.[1]

$$\log N_f = 8.59 - 3 \log \Delta\sigma$$

Solution

The calculation utilizes columns 2 through 5 of Table 5.4. From Eq. (5.34)

$$\Delta\sigma_{rms} = \left\{ \sum_i \alpha_i (\Delta\sigma_i)^3 \right\}^{1/3}$$

$$= (19.21)^{1/3} = 2.68$$

and

$$\log N_f = 8.59 - 3 \log 2.68$$
$$= 8.59 - 1.284$$
$$= 7.306$$

This gives

$$N_f = 20{,}230{,}192 \text{ cycles}$$

Hence, from Eq. (5.36)

$$N_y = \frac{20{,}230{,}192}{5{,}582{,}400} = 3.6 \text{ years}$$

This value is consistent with the requirement of a three-year service life specified by the design problem. The log–log equation (N_f vs. $\Delta\sigma_{rms}$) used in this example corresponds roughly to the category of fatigue life design curves dealing with welded thick beams and plate girders. The problem falls into the highly complex regime of fatigue of welded components,[1] of special interest to the automobile, shipbuilding, and other industries. Further comments related to crack mechanics and welding technology are included in Chapter 6.

SYMBOLS

A	Area of cross-section, in.^{2} (mm^2)
A	Constant (Paris equation)
A	Constant, Eqs. (5.20) and (5.21)
a	Crack length (also minor half-axis of ellipse), in. (mm)
a_{CR}	Critical crack length, in. (mm)
a_{ave}	Average crack length, in. (mm)
a_f	Final crack size, in. (mm)
a_i	Initial crack size, in. (mm)
B	Thickness of plate, in. (mm)
C	Reduction of area factor
c	Major half-axis of ellipse, in. (mm)
E	Modulus of elasticity, ksi (MPa)
$f(g)$	Geometrical parameter
K_{eff}	Effective stress intensity factor, ksi (in.)$^{1/2}$ [MPa (m)$^{1/2}$]
K_I	Stress intensity factor (Mode I), ksi (in.)$^{1/2}$ [MPa (m)$^{1/2}$]
K_{Ic}	Plane-strain fracture toughness, ksi (in.)$^{1/2}$ [MPa (m)$^{1/2}$]
K_{Id}	Dynamic fracture toughness, ksi (in.)$^{1/2}$ [MPa (m)$^{1/2}$]
K_T	Stress intensity factor range, ksi (in.)$^{1/2}$ [MPa (m)$^{1/2}$]
k	Conventional stress concentration factor
k_f	Stress factor from fatigue tests
$k_t(a)$	Stress factor as a function of crack length
M	Bending moment, lb-in. (N-mm)
M_K	Front free-surface correction factor
m	Constant (Paris equation)
m	Constant, Eq. (5.21)
N	Number of fatigue cycles to failure
N_f	Total number of cycles
N_i	Number of cycles at a given interval
N_o	Number of applied cycles
N_y	Number of years of fatigue life
n	Strain-hardening exponent
P	Concentrated load, lb (N)
Q	Flaw shape parameter
q	Notch sensitivity factor
R	Stress ratio ($\sigma_{min}/\sigma_{max}$)
$R.A.$	Reduction of area (percent)
r	Arbitrary radius, in. (mm)

S_e	Endurance limit, ksi (MPa)
S_u	Ultimate strength, ksi (MPa)
S_y	Yield strength, ksi (MPa)
Z	Section modulus, in.^{3} (mm^3)
α_i	Special parameter, Eqs. (5.34) and (5.35)
Δa	Increment of crack size, in. (mm)
ΔK	Stress intensity fluctuation, ksi (in.)$^{1/2}$ [MPa (m)$^{1/2}$]
ΔK_{eff}	Range of effective stress intensity, ksi (in.)$^{1/2}$ [MPa (m)$^{1/2}$]
ΔK_I	Stress intensity factor range, ksi (in.)$^{1/2}$ [MPa (m)$^{1/2}$]
ΔK_{th}	Threshold stress intensity range, ksi (in.)$^{1/2}$ [MPa (m)$^{1/2}$]
$\Delta \sigma_i$	Stress range at N_i, ksi (MPa)
$\Delta \sigma_{max}$	Maximum stress amplitude, ksi (MPa)
$\Delta \sigma_{rms}$	Root-mean-square stress range, ksi (MPa)
θ	Arbitrary angle, degrees
ρ	Notch tip radius, in. (mm)
σ	Applied stress, ksi (MPa)
σ	Amplitude of nominal stress, also stress range, ksi (MPa)
σ_a	Stress amplitude, ksi (MPa)
σ_{alt}	Alternating stress, ksi (MPa)
σ_{max}	Maximum fatigue stress, ksi (MPa)
σ_{mean}	Mean fatigue stress, ksi (MPa)
σ_{min}	Minimum fatigue stress, ksi (MPa)

REFERENCES

1. Barsom, J.M.; Rolfe, S.T. *Fracture and Fatigue Control in Structures*, 2nd Ed.; Prentice-Hall: Englewood Cliffs, NJ, 1987.
2. Broek, D. *The Practical Use of Fracture Mechanics*; Kluwer Academic Publishers: Dordrecht, 1988.
3. Broek, D. *Cracks at Structural Holes*, Research Report MCIC-75-25; Battelle Columbus Laboratories: Columbus, OH, 1975.
4. Bowie, O.L. Analysis of an infinite plate containing radial cracks originating at the boundary of an internal circular hole. J. Mat. Phys. **1956**, 25.
5. Isida, M. On the determination of stress intensity factors for some common structural problems. Eng. Fract. Mech. **1970**, 2, 61.
6. Van Oosten Slingeland, G.L.; Broek, D. *Fatigue Cracks Approaching Circular Holes* (in Dutch); Delft University: The Netherlands, 1973.
7. ASTM. *Manual on Fatigue Testing*; ASTM STP No. 91; ASTM: Philadelphia, 1949.
8. Tall, L. *Structural Steel Design*, 2nd Ed.; Ronald Press: New York, 1974.
9. Neuber, H. *Theory of Notch Stresses*; J.W. Edwards: Ann Arbor, MI, 1946.
10. Peterson, R.E. *Stress Concentration Design Factors*; John Wiley: New York, 1953.

11. Field, M.; Kahless, J.F. *The Surface Integrity of Machined-and-Ground-High Strength Steels*, DMIC Rep. 210; Defense Metals Information Center, Battelle Memorial Institute: Columbus, OH, 1964.
12. Moore, H.F. *Shot Peening and the Fatigue of Metals*; American Foundry Equipment Co.: Tulsa, OK, 1944.
13. Tapsell, H.J. *Fatigue at High Temperatures, Symposium on High Temperature Steels and Alloys for Gas Turbines*; Iron and Steel Institute: London, 1950.
14. Travernelli, J.F.; Coffin, L.F. *Experimental Support for Generalized Equations Predicting Low Cycle Fatigue*; Appl. Ser. M-3; Instron Engineering Corp.: Canton, MA, 1959.
15. American Society for Testing and Materials. *Annual Book of ASTM Standards*, E466-82; ASTM: Philadelphia, 1983; Sect. 3, Vol. 03.01.
16. American Society for Testing and Materials. *Annual Book of ASTM Standards*, E606-80; ASTM: Philadelphia, 1983; Sect. 3, Vol. 03.01.
17. *Metals Handbook*, 8th Ed.; American Society for Metals: Metals Park, OH, 1975; Davis, J.R., Ed. Vol. 10.
18. Barsom, J.M. Fatigue-crack propagation in steels of various yield strengths. Trans. ASME J. Eng. Ind. Ser. B **1971**, *93* (4).
19. Faupel, J.H. *Engineering Design*; John Wiley: New York, 1964.
20. Raske, D.T.; Morrow, J. Mechanics of materials in low cycle fatigue testing. In *Manual on Low Cycle Fatigue Testing*, ASTM STP 465; American Society for Testing and Materials: Philadelphia, 1969.
21. Graham, J.A., Ed. *Fatigue Design Handbook*; Society of Automotive Engineers: Warrendale, PA, 1968; Vol. 4.
22. Gordon, J.E. *Structures, or Why Things Don't Fall Down*; Plenum Press: New York, 1978.
23. Creager, M. *The Elastic Stress-Field Near the Tip of a Blunt Crack*, Lehigh University: Bethlehem, PA, 1966; MSc Thesis.
24. Dowling, N.E. *A Discussion of Methods for Estimating Fatigue Life*; SAE Technical Paper Series, Paper No. 820691; Society of Automotive Engineers: Warrendale, Pennsylvania, 1982.
25. Mowbray, D.F.; McConnelee, J.E. Applications of finite element stress analysis and stress–strain properties in determining notch fatigue specimen deformation and life. In *Cyclic Stress–Strain Behavior—Analysis, Experimentation, and Failure Prediction*, ASTM STP 519; American Society for Testing and Materials: Philadelphia, May 1973.
26. Barsom, J.M.; McNicol, R.C. *Effect of Stress Concentration on Fatigue Crack Initiation in HY-130 Steel*, ASTM STP 559; American Society for Testing and Materials: Philadelphia, 1974.
27. Frost, N. Notch effects and the critical alternating stress required to propagate a crack in an aluminum alloy subjected to fatigue loading. J. Mech. Eng. Soc. **1960**, *2* (2).
28. McClintock, F.A. On the plasticity of the growth of fatigue cracks. In *Fracture of Solids*; Wiley-Interscience: New York, 1963.
29. Paris, P.C. Testing for very slow growth of fatigue cracks. MTS Closed Loop Mag. **1970**, *2* (5).
30. Various Authors. In *Fatigue Thresholds*, Eng. Mat. Adv. Serv. (EMAD); Warley Engineering Materials Advisory Services, 1982; 2 Vols.

31. Clark, W.G., Jr. How fatigue crack initiation and growth properties affect material selection and design criteria. Met. Eng. Q. **1974**, August.

32. Ramberg, W.; Osgood, W.R. *Description of Stress–Strain Curves by Three Parameters*, NACA TN 902; NACA, July 1943.

33. Clausing, D.P. Tensile properties of eight constructional steels between 70 and −320°F. J. Mater. **1969**, *4* (2) June.

34. Royer, C.P.; Rolfe, S.T.; Easley, J.T. Effect of strain hardening on bursting behavior of pressure vessels. *Second International Conference on Pressure Vessel Technology: Part II — Materials, Fabrication and Inspection*; American Society of Mechanical Engineers: New York, 1973.

35. Roberts, R.; Barsom, J.M.; Rolfe, S.T.; Fisher, J.W. *Fracture Mechanics for Bridge Design*, Report No. FHWA-RD-78-69; Federal Highway Administration: Washington, DC, 1977.

36. Taylor, M.E.; Barsom, J.M. Effect of cyclic frequency on the corrosion-fatigue crack-initiation behavior of ASTM A517 Grade F steel. In *Fracture Mechanics: Thirteenth Conference*; Richard Roberts, Ed.; ASTM STP 743; American Society for Testing and Materials: Philadelphia, 1981.

37. Paris, P.C.; Erdogan, F. A critical analysis of crack propagation laws. ASME Trans. J. Basic Eng. **1963**, 528–534.

38. Barsom, J.M. Fatigue behavior of pressure vessel steels. WRC Bull. (Welding Research Council, New York) **1974**, *194*.

39. Richie, R.D. Near-threshold fatigue crack propagation in steels. Review no. 245. Int. Met. Rev. **1979**, *5* and 6.

40. Barsom, J.M. *Investigation of Subcritical Crack Propagation*, University of Pittsburgh: Pittsburgh, 1969; DPhil Dissertation.

41. ASTM. *Annual Book of ASTM Standards*, E647-83; American Society for Testing and Materials: Philadelphia, 1983; Sect. 3, Vol. 03.01.

42. Barsom, J.M.; Imhof, E.J., Jr.; Rolfe, S.T. Fatigue-crack propagation in high-yield-strength steels. Eng. Fract. Mech. **1971**, *2* (4).

43. James, L.A. The effect of elevated temperature upon the fatigue-crack propagation behavior of two austenitic stainless steels. In *Mechanical Behavior of Materials*; Society of Materials Science: Tokyo, 1972; Vol. 3.

44. Novak, S.R.; Barsom, J.M. *Brittle Fracture (K_{Ic}) Behavior of Cracks Emanating from Notches, Cracks and Fracture*, ASTM STP 601; American Society for Testing and Materials: Philadelphia, 1976.

45. Crooker, T.W. *Crack Propagation in Aluminum Alloys Under High-Amplitude Cyclic Load*, Report 7286; Naval Research Laboratory: Washington, DC, 1971.

46. Shahinian, P.; Smith, H.H.; Hawthorne, J.R. Fatigue crack propagation in stainless steel weldments at high temperature. Weld. J. **1972**, *51* (11).

47. Maddox, S.J. Assessing the significance of flaws in welds subject to fatigue. Weld. J. **1974**, *53* (9).

48. Leve, H.L. Cumulative damage theories. In *Metal Fatigue: Theory and Design*; Madayag, A.F., Ed.; John Wiley: New York, 1969.

49. Klippstein, K.H.; Schilling, C.G. Stress spectrums for short-span steel bridges. In *Fatigue Crack Growth Under Spectrum Loads*, ASTM STP 595; American Society for Testing and Materials: Philadelphia, 1976.

50. Von Euw, E.F.J. *Effect of Single Peak Overloading on Fatigue Crack Propagation*, Lehigh University: Bethlehem, PA, 1968; MSc Dissertation.

51. Barsom, J.M. *Fatigue-Crack Growth Under Variable-Amplitude Loading in ASTM A514 Grade B Steel*, ASTM STP 536; American Society for Testing and Materials: Philadelphia, 1973.

52. Barsom, J.M.; Novak, S.R. *Subcritical Crack Growth and Fracture of Bridge Steels*, NCHRP Report 181; Transportation Research Board: Washington, DC, 1977.

53. Hertzberg, R.W. *Deformation and Fracture Mechanics of Engineering Materials*, 3rd Ed.; John Wiley: New York, 1989.

54. Anderson, T.L. *Fracture Mechanics Fundamentals and Applications*; CRC Press: Boca Raton, FL, 1991.

6

Elements of Structural Integrity

DUCTILE AND BRITTLE FRACTURE

Recognition of the nature of fracture modes is tied to knowledge of the basic elements of ductile and brittle behavior. For instance, the effects of temperature and the state of stress on brittle behavior are related directly to the various modes of fracture. This information is intended to provide the designer with additional confidence in developing adequate safeguards against structural failures. This is particularly timely because of the maturing processes and procedures in linear elastic fracture mechanics (LEFM) and because of the increasing need for special and higher strength engineering materials. It is well known that the increasing material strength is attained at a sacrifice in ductility and toughness, and we need all possible skills in designing around such a sacrifice.

The mechanism of ductility can be described as a process of formation of free surfaces around any inclusions and as a growth of plastic strains, in addition to the coalescence of microscopic voids. Although this action is confined to a small tear zone, it results in a significant plastic deformation inclined at about 45° to the applied stress. In this manner we arrive at a full shear consistent with the plane-stress condition and brittle-to-ductile transition. The mechanism of ductility was always considered as a way of distributing the loading around the stress concentrations caused by structural holes, smooth fillets, and similar geometric discontinuities. However, it soon became obvious that the role of ductility in stress redistribution around the sharp cracks was rather limited.

The increased use of modern high-strength materials in many applications critical to various industries has resulted in accelerated programs of research aiming at solving the materials problems. Unfortunately, in unison with the entire effort, many multifaceted approaches were undertaken. This process became confusing and bewildering to the practical designers in the field because a maze of results and conclusions not always compatible with one another were reported during the late 1960s and beyond. And even today few research results lead to quantitative design procedures in the area of fracture resistance of a material. It appears that the forces driving research are quite different from those required to bring order and utility of design information for practical purposes.

Before making any comments regarding the entire topic of ductile and brittle behavior, short of venturing into the area of dislocation theory, it may be helpful to examine the basic characteristics of two of the simplest structures: the unnotched cylindrical tensile specimen and the notched slow-bend test piece. These two cases appear to highlight the nature of structural failure in general.

The classical case of an unnotched cylindrical test piece in tension at ambient temperature can be represented by the load deformation curve in Fig. 6.1. After the initial elastic deformation, the test piece elongates until the maximum load is exceeded, with local necking taking place. The existing tension, in combination with the presence of inclusions and defects, leads to incipient cracking, void formation, and continued extension. At this stage a critical crack is formed, which becomes unstable and propagates rapidly. This process of crack growth is governed by either brittle or ductile response. The brittle behavior is normally described by the Griffith–Irwin criteria,[1-3] while the ductile response requires

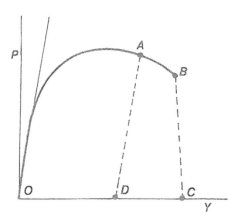

FIGURE 6.1 Load deformation curve for unnotched specimen (*P*, load; *Y*, deformation).

an Orowan correction.[4] The free surface experiences a gross shear failure result-ing in shear lip formation.

The region of the local contraction (necking) has the energy required for crack propagation. Since the neck itself is not inhibited by the surrounding material, rapidly developing brittle fracture cannot be suppressed. This condition is difficult to change unless the specimen is rather short and rigidly supported. In addition, brittle fracture in general can develop and proceed under a small fraction of the yield strength. This particular characteristic is most troublesome, and it is quite contrary to the ductile behavior under which ductile fracture entails stresses above the yield.

At the point at which a crack becomes unstable and grows rapidly, such as point A in Fig. 6.1, the crack growth is related to the mode of fracture and the stiff-ness of the test piece. The length of the curve between points A and B represents crack propagation. Also, the area under the appropriate portion of the curve relates to the fracture resistance of a material. It has been convenient to link the area under the curve with the amount of the energy absorbed. In terms of the illustration of the unnotched specimen in Fig. 6.1, the initiation energy is given by the area OAD, while the propagation energy is bounded by the region ABCD. The initiation energy is certainly the dominant factor in an unnotched specimen.

A further point of interest is concerned with the effect of thickness (size) on the mechanism of fracture,[5] examined using cylindrical tensile specimens. The experiment involved a large-diameter test piece and an order of magnitude smaller diameter specimen subjected to the same type of loading, causing tensile failure in both cases. Both fractured specimens showed essentially the same load deformation characteristics of typical tensile tests. An important difference, however, was discovered in the manner of a load drop following the formation of the neck contour. In the case of a large-diameter specimen there was an uncontrollable and spontaneous load drop, with the net stress level well below the yield strength. On the other hand, the small specimen did not experience the catastrophic drop in load, and the stresses continued to rise in uniform fashion.

The difference in crack mechanics between the two cases was matched by equally contrasting details of fractured surfaces. Although both fractures developed slowly in a flat, fibrous mode, the smaller specimen was unable to reach the point where the critical size of the crack for rapid growth could be attained. The critical size for the large specimen was of the order of 0.75 in. (according to the Griffith criterion), while the entire diameter of the small specimen was only 0.35 in. Hence the large specimen, by reaching the critical dimension, progressed rapidly and in a brittle manner to fracture, driven by the strain energy available within the necked portion of the specimen. Since the small specimen could not accommo-date the critical crack length of 0.75 in., only slow crack extension could take place.

The experiments with large- and small-diameter, unnotched specimens also show that, as the front of the crack gets closer to the free surface, a shear lip develops that is of equal size for both large and small specimens. This finding was confirmed by many experiments with plates (rather than cylinders) of varying thicknesses, paving the way toward the concept of critical thickness.

When the experimentation was extended to the so-called slow-bend, notched specimens, some features of fracture were similar to those observed in unnotched samples. Hence the initiation phase, such as OAD in Fig. 6.1, became smaller, while the propagation phase, ABCD in Fig. 6.1, was enlarged.

The shape of the curve shown in Fig. 6.1 is expected to change with temperature. This development provides yet another slant on the theory of ductile and brittle behavior of engineering materials.

An example of early slow-bend data[6] on load deformation curves at various temperatures is given in Fig. 6.2. Clearly, the general shape of the curves changes as the temperature decreases, with the specimens failing in a brittle manner at stresses well below the yield. The crack propagation is driven by the elastic strain energy stored near the crack front and the fractured surface shows little or no shear lip. The approximate characteristics shown in Fig. 6.2 as curves A, B, and C cover the temperature range from −20°F to about −240°F, with the C curve corresponding to the lowest temperature. The curves shown in Fig. 6.2 are not to scale and are intended only to illustrate the trends for the purpose of general discussion. In the range of temperatures closer to normal working conditions, the test piece is likely to undergo gross plastic deformation before the onset of crack initiation at the base of the notch. In this type of ductile mode of fracture, the nominal stresses during the test exceed the yield strength, and the fractured material surface will be a flat fibrous area bordered by the shear lips at the free surfaces. This type of fracture with the mechanism of a transient buildup of shear lips is consistent with the minimum thickness criteria and the

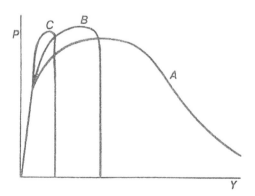

FIGURE 6.2 Effect of temperature on load deformation curves (P, load; Y, deformation). See text discussion of curves A–C.

original idea of the Naval Research Laboratory (NRL) to develop the drop-weight tear test (DWT) methodology discussed in Chapter 4.

At other levels of loading, ductile tearing mixed with brittle response will result in an intermediate degree of fracture resistance. The various forms and modes of fracture in relation to the plane of the fracture can be summed up as shown in Fig. 6.3. It also appears that both unnotched and notched tensile specimens can develop slow crack growth prior to a rapid separation. It is, therefore, possible that an occasional overload may go unnoticed in dealing with a complex structural configuration, and this overload may start a low-stress fracture. Brittle fracture can take place with or without, so to speak, a precursor created by gross plastic flow.

In referring to the several modes of fracture illustrated in Fig. 6.3, the following designations explain details of fractured surfaces.

C	Cleavage
F	Fibrous appearance (microshear mode)
S&C	Gross shear and flat cleavage
S	Gross shear
S&F	Gross shear and flat fibrous surface

Flat fracture defines a plane normal to the nominal tensile stress consistent with the direction of loading as shown by the two-way arrow in Fig. 6.3. Gross shear acts at about 45° to the direction of loading. Flat fracture consists of brittle cleavage facets or dimples characteristic normally of ductile behavior. Some of these modes are also shown in Fig. 4.26. While the ductile dimple mode is a randomly oriented, local microshear phenomenon, the shear lip configuration is definitely a well-defined inclined surface. There is also a likelihood of the two additional mechanisms of shear lip formation existing under unique conditions. Here the term VC, Fig. 6.3, refers to one of the mechanisms controlled by the volume of the shear lip or by the so-called internal process. Finally, the term SC in

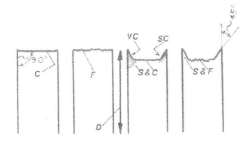

FIGURE 6.3 Orientation and modes of fracture (*D*, direction of loading; *C*, cleavage; *F*, fibrous; *S&C*, gross shear and flat cleavage; *S&F*, gross shear and flat fibrous).

Fig. 6.3 implies control by the surface area or the "external process." For practical purposes, however, only flat fibrous, flat cleavage, and gross shear are considered in design. This is convenient because the cleavage provides a minimum resistance to fracture, while the gross shear implies the relevant maximum. This also leaves the flat fibrous mode as the intermediate characterization of fracture.

In addition to metallurgical and environmental factors, the fracture modes of brittle and ductile type are most affected by temperature and the state of stress.

The entire topic of ductile and brittle fracture has been debated for many years, particularly because of frequent structural failures encountered during the past hundred years. Brittle fracture, characterized as the most insidious mode of a material's behavior, can occur without prior plastic deformation and can grow at speeds as high as 7000 ft/s. In view of such a constant threat, the choice of engineering materials contains negative aspects, and early investigators[5] utilized the concept of notch strength as one of the tools for judging structural integrity. In terms of consistent units, the notch strength can be related to tensile strength, as illustrated roughly in Fig. 6.4.

According to Fig. 6.4 the material's behavior in the area of lower tensile strength is acceptable as long as the strength ratio (notch over tensile) is higher than 1.0. This condition appears to be consistent with the theory of plasticity. However, as the selected strength of the material is gradually increased, the notch strength attains a maximum value, and then it begins to fall off at a rapid rate. It becomes obvious at this point that specifying higher tensile strength to correct the design efficiency can only cause further degradation of structural integrity because of the natural tendency of the material of higher strength to invite brittle behavior.

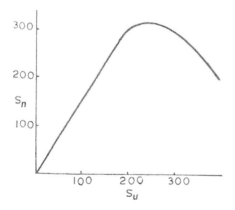

FIGURE 6.4 Notch vs. tensile strength (S_n, notch strength (ksi); S_u, tensile strength (ksi)).

Over the years the traditional mechanical tests of strength — elongation and elastic modulus — had to be supplemented by some form of experiments on notch strength and toughness. By definition, "notch toughness" is the ability of a given material to absorb energy in the face of a flaw. At the same time, "material toughness" is the ability of an unnotched specimen to absorb energy under a slow rate of loading. Notch toughness, then, is well represented by standard tests such as Charpy V-notch (CVN) impact, dynamic tear (DT), or K_{Ic} crack tip opening, to mention a few. Material toughness, on the other hand, is characterized by the area under a stress–strain curve developed in a more common tensile test. Hence the presence of a discontinuity such as a notch, groove, or crack invites brittle behavior.

It is now well established that the majority of structural steels are open to ductile or brittle fracture depending on the loading rate, temperature, and the degree of constraint. Ductile and brittle fractures may be involved in various forms of behavior such as those shown in Fig. 6.3, constituting extreme and mixed modes. Hence the designer has to decide which criteria apply to the case at hand, in view of interrelationships such as those depicted in Fig. 6.5. The vertical axis of this diagram represents performance in terms of absorbed energy, which can be derived from standard tests on notched specimens. The individual curves have the following designations indicating type of external loading:

A   Static
B   Intermediate rate of application
C   Impact

In order to be on the conservative side, curve C is used for design purposes because, in the real world, the actual rates of loading on structures (existing or in the planning stages) are poorly defined. The specific design value of a notch

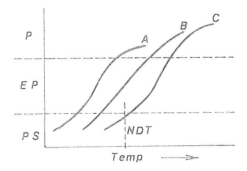

FIGURE 6.5 Areas of structural performance (*P*, plastic; *EP*, elastic–plastic; *PS*, plane strain).

toughness corresponds to the nil-ductility transition (NDT) temperature, defined as the upper limit of plane-strain response.

The majority of notch toughness tests have been developed for specific purposes. However, they all have a common goal: to produce test fractures in steels in such a manner as to assure a good correlation with service performance in the field. Furthermore, to supplement this effort, notch toughness can be characterized using fracture mechanics methodology and involving structural parameters such as stress and crack size.

The lateral constraint ahead of a sharp crack increases with an increase of plate thickness, and it creates a triaxial state of stress. In turn, this state of stress reduces the apparent ductility of the material (such as steel) without changing the metallurgical properties. This process then decreases the notch toughness.

It is not very easy to visualize the various fracture modes in a notched plate in tension or to understand the relationship between the changing mode of fracture and the notch strength. Typically the onset of propagation starts with a flat-fractured surface in the mid thickness and it gradually changes (partially or totally) into a shear mode. And, since the flat mode absorbs less energy than the shear mode, there must be a gradual increase in the resistance to fracture. The feasible changing patterns of a fractured surface are illustrated in Fig. 6.6. The appropriate designations of these patterns are:

A   Machined crack (notch) surface
B   Fatigue crack
C   Flat fracture ("pop-in" type)
D   Steady-state mixed pattern
E   Shear lip
F   Flat fracture (normal)

There is still some question about the extent to which the brittle behavior is affected by an increase in thickness. Experience shows that thin specimens fail

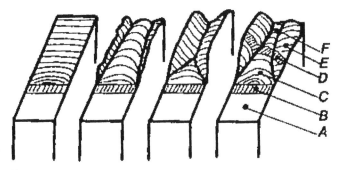

FIGURE 6.6

in a ductile mode at stresses beyond the yield strength of the material, while thick specimens show considerable tendencies toward brittle failure at low stresses. The general features of this situation are best described as shown in Fig. 6.7, where fracture toughness is plotted as a function of thickness of the test pieces.[5] The left side of the diagram concerns the shear-controlled condition, and therefore this area is assigned to plane stress. Beyond the vertical dotted line denoted as C_t (critical thickness), the fracture is of a mixed pattern of pure shear and flat appearance. The horizontal dotted line corresponds to a plane-strain fracture toughness normally expected near the crack tip of thick plates. The other extreme is noted in thin sections with a totally developed shear appearance. In Fig. 6.7 this area is found to the left of the vertical dotted line.

In the case of a significant plastic deformation the fracture behavior of notched components has a number of special characteristics. It appears that with the increasing depth of a notch, the plastic constraint increases and therefore a higher axial stress is required to further deform the specimen. Hence, the yield strength of a notched specimen may be higher than the yield strength obtained from a conventional tensile test on a smooth bar. We observe here the case of "notch strengthening," which can be related to reduction of area in a notched specimen.

An example of such a relationship is given in Fig. 6.8 for the laboratory data on 1018 steel.[7] In this diagram the yield strength ratio (notched sample/smooth sample) is plotted against the reduction of the cross-sectional area, in percent, of the notched sample. Experiments indicate that materials with limited capacity to deform will "notch weaken," while highly ductile materials will "notch strengthen." According to Hertzberg,[7] the net section stress (notched specimen) in a highly deformable material may be two to three times higher than that in a smooth (unnotched specimen), in full compliance with the theory of plasticity.

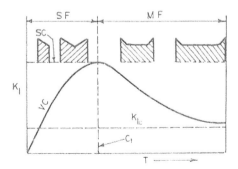

FIGURE 6.7   Variation of toughness (SF, shear fracture; MF, mixed pattern; SC, surface control; VC, volume controlled; C_t, critical thickness; T, temperature).

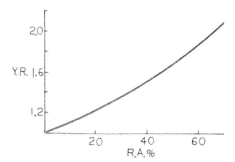

FIGURE 6.8 Notch strengthening (Y.R., yield strength ratio; R.A., reduction of area).

The topic of ductile and brittle fracture is so vast that any treatment of this subject within the confines of this section, or even the entire book, barely scratches the surface. The reasons for doing basic research or developing practical information for design purposes must be driven by the need to establish strategies for failure avoidance.[8] This book only attempts to select the basic notions and elementary formulas useful to practicing engineers in making design decisions at those times when trial-and-error approaches are no longer economically possible. It is only left to sum up a few comments on the mechanics of ductile and brittle fracture.

Irrespective of the manner of crack initiation, brittle fracture has a built-in feature of instability that can be triggered with little or no plastic deformation at applied stresses well below the traditional yield strength of the material. Even the smallest internal or external defects can lead to the failure of large structural members. And for many years the explanations, even at the level of microscience invoking the crystalline appearance, provided no guidance on how to avoid the failures. The problem is, in a nutshell, that under special conditions a metal, such as steel, can be "glass brittle." It appears now that history is full of examples of insidious brittle behavior, including the *Titanic*.

Examinations of the various brittle failures prompted the development of the crack-starter techniques that led to the concept of the nil-ductility transition (NDT) temperature. This determines the location on the curve of the ductile–brittle transition temperature that can be described as the point of no observable ductility. Also, studies of catastrophic propagation of cracks have established that the mere presence of sharp notches makes structural materials susceptible to brittle fracture. The tests widely used for the determination of ductile-to-brittle transition include CVN, DT, and DWT, as well as notched tensile tests. However, certain metals and alloys do not show a ductile-to-brittle transition.

Although the main task for the design engineer is to avoid brittle character-istics in selecting structural materials and configurations, complete elimination of defects is impractical. Even with the best detection of any original defect there can be at least two or three dozen material, environmental, and operational degradation mechanisms that, over the lifetime of the component or a structure, can contribute to failure.

TRANSITION TEMPERATURE APPROACH

For many years after the publication of early classical papers in fracture mechanics,[1,9] pragmatic users and students of this new branch of engineering science had to flounder on the impossibility of coping with a flood of new information with an apparently narrow span of interest and a rather wide field for contradictions. But, above all, the fracture-oriented literature at the time was not useful for the solution of practical problems in the area of brittle failures of ships and tankers. Industry and the government needed new engineering approaches to achieve the design of the high-performance ship fleet of the future and to form a structural integrity technology guide for applications in other areas. This task was accomplished by the Naval Research Laboratory between the 1950s and 1970s as it developed the new engineering design procedure for fracture-safe steel structures referred to since then as the "fracture-safe design" or the "transition temperature approach." In a sense this was an independent school of thought existing in parallel with the branch of theoretical fracture mechanics. The leader of the fracture-safe design school was Pellini.[10]

Although the transition temperature approach is not necessarily a preferred, generic topic for textbooks on fracture mechanics, it does fit the category of practical applications of the relevant fundamentals and design philosophy selected for this book.

The new procedure is intended for qualifying structures for service by focusing on the effects of sharp cracks and metallurgical defects. The method is aimed at quantifying the results in order to minimize the (often costly) retrofit. It is based on accurate measurement of fracture and crack growth properties. The specimens are expected to have sharp notches and cracks that model conditions under which the real structure may fail. Accurate numerical results then provide the basis for a formal certification of the structural component. The qualitative assessments should not be used as a basis for a design standard.

The main objections to use of fracture mechanics analytical methods in quantifying fracture may be that LEFM is limited to brittle metals, and the assumption that all structures are defective. The transition temperature approach, on the other hand, suggests engineering trade-offs that must be formalized to meet the certification needs.[11]

The new approach applies fracture-safe principles and it is used to circumvent the fracture problem by designing the structures to operate in the ductile region, that is, above the transition temperature. It should be stated for the record that a similar approach was developed earlier, during and shortly after World War II. In essence the old method involved load and stress analysis of the prototype component and determination of the equivalent Charpy transition temperature.

The concept of fracture-safe design[11] covers essentially the intermediate- and low-strength steels. These are known as the structural grades in the form of plates, forgings, weldments, and castings. The procedure involves practical tests and simple analysis applicable to structures starting from 5/8 in. thick plate weldments and progressing to castings and forgings thicker than 1 ft.

The nature of crack initiation and fracture under brittle and ductile conditions is related directly to the concept of the "fracture analysis diagram," known as FAD. With decreasing temperatures the ductile behavior at crack tips transforms rather sharply into cleavage (or brittle) behavior over a narrow temperature range. This feature meshes rather well with the various boundary curves in a generalized stress–temperature field. A reference transition temperature is designated as the NDT temperature, as alluded to previously in connection with other topics and noted in Figs. 4.6, 4.12, and 6.5. This is an important parameter[12,13] and very basic to the development of improved materials such as structural steel. The significance of the NDT temperature is further illustrated in Fig. 6.9.

Curve A in Fig. 6.9 represents fracture stress in steel, as a part of the FAD derived from consolidation of the available test results and involving flaw size,

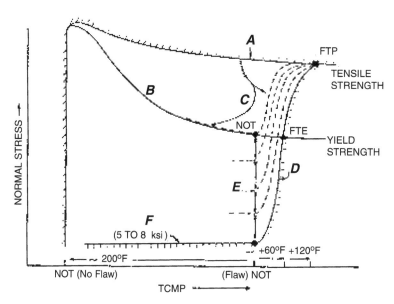

FIGURE 6.9 Concept of transition temperature (from Ref. 11).

stress, and temperature. This curve corresponds to a fracture stress without a flaw at the maximum level, that is, tensile strength. As the temperature drops, the tensile strength shows some increase, corresponding finally to the maximum value at a "no-flaw" condition, indicated by the vertical line to NDT at an approximate temperature of $-200°F$.[11] At the same time, curve B represents the change in yield strength with decrease in temperature. At the location where the A and B curves coincide, the steel ductility as determined by elongation or reduction of area is essentially reduced to the zero (or nil) value. When a small, sharp flaw is machined into the test specimen, fracture stress is decreased, as indicated by the arrows C pointing at the dashed curves. The letters A–F in Fig. 6.9 relate to the following characteristics:

A   Fracture stress without flaw (tensile strength).
B   Variation of yield stress (solid line); note that the B curve is more sensitive to temperature drop than curve A.
C   Fracture stress decreases due to small, sharp flaws (dashed line).
D   CAT (crack arrest temperature) curve (arrest of propagating brittle fractures).
E   Additional decreases in fracture stress due to increasing flaw sizes.
F   Lower stress limit for fracture propagation.

Below the temperature defined by the vertical line and denoted by NDT (flaw condition), the fracture stress curve for the small flaws coincides with the yield strength curve. The arrows coming down along the NDT line signify the increases in the flaw size. Here the approximate fracture stress is inversely proportional to the square root of the flaw size. The crack arrest temperature, popularly known as the CAT curve, represents the temperature of arrest of a propagating brittle fracture at the various levels of applied nominal stress. The CAT curve intersects the yield strength curve at the point known as FTE, which stands for "fracture transition elastic." It signifies the highest temperature of fracture propagation for elastic loading. The point of intersection between the CAT curve and the tensile strength curve is called FTP, the "fracture transition plastic." This point determines the temperature beyond which fracture is entirely in the shear mode. The fractured surface is not of a cleavage type and the stress level required for the fracture is the tensile strength of the material. The original concept of temperature transition was intended for steel products.

The lower portion of the CAT curve, identified by the letter F in Fig. 6.9, is referred to by Pellini and Puzak [11] as the "lower stress limit for fracture propagation." This limit defines the lower shelf value in the range 5–8 ksi, below which fracture propagation is considered to be next to impossible. The reason for this structural limit is generally given as insufficient source of elastic strain energy for propagation of a brittle fracture. It is shown below that this lower shelf stress criterion can be used as a practical design tool.

A rather remarkable variety of structural tests can be employed in development of the transition temperature details for the fracture analysis diagram. The first type of experiment involves a composite plate consisting of a brittle plate welded to a test plate and employing the technique of a forced crack initiation applied either to the composite or to a single test plate, as shown in Fig. 6.10. The idea here is to develop a rapidly running "brittle fracture, force initiated" by means of a wedge impact. The crack can either propagate through a test plate or be arrested, depending on the temperature and the specific level of stress. In this manner a series of tests is made to establish the boundary between the "run" and "stop" temperatures. This type of test procedure, using the test samples illustrated in Fig. 6.10, is appropriate for the portion of the CAT curve below the FTE point shown in Fig. 6.9.

For the CAT curve above the FTE point the procedure includes the use of crack-starter explosion tests, indicated roughly in Fig. 6.11. Plastic response in this case can be obtained by the development of a deep hemispherical bulge with emanating short tears in a complete shear mode.

The last series of tests for the completion of the transition temperature diagram involved large-scale specimens using fatigue and brittle weld fracture initiation of cracks of various dimensions. Loading of the test plates in a stress–temperature field is shown in Fig. 6.12. The family of dashed curves that represents the initiation of fracture is bounded by the CAT curve.

The NDT temperature constitutes a reference point for the generalized diagram and it is determined by the drop weight test (DWT), which is relatively simple and reliable. The test shows a sharp transition from "break" to "no break," making the test work easier. Also, analysis of experimental results from DWT

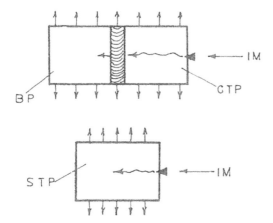

Figure 6.10 Wedge impact tests (IM, impact; CTP, composite test plate; STP, single test plate; BP, brittle plate).

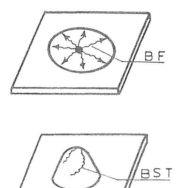

FIGURE 6.11   Explosion crack starters (BF, bulge fracture; BST, bulge shear tears).

indicates that this procedure is highly reproducible, and that it is relatively insensitive to orientation with respect to rolling or forging direction. It should also be noted that there is a dramatic change in fracture toughness of steels over a narrow temperature range and that the interpretation and terminology of NDT, FTE, and FTP developed during the early 1950s are still valid.

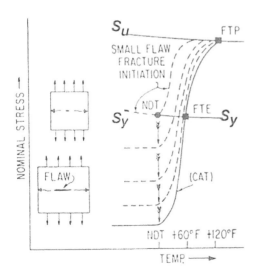

FIGURE 6.12   Large-scale fracture initiation and arrest.

The generalized fracture analysis diagram given in numerous older publications is shown in Fig. 6.13. The approximate range of flaw sizes noted in Fig. 6.13 has been obtained from a variety of small flaw tests, from studies of large flaws, and from interpretations of theoretical fracture mechanics.[14] Validation of this aspect of FAD has been provided by many experiments and analysis of service failures, covering the range of crack and flaw sizes between 1 in. and 2 ft.

As far as practical design is concerned, the FAD defines four major index points, which can be summed up as follows:[10]

- When service temperature is restricted to slightly above the NDT, the design is protected against the initiation of fracture due to small cracks in the regions of high local stresses.
- When service temperature is a little above the point on the CAT curve corresponding to $0.5S_y$, and designed structure is protected against the fracture if the nominal stress does not exceed $0.5S_y$.
- Restricting service temperature above the FTE temperature provides fracture arrest protection provided nominal stresses are not higher than the yield strength of the material.
- When service temperature is above the FTP temperature, the designer has the assurance that only fully ductile fracture is feasible.

It appears that in most practical situations, finer divisions than the four design points of the FAD stated above are not required. Considerable increase

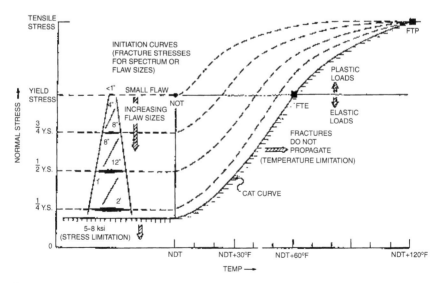

FIGURE 6.13   Fracture analysis diagram.

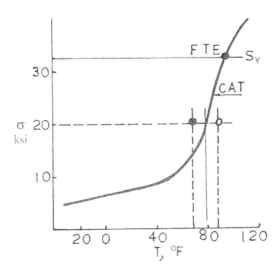

FIGURE 6.14 Example of CAT curve use.

in fracture resistance at NDT $+30°$F reduces the problems of fracture-safe design to a temperature reference system of great simplicity.

The full meaning of the crack arrest features can be summed up with the aid of Fig. 6.14. Suppose we have a working stress of 20 ksi and the horizontal line crossing the CAT curve at a temperature of approximately 80°F. The structural member designed for 20 ksi should be fracture-safe at a service temperature higher than 80°F. Here the filled circle signifies crack propagation with a total fracture of the specimen, while the open circle corresponds to complete fracture arrest.

The experimental basis for the location of the CAT curve within the stress–temperature coordinates was derived from all available crack arrest data for steel plates some time ago.[16] A brittle fracture propagated across the starter plate was

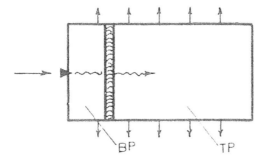

FIGURE 6.15 Test assembly for indexing the CAT curve (BP, brittle plate; TP, test plate).

used as an indicator in locating the CAT curve. Depending on the temperature, the brittle fracture either stopped on entering the test plate or continued through to a complete fracture as shown in Fig. 6.15. The test assemblies were loaded to one-half the yield strength of the steel and a hammer blow was applied to a wedge-in notch at the edge of the brittle "crack-starter" plate. The NDT temperature determined by the DWT indexes the temperature at which the CAT curve begins to rise. For a complete discussion of this process, the reader may wish to consult the basic text dealing with the principles of structural integrity technology.[10]

Practical use of crack arrest diagrams can be summarized with the help of Fig. 6.16. In this case YS stands for yield strength of the material, σ_W denotes working stress, σ_{CA} is the limiting design stress, and ΔT_W is the temperature increment above the NDT. The continuous line FABC represents the crack arrest temperature and it is the important element of the fracture analysis diagram compatible with conservative design philosophy. The CAT curve divides the two main regions of material's behavior. The area below the CAT curve represents a "safe" region, free of initiation and propagation of cracks. On the premise that the CAT curve and NDT have been determined for the specific material at hand, the limiting design stress corresponds to a point B on the CAT curve defined by the temperature increment ΔT_W. For a working stress σ_W and $\sigma_{CA} > \sigma_W$, a positive margin of safety is established. When the actual service temperature is lower than (NDT + ΔT_W), corresponding, say, to point B', a fracture may or may not propagate at a working stress level of σ_{CA}. The region above the CAT curve can also be analyzed in terms of crack length and stress, as shown

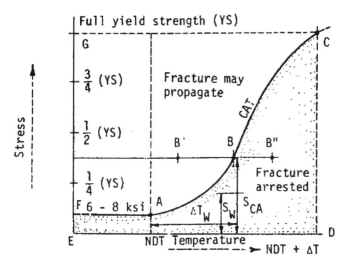

FIGURE 6.16   Crack arrest temperature (CAT) curve in design.

in Fig. 6.13. However, any increase in ΔT_W, consistent, say, with point B′; in Fig. 6.16 provides assurance of crack arrest.

It should be noted that Figs. 6.9, 6.12, 6.13, 6.14, and 6.16 involve the use of stress and temperature. These two variables, in combination with crack size, constitute quantitative information of direct interest to designers. It is also well known that the main scientific issue in the early 1960s was the effect of section thickness on the range of transition temperature. The main question was whether the temperature transition from plane strain to elastic–plastic fracture could be put down by rather thick sections. Several years went by, producing controversial results, until heavy plates of reactor-grade steels were tested in 1969. The current engineering significance of this research is that in thick sections, the plane-strain constraint is essentially lost and elastic–plastic fracture starts at higher temperatures and lower strength than with thin sections. The approximate dividing line between thin and thick sections can be taken as 1.0 in. In metallurgical terms, increasing metal grain ductility in the plane-strain transition region is, according to Pellini,[10] difficult to suppress.

The general diagram in Fig. 6.13 was supplemented by comments on four major points regarding the role of the test temperature in relation to service temperature and nominal stresses. The FAD procedures, however, only apply to structures that contain flaws. If it can be determined that our structure is, or will be, free of flaws, then there is no need to continue with establishing fracture characteristics. Nevertheless, all items requiring engineering judgment in a variety of technical areas should be reviewed with the following guidelines in mind.[11]

1. *NDT Temperature:* Service above the NDT is needed for structural members not thermally or mechanically stress relieved. It should also be noted that some members may be expected to develop points of local yielding, and that even very small flaws, under these conditions, may act as initiators below the NDT. Fracture protection, however, is available above the NDT.

2. *NDT + 30°F Criterion:* This relates to CAT at stresses equal to one-half the yield strength of steel. If the service temperature is higher than NDT + 30°F, the fractures can neither initiate nor propagate.

3. *NDT + 60°F Criterion:* This relates to the level of general stress equal to yield strength of steel. The criterion is intended for special cases of high-pressure testing, reactor pressure vessels, and nozzles under severe thermal conditions.

4. *NDT + 120°F Criterion:* This criterion applies to plastic overload of structures in accidents or military activities. Hence the structural members are analyzed or designed for utmost fracture resistance at a temperature level allowing full shear fracture.

The successful use of FAD in design, described so far, depends on knowl-edge of the NDT properties of steel. Once the NDT temperature is established the most important role of the FAD technique is to provide quantitative information on the crack size that can be tolerated throughout the transition region.

So far the transition temperature criteria discussed in this section are con-cerned with the stress–temperature field as the more convenient parameters for design. The FAD has not been established as a parameter of the material but rather as a correlation technique between the laboratory test and structure behavior. While the use of FAD has obvious pragmatic values, other factors must also be taken into account such as toughness-to-strength ratio and the relationship between the crack size and the stress-to-yield ratio. After the World War II ship failures, there was a special degree of urgency in dealing with the practical problems with low- and intermediate-strength steels in terms of the transition from brittle to ductile beha-vior. The FAD methodology was followed by the ratio analysis diagram, known as RAD, also as a tool of fracture-safe design. The RAD technique combined the metal data-bank summary of the time with the critical fracture conditions defined in terms of K_{Ic}/S_y ratios. Essentially, RAD relates K_{Ic} and DT parameters to the yield strength of the material, bounding the ductile zone between the lower and maximum technological limits of the day.[10] The upper boundary represented the state of the art in metallurgical research.

The RAD was divided into three areas:

1. Plastic region
2. Elastic–plastic regime
3. Plane-strain area

The K_{Ic}/S_y ratio lines also involved crack size for a given S_y/σ ratio where σ was the nominal stress. The lower portions of the straight lines included limits of detectable crack sizes. These diagrams were used for some steels and other materials such as aluminum and titanium alloys. The RAD approach was strictly empirical and required careful interpretation of the various parameters when correlating RAD and LEFM numerical results. The discrepancy may often be due to metallurgical vari-ations between RAD and LEFM sources of material properties. There are, of course, certain limitations of both the fracture-safe design and linear elastic fracture mech-anics in solving practical problems where so many variables and constraints are involved. The ideal appears to be a blend of fracture-safe philosophy and LEFM on an individual-case basis. Unfortunately for the designer, current papers and textbooks seldom recognize the FAD and RAD techniques as tools of special merit in design.

LOWER-BOUND STRESS IN DESIGN

The preceding section provides a brief introduction to the concept of transition temperature approach to fracture avoidance. Instead of evaluating the critical crack size for a given structure, the load-carrying capacity can be calculated using conventional stress analysis and disregarding the presence of potential small cracks provided the selected material has sufficiently high shelf energy at low temperatures. This technique makes sense if the designer first establishes the service temperature and then proceeds to obtain an energy vs. temperature design curve for the candidate material. As stated in other sections of this book, various tests can be used in developing the transition temperature curve for the case at hand. These include Charpy V-notch, drop-weight tear, explosion tear, dynamic tear, and Robertson crack arrest techniques. Needless to say, the Charpy V-notch test is still preferred in many situations because it provides a rather severe test of material toughness loaded at very high strain rates. Also, the amount of energy absorbed by the notched Charpy bar can be measured with good accuracy.

The Charpy test results can be presented as the absorbed energy vs. temperature with three points of interest, indexed as fracture appearance A, transition B, and midpoint temperature C, in Fig. 6.17. The fracture energy level often used in such interpretations is 15 ft-lb. The 50% cleavage number indicates that the fractured surface is 50% cleavage and 50% fibrous. This illustration is based on some older results obtained from steel plates and the CVN data of small specimens; this illustration may differ from an actual structure response because of the difference in thickness, as indicated previously in Fig. 4.9. It also follows from Fig. 4.9 that, if the designer uses the Charpy curve for establishing a temperature operating criterion, it is possible that the structure will fail in a brittle fashion. It is prudent at this point to restate the following criterion:

> When using the transition temperature approach in design, it is important to make sure that the selected transition temperature for the relevant material is below the service temperature of the structure.

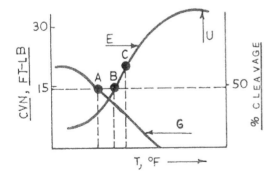

FIGURE 6.17   Charpy test results (E, energy; U, upper shelf; G, cleavage curve).

The lower-bound stress idea for design purposes given in this section is based on the Robertson crack arrest criteria[15] and experience with fracture control technology of mechanical and structural systems developed and fielded in support of underground nuclear tests. This work was driven by the concerns about the existence of unqualified steels in various branches of industry affecting certain new technologies and materials in fracture-critical applications. Several details of interpretation, verification, and the general approach to the lower-bound stress problem are given in Chapter 11.

In developing the procedure for utilizing the concept of lower-bound stress it is essential to face a number of constraints and simplifying assumptions. The problem also demands a degree of blending of the transition temperature approach and the rudimentary elements of LEFM.

Since the procedure is to deal with a material that is not certified, it is necessary to provide assurance that a fracture-critical component is loaded to stresses below a certain allowable level. Hence the nominal working stress must be subject to a number of constraints. One of these is concerned with the minimum practical value of plane-strain fracture toughness. For virtually all structural steels the lower-bound toughness K_{Ic} should be of the order of 25 ksi (in.)$^{1/2}$ or 27.5 MPa (m)$^{1/2}$. The corresponding values for the titanium and aluminum alloys can be taken as 19 ksi (in.)$^{1/2}$ or 21 MPa (m)$^{1/2}$ and 10 ksi (in.)$^{1/2}$ or 11 MPa (m)$^{1/2}$, respectively.[7] These values include the effects of welding, alloying, low temperatures, and high loading rate (for structural steels), and the effects of optimization of microstructure (for titanium and aluminum alloys).

The next assumption is that of a larger, undetected crack residing in the structure. Here the flaw depth is sometimes taken to be equal to half the section width or thickness. In some instances crack arrest due to increase of toughness can be followed by gross plastic yielding with potential specimen failure. It is quite easy to overlook this point when our primary interest lies in analyzing crack arrest. The problem is that even with full arrest of a brittle crack, the structure may be damaged due to the reduction of loaded area. The consequences can still be severe in spite of the likely change in the failure mode, say, from brittle to ductile.

Although the transition temperature approach has in the past met with various successes, it is not altogether obvious to design practitioners or even LEFM specialists which stress — local or gross — should be considered.[7] When on top of this question we dare to add descriptive terms such as reference, gross, nominal, primary, maximum nominal, average, uniform, or local stress, it is no wonder the questions and debates continue. In the context of lower-bound criteria we have little choice but to consider the stress component (away from the crack tip), which we can predict with some degree of confidence, and which is likely to open the crack. It must therefore be the product of experience and conservatism.

In line with maintaining some degree of conservatism, single-edge notched geometry is assumed for the panel of finite width under uniform tension shown in Fig. 6.18.

The stress intensity factor for a single-edge panel of finite width is

$$K_I = \sigma(\pi a)^{1/2} f(w) \tag{6.1}$$

where $f(w)$ is given by Fig. 3.6 in terms of the a/w ratio. Assuming $a/w = 0.8$, the exact calculated value (including a 1.12 correction) is

$$f(w) = 2.139$$

The approximate readout from Fig. 3.6 is 2.1. Taking the calculated number and making $K_I = K_{Ic}$, we have from Eq. (6.1)

$$K_{Ic} = 2.139 \times 1.7725\sigma(a)^{1/2}$$

or

$$K_{Ic} = 3.8\sigma(a)^{1/2} \tag{6.2}$$

This form of the equation pertinent to the assumed model is intended for Fig. 6.18. Next, making $K_{Ic} = 25$ ksi (in.)$^{1/2}$ and solving for stress, Eq. (6.2) yields

$$\sigma = 6.6(a)^{-1/2} \tag{6.3}$$

The design curve for the lower-bound stress, which represents the maximum nominal stress level for unqualified structural steels, is shown in Fig. 6.19.

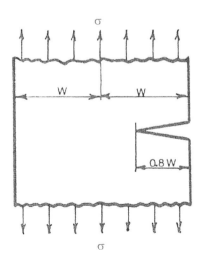

FIGURE 6.18   Model for lower-bound stress.

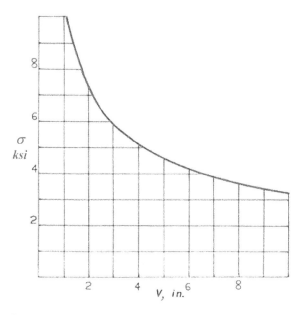

FIGURE 6.19 Lower-bound design stress for unqualified steel (V, total width of plate $= 2w$ (Fig. 6.21)).

Since the minimum plane-strain fracture toughness for the titanium alloys is about 19 ksi (in.)$^{1/2}$,[7] and assuming the crack length to panel width ratio to be 0.8 as in Fig. 6.18, we obtain

$$19 = 3.8\sigma(a)^{1/2} \tag{6.4}$$

or, solving for stress, Eq. (6.4), gives

$$\sigma = 5(a)^{-12} \tag{6.5}$$

which represents the maximum nominal stress level for unqualified titanium alloys. The design curve for this case is given in Fig. 6.20.

The conservative minimum plane-strain fracture toughness for the aluminum alloys is about 10 ksi (in.)$^{1/2}$.[7] Hence assuming the a/w ratio to be 0.8, as in Fig. 6.18, we have

$$10 = 3.8(a)^{1/2} \tag{6.6}$$

or, solving for stress again, we obtain

$$\sigma = 2.6(a)^{-1/2} \tag{6.7}$$

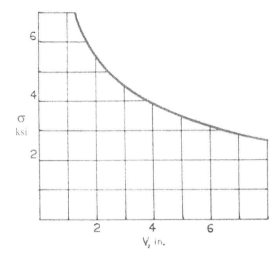

FIGURE 6.20 Lower-bound design stress for unqualified titanium alloys (V, total width of plate $= 2w$ (Fig. 6.21)).

which defines the maximum nominal stress level for unqualified aluminum alloys. The design curve for this material is shown in Fig. 6.21, obtained by plotting Eq. (6.7).

The lower-bound stress design methodology described in this section falls into a special category of safety requirements and fracture control. It pertains to test or temporary equipment, generally expected to be of the "one-time usage" type. This may include the components designed for a remote control and/or support of loaded test pressure vessels or explosive devices. Safety also must be ensured in cases of remote operation and containment of hazardous and radioactive materials. The control and handling systems in this special category are often expensive to design and fabricate but they are also indispensable. There are also special areas of components and equipment in use for which material properties are not known and for which there is no certification that fracture-critical parts are loaded below a maximum allowable stress. These can include mechanical handling equipment, pressurized components, laboratory apparatus, or welded support structures, to mention a few. Another problem with this type of equipment or hardware is that the original design calculations are either lost or not available for proprietary reasons. In the case of the much older handling equipment still in use, the criterion of a lower-bound design stress can be applied to derating, recertification, or decommissioning purposes.

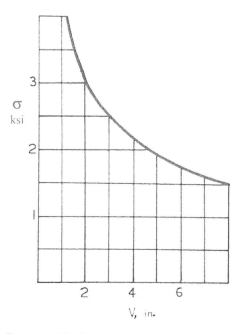

FIGURE 6.21 Lower-bound design stress for unqualified aluminum alloys (V, total width of plate $= 2w$ (Fig. 6.18)).

CRACK BEHAVIOR IN WELD REGIONS

Considering all the variables in analysis, design, materials, and certification of welded structures, it is not surprising that welds are prime targets in most failures. Volumes have been written on this subject, with the primary sources for welding information being the institutes and professional societies in this country and elsewhere. This section barely scratches the surface of this topic and it is only intended to instill some awareness in the mind of the reader as to the basic characteristics of crack behavior in a welded region. Subtle peculiarities lie at the root of many weld problems, which can be put in a number of categories of weld zone cracking.[8,10] Many of these relate to the heat-affected zone (HAZ), weld proper, and the fusion line. Here HAZ may be characterized by plane-strain criteria, while the weld and base metals can also be involved. The fusion line seldom propagates fracture, unless it becomes very brittle.

Pellini[10] describes several principal types of cracking that can occur during or after welding. Brief descriptions of the generic problems of weld zone cracking are given below. The main causes can be traced to inherent metallurgical sensitivities and stress–strain fields.

Solidification. Solidification is triggered by the process of cooling of the pool of weld metal, with the liquid film becoming the site of cracks arranged according to the direction of restraining forces. Impurities consisting of small amounts of sulfur and phosphorus are critical to hot cracking of steels. The mechanism of cracking in this case is called "hot tearing" because it appears at high temperatures before the metal develops sufficient strength. This mode is shown in Fig. 6.22.

Liquation. While the mechanism of liquation cracking is similar to strain-controlled cracking during solidification, the temperature of a liquid film is not sufficiently high to melt the HAZ of the base metal. Hence the tensile strain develops from shrinkage of the HAZ and the liquation crack is formed, as shown in Fig. 6.23. Cross-hatched areas in Fig. 6.23 represent pools of weld metal.

Cold cracking. In metallurgical terms, cold cracking occurs after hard products such as bainite and martensites are formed. This process of cracking is accelerated with the help of tiny liquation cracks. Hardness can be reduced by preheating or slower cooling. The best overall remedy is to lower the carbon and sulfur contents.

Stress relief. Stress relief is a complex situation from the point of view of material science. During the welding process the HAZ area and the weld bead are subject to high-temperature solution treatment. Under this condition carbides are dissolved and put back into solution. If, next, a high rate of cooling does not permit all the carbides to be precipitated, some of them will remain in the solution. However, when a weld joint is reheated for the purpose of stress relief, carbon precipitation is continued, creating fissure-type cracks at grain boundaries. The crack pattern in this case may be in the form illustrated in Fig. 6.24. While precipitation strengthens the grains, it also leaves the grain boundaries weaker.[10] It is clear from this simple example that welding and heat treatment processes create opportunities for originating and enlarging defects in the form of complicated crack geometries. Such processes involving thermal inputs are

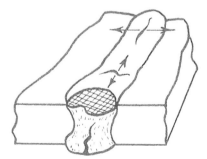

FIGURE 6.22   Direction of hot cracking.

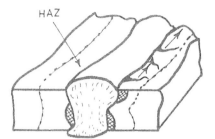

FIGURE 6.23 Liquation cracking.

usually complex and demand strict control of heating and cooling rates, holding intervals, and quenching operations with a follow-up of reheating and cooling. These activities have a very significant effect on fracture toughness and ductility of the great majority of engineering materials. These comments should apply to most welding, brazing, and soldering operations. And it should be stressed from a practical point of view that metal-joining data developed for one set of service conditions may not be transferable to another.

Lamellar tearing.   Lamellar tearing applies to highly restrained weldments and fillet welds indicating the ideal conditions for originating cracks, as shown in Fig. 6.25. Separation of the base metal along the planes of nonmetallic inclusions is referred to as "lamellar tearing." This tearing process develops in the through-thickness direction due to low ductility, and it normally follows the onset of cracking caused by low-quality melting practices.

Hydrogen effect.   The negative role of hydrogen in welded joints is to encourage fissuring in the plastic zones at crack tips and to extend the existing cracks in the metallic parts generated by other means. Hot liquation or cold cracking can provide the right conditions for this type of crack extension. Even the smallest cracks can be extended if the stress fields and hydrogen contents are sufficiently high. The directions of hydrogen-assisted crack growth follow paths normal to the directions of applied stresses σ such as, for instance, those shown in Fig. 6.26. The cracks in this case are located in underbead toe regions.

FIGURE 6.24 Crack patterns under stress relief.

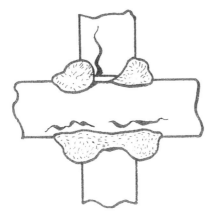

FIGURE 6.25 Example of lamellar tearing.

It also appears that both short- and long-range, high-intensity stress systems are critical in hydrogen presence.

   Restraint effect. Several metallurgical mechanisms and modes of cracking can be combined in so-called restraint cracking,[10] as illustrated in Fig. 6.27. These can include hot tearing, cold cracking, and lamellar tearing assisted by hydrogen. The sketch in Fig. 6.27 highlights specifically underbead hot tearing and lamellar cracking combined. The initial fissuring by hot tearing and crack extension (induced by hydrogen) may be difficult to identify and will always be subject to differing opinions among experts unless we can develop the appropriate experiments.

   The entire process of crack behavior in a weld regime involves as many facets of imperfections as there are geometrical, materials, and fabrication characteristics. Just a handful of the possible types of imperfections and cracks is given in Fig. 6.28 for a common weld joint. The various forms of flaws and

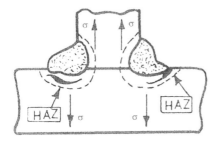

FIGURE 6.26 Location of hydrogen-assisted cracks.

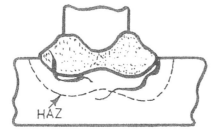

FIGURE 6.27 Combined effects.

crack shapes are identified here for general purposes, on the premise that only some of them will be found in a particular joint at hand:

1. Toe crack
2. Slag inclusion
3. Gas pocket
4. Porosity
5. Underbead crack
6. Lack of fusion here constitutes a crack
7. Solidification crack
8. Fillet overlap
9. Root crack
10. Undercut condition
11. Throat area crack

Definition 6 is concerned with a geometrical discontinuity falling in the area of a stress concentration caused by fabrication. Theoretically we can have

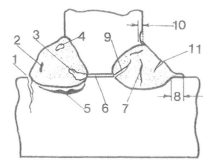

FIGURE 6.28 Potential faults and cracks in a welded joint. See text for identification of features 1–11.

several regions of stress concentration that can eventually become the initiation sites for fatigue and fracture of a welded joint. This applies equally to various full- and partial-penetration welds in butt- and fillet-types of welded connections.

Several aspects of mechanical cracking have been illustrated in Figs. 6.22 to 6.28. This problem, however, should also be analyzed from the point of view of metallurgical characteristics.[8,10] The appropriate comments are included here because the nature of weld cracking is highly complex and can only be assessed with the help of a special blend of material science, stress analysis, and fracture mechanics.

In certain applications it is required to have welded joints of dissimilar materials. Experience shows that such joints are likely to have cracks, and considerable effort was expended in the past to find a solution. It appeared that a dissimilar combination of metals was naturally susceptible to cracking. With time the problem was minimized using weld-overlay buildups, but the new technique was not 100% reliable and it was a very time-consuming operation. A series of metallurgical studies followed involving a nickel–copper alloy and alloy steel combinations with special emphasis on the amounts of phosphorus and other contaminating elements, which could make the alloys susceptible to cracking. The final outcome of the study was that it was necessary to monitor the composition of the pertinent alloy in order to assure a lower threshold for residual phosphorus.[8] The lesson here is that assuring the quality of a welded joint demands an appropriate composition and microstructure proven by pedigree documentation. In spite of this, the process of welding unfortunately still creates conditions particularly susceptible to cracking even in well-engineered structures. The reason for this is that fracture mechanisms and failure modes can be insidious, with virtually no outward signs of imminent disaster. The critical crack can propagate under normal service conditions and unusually low operating stresses.

Because of the complexity of fracture mechanisms in a welded joint, it is difficult (or next to impossible) to provide a single unique formula, tip, or advice to assure high-quality, failure-resistant welded structure without a long and tedious process. However, improved understanding of several features and conditions responsible for, or at least implicated in, failure analysis of welds should help to focus on the root of the problem.

The lack of fused regions (Figs. 6.25 and 6.28) as well as the undercut condition (Fig. 6.28) fall in the areas of geometric transitions and discontinuities together with notches, pockets, porosities, and inclusions that are involved in the process of load transfer and stress concentration. Welds are often subject to external loads and deformations required to assure a tolerance or configuration of the welded structure. Such a procedure leads to additional residual stresses, with further influence on microstructure and mechanical properties. Also, partially fused regions, as well as sharp notches, are known locations (Figs. 6.25 and

6.28) for crack initiation. Weld bead areas include ripples and ridges that can act like stress concentrations where hydrogen and restraint effects already drive cracking, as shown in Figs. 6.26 and 6.27. It is also difficult to say whether surface irregularities can be removed by mechanical means without introducing new discontinuities in the form of grinding marks and scratches.

There are certainly various geometrical discontinuities and defects that trigger cracks in the weld itself or the HAZ (Figs. 6.24 through 6.27). However, certain metals and alloys are more tolerant of the welding effects. In general, weldability varies inversely with the conventional strength properties, and more exotic materials may have less tolerance for errors during the welding process. For other aspects of metallurgical nature including compositional and microstructural heterogeneity, residual and thermal effects, and steps to minimize hydrogen-induced cracking, the reader may find the practice-oriented material of further interest.[8,10]

While welding technology has had a considerable influence on production and fabrication costs in various industries, gaining knowledge of the process is still very complex because of the nature of residual stresses, imperfections, and stress concentrations. Some of these characteristics have already been illustrated in rough sketches (Figs. 6.22 through 6.28).

The concept of "residual stress," often referred to in welding technology, should not be confused with "residual strength." The first term relates to a stress existing in a structure while there are no loads acting on the structure. The second definition refers to the remaining strength under the presence of cracks, and therefore it is based on the size of the ligament. Local stresses at the crack tip are normally expressed as a function of the applied stress field. Such a function can be established provided we know the mode of loading. The so-called opening or tension mode used in this book is, by far, the most important. It should be noted that the modes of loading are not modes of cracking. However, one of the important reasons why the mode designation "roman numeral I" is selected stems from the fact that the majority of cracks result from Mode I loading. It is also generally accepted and recognized that although the residual stresses themselves cause a certain amount of stress intensity, which can be superimposed on the stress intensity due to the applied loads, the result of the superposition can only be judged qualitatively. The quantitative evaluation of the residual stress field and its effect on the stress intensity factor K is, for all practical purposes, very difficult and can only be estimated. Residual stresses are always found in the HAZ, and finite-element techniques have been used to estimate residual stresses at welds in piping systems.[19,20] Fracture mechanics principles apply to the weld metals and HAZ areas using the K_{Ic} and da/dN parameters, provided HAZ data can be obtained. The latter procedure, however, may be time consuming and costly. The principles of LEFM are useful because the stress intensities are additive. Unfortunately, the magnitude of the residual stress field is very often unknown.

A rather interesting case of a triaxial field of residual stresses with large gradients in the railhead is quoted by Broek.[21] This case raises the question of why fatigue cracks occur at all if the stresses in the railhead are compressive.

Fabrication imperfections in welding may be found in the filler and base metals. The cracks in welded regions are generally caused by a number of reasons, which can be summed up as follows:

- Improper design configuration, which restricts the use of correct electrode angle.
- Poor selection of a welding process and the relevant welding parameters for the material and joint configuration.
- Incorrect fabrication procedures.
- Questionable handling and care of the electrode and flux.

These general observations are noted because so many factors are responsible for the way the cracks appear and behave in welded connections.

The majority of welded structures are designed and built in compliance with codes and standards prepared by technical and professional organizations in the United States and elsewhere. The regulatory documents accept tolerable fabrication-induced imperfections, and care should be taken not to remove the allowable flaws and cracks without a very sound reason. This specifically applies to any repair welding, which can lead to dire consequences. The problem of fatigue behavior of welded components is particularly important and the welded steel structures should be designed, fabricated, and inspected according to the rules of the American Association of State Highway and Transportation Officials, known as the AASHTO specifications.

It is stated previously in this book that the maximum stress intensity develops when the plane of the crack is normal to the direction of the primary tensile stress. However, as the plane of the crack tilts further away from the Mode I tensile loading, the structural member, in the form of a plate or similar configuration, will lose its ability to support the external load. For further comments on the problem of a slanted crack mode, see Fig. 3.32 and Eq. (3.42).

Plate laminations parallel to the surface of the plate, which supports the in-plane tensile stress, rarely degrade the fatigue resistance, provided there are no stress fluctuations of a tensile nature in the through-thickness mode.

There is a distinct difference between the effect of the embedded and the surface imperfections. For a given shape, an embedded flaw must be twice the size of a surface defect to cause the same stress intensity. A similar condition exists in judging the effect of planar vs. volumetric imperfections. Hence fatigue cracks at the weld surface can initiate and grow faster than the gas pocket or porosity-induced cracking, on the premise of equal projected size and shape. By "projected" we mean to compare the dimensions of the defects on the plane perpendicular to the direction of stress. In general, fatigue cracks may originate from internal or external weld areas, as shown, for instance, in Fig. 6.28. The majority of weld cracks, however, originate at the surface, at weld toes or terminations, where geometrical discontinuities of the joint reside. The maximum depths at the fatigue crack initiation points were found to be less than 0.016 in., or 0.4 mm. The maximum distance from the surface of the weld to the embedded defects was on the order of 0.08 in., or 2 mm.[22,23]

An example of a fatigue crack initiated at a gas pocket in a web-to-flange fillet weld is shown in Fig. 6.29. The crack propagated, continually changing its shape, until it broke through the fillet-weld surface and became a penny-shaped crack. Normally the cracks are expected to propagate in all directions along the plane perpendicular to the direction of the applied tensile stress. The percentages given in Fig. 6.29 relate to the various levels of the exhausted fatigue life of the weld joint. For instance, the crack is shown to break through the back surface of the tension flange after 92% of the fatigue cycles. The dashed lines indicating the crack boundaries at various stages of fatigue crack propagation are only approximate and for the purpose of general comments. In this and similar cases of welded connections the cracks tend to propagate first as through-thickness defects and then as edge cracking, while the frequently encountered areas of crack origins in other situations can be traced to the toe of the weld. Experience also seems to indicate that in the case of flanged connections, about 90% of the fatigue life is used up prior to the crack breaking through the back surface of the flange.[24] It appears then that the major part of the fatigue life of the welded joints would be expended while the fatigue crack is still rather small. Since the K_{Ic} parameter varies as the square root of crack size, a more significant increase in fracture toughness would have only a moderate influence on the fatigue life of the weldment.

In discussing the topic of residual stresses in this section, the question of initiation and propagation of cracks in a compression stress field arose. In the

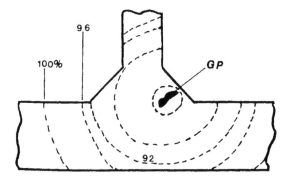

FIGURE 6.29 Approximate pattern of crack propagation in a fillet weld. Key: GP, gas pocket.

case of welded flanges and girders in compression, the cracks are likely to develop in regions of tensile residual stresses, which can be of the order of the yield strength of steel. However, these cracks are expected to grow out of the field of residual tensile stresses and arrest.

Prediction of the fatigue life of welded components depends on a number of variables that are not easy to quantify in spite of the progress of fracture mechanics technology.[25,26] It becomes necessary to characterize the behavior of small imperfections, which is a very difficult and costly process, as one can imagine. It is a practical consideration. The probability of finding and measuring small cracks in complicated weldments is, unfortunately, very low. The magnetic particle and ultrasonics techniques are probably the best for detection of smaller flaws, while the x-ray method is at the bottom of the list in this regard. The dye penetrant detection technique is likely to represent the average values of the reliability index. The reliability index varies between 0 and 1.0. In general, the choice of the instrumentation for crack detection in complex welding of bridges, ships, offshore platforms, and similar important structures should be conservative. Alas, "conservative" does not mean "economical."

While the engineering world is waiting for more research results on characterization and prediction of small initial imperfections, it remains to use the existing LEFM techniques and conservative assumptions. In relation to welded structures, the work of AASHTO resulted in establishing five basic fatigue categories and one special category E' for thick flanges and covers, as shown in Fig. 6.30.[27] The design categories A, B, C, D, and E apply to various cases as follows.

A	Plain plates and rolled beams
B	Plain welds and welded beams and plate girders
C	Stiffeners and attachments less than 2 in. long
D	4 in. long attachments
E	Cover-plated beams

Design Problem 5.10 (Chap. 5) illustrates the calculation procedure of fatigue life corresponding to the AASHTO design curve E' dealing with thick flanges and covers. The AASHTO curves have been derived from fatigue tests and field experience. These curves represent the 95% confidence limit for all design categories and are essentially independent of steel strength. However, under static conditions the design stress is proportional to the strength of the steel.

The testing and analysis of fracture toughness of metals is reviewed in Chapter 4 with special reference to the geometry of test specimens and LEFM parameters. The behavior of toughness and imperfections in weldments represents a unique topic because of a heterogeneous microstructure. This is a complex problem that is certainly beyond the scope of a practice-oriented

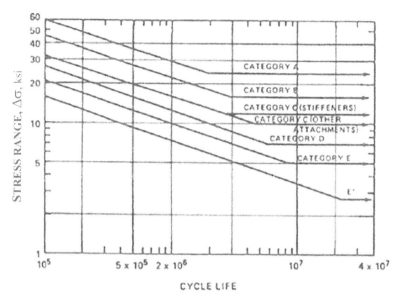

FIGURE 6.30 AASHTO design curves for fatigue of welded structures.

book. Only a few general comments can be made concerning fracture toughness of weldments.

By analogy to conventional steel products, the properties of filler metals and fluxes can be modified by adding the appropriate chemical elements. The steel can be affected by the presence of microstructural constituents. These include ferrite, pearlite, bainite, and martensite. For instance, the strength of steel decreases and the fracture toughness transition temperature goes up as the tempered martensite is changed to other microstructural constituents. The actual role of the various constituents depends on the composition, processing, and heat treatment of the steel. For other details, as required, the design engineer should consult the metallurgist because no single process can be utilized for all weldments. This fact, in itself, shows the overall complexity of the process of selection of the welding procedure.

The most frequently used welding process is known as arc welding, which creates a higher temperature for the weld metal than the base metal. The size of the formed HAZ depends on a number of factors that determine grain size and microstructure responsible for the level of HAZ fracture toughness. The complexity of this phenomenon does not end here, because the variations in temperature gradient and cooling rate can, in turn, change the grain size. The experience shows that in the case of carbon and low-alloy steels, the regions next to the weld have coarse grains, resulting in the lowest toughness. The entire picture of causes and effects within the HAZ regions becomes even more involved when the weld metal is deposited in a multipass fashion. It is fortunate, however, that the zones of low toughness are surrounded by the regions of higher toughness.

CRITERIA OF LEAK-BEFORE-BREAK

Quantitative assessment of toughness requirements is possible because of the developments in LEFM.[28,29] The requirement is that in the presence of a large, sharp crack in a large plate, through-thickness yielding should develop prior to fracture. The criterion here is based on the ratio of plate thickness to plastic zone ahead of a large and sharp crack. The yielding takes place when the plate thickness is decreased, causing the change of stress from plane strain to plane stress. Here a ductile fracture occurs along a 45° plane through the thickness. In most materials a failure of this kind is preceded by a through-thickness yielding and it is not catastrophic. The situation is different in thicker materials, where we have plane-strain conditions and where the fracture is oriented normal to the direction of loading. The through-thickness yielding may not occur and the fracture may become unstable, unless we are dealing with a recognized ductile material. However, if plane stress can be assured, a plastic overload would be required for the structure to fail. Unfortunately, the majority of structural members may fall between the plane-stress and plane-strain conditions.

    In order to be able to judge and control the through-thickness deformation, the following relation is recommended,[30] which follows from Eq. (4.4).

$$\left(\frac{K_{Ic}}{S_y}\right)^2 \frac{1}{B} \geq 1 \tag{6.8}$$

In Eq. (6.8) K_{Ic} = plane-strain fracture toughness, ksi (in.)$^{1/2}$, and S_y = yield strength, ksi. Also the condition of plane-strain behavior follows from Eq. (4.1):

$$\left(\frac{K_{Ic}}{S_y}\right)^2 \frac{1}{B} \leq 0.40 \tag{6.9}$$

where B in these formulas denotes the plate thickness in inches. Using Eq. (6.8) as a basis, Irwin[14] proposed the minimum fracture toughness K_{Ic} for a "leak-before-break" criterion to be as

$$\frac{1}{B}\left(\frac{K_{Ic}}{S_y}\right)^2 = 1.5 \tag{6.10}$$

from which

$$B = \frac{2}{3}\left(\frac{K_{Ic}}{S_y}\right)^2 \tag{6.11}$$

This criterion constitutes a practical goal for establishing the proper toughness for large, thin plates containing through-thickness cracks. Hence for a fixed value of yield strength S_y, the toughness must increase as a function of thickness. However, Eqs. (6.8) through (6.11) become rather conservative when the plate thickness exceeds 2 in. For plate thicknesses lower than 2 in., Barsom and Rolfe[24] suggest the following expression:

$$K_{Ic} = S_y(T)^{1/2} \tag{6.12}$$

Equation (6.12) determines K_{Ic} values required for through-thickness yielding before fracture. However, current ASTM practice still uses a slightly different formula intended for the maximum valid K_{Ic} parameter as

$$K_{Ic} = 0.63S_y(T)^{1/2} \tag{6.13}$$

For a K_{Ic} value higher than the maximum, the test specimen is likely to indicate some ductility, and the pure plane-strain conditions will not be met.

The important leak-before-break criterion was first established by the Naval Research Laboratory[31] almost 30 years ago. The intent was to estimate the proper level of fracture toughness of pressure vessel steels, so that a surface crack could grow through the wall and cause the vessel to leak prior to catastrophic fracture. In other words, the critical size of the crack at the design stress level would have to be larger than the wall thickness of the vessel. Ideally, a detectable leak is desirable because it gives the opportunity to repair the vessel prior to any violent fracture. It requires little imagination to see that all pressure vessels contain large amounts of stored energy, which have very serious consequences during catastrophic bursts. This situation is even more disturbing in the nuclear industry. There is a dire need to understand the scale of a potential nuclear accident, and every effort should be made during the design phase to quantify the fracture mechanics parameters and stress analysis results to assure leak-before-break conditions. Knowledge of stress intensity, materials' resistance to fracture, and morphology of growing cracks comes to the forefront of the assessment of structural integrity.

The concept of leak-before-break criterion is illustrated in Fig. 6.31. This shows the initial crack (1), crack growth through the wall (2), break-through portion (3), and a local section of the wall (4). As the surface crack begins to grow, the stress intensity increases, and as long as the stress intensity factor K does not exceed the material's resistance to fracture K_{Ic}, before the crack becomes rather large it is quite feasible that the remaining thin ligament will

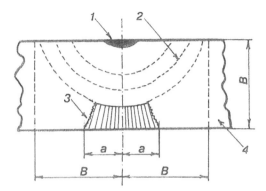

FIGURE 6.31   Model of crack growth for leak-before-break. See text for explanation of 1–4.

be ruptured and a through-wall crack of length $2a$ will be created. This happens when the net section stress in the ligament exceeds the tensile strength of the material. When the length of the through-thickness crack $2a$ shown in Fig. 6.31 does not become greater than the critical, through-wall crack length, then we have compliance with the leak-before-break criterion. However, if the through-thickness crack created at the time of the breakthrough is longer than the critical size for failure, it will certainly continue to propagate along the wall length, leading to a large break condition.

The entire process of crack growth consistent with the leak-before-break criterion (Fig. 6.31) indicates how a surface crack is transformed into a through-thickness crack. It is customary to assume here that a crack of twice the wall thickness in length (shown as $2B$ in Fig. 6.31) is likely to be stable at a stress level equal to the nominal design stress. So much can be said at this point regarding the morphology of the growing crack.

The final geometry, dimensions, and loading of the through-thickness crack are given in Fig. 6.32.

The stress intensity factor K_I for a through-thickness crack in a large plate can be defined in terms of a stress ratio σ/S_y, as stated by Barsom and Rolfe.[24]

$$K_I^2 = \frac{\pi\sigma^2 a}{1 - \frac{1}{2}(\sigma/S_y)^2} \tag{6.14}$$

When the design stress σ is relatively low, Eq. (6.14) can be reduced to the simplest form: $K_I = \sigma(\pi a)^{1/2}$, which resembles Eq. (2.14) and a standard formula for stress intensity taken from Fig. 6.33. At the time of fracture and plane-stress

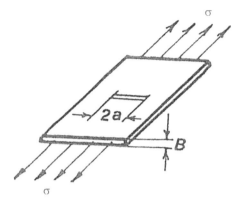

FIGURE 6.32   Symbols for through-thickness crack.

conditions, the stress intensity becomes

$$K_c^2 = \frac{\pi \sigma^2 a}{1 - \frac{1}{2}(\sigma/S_y)^2} \qquad (6.15)$$

The relationship between the parameters K_c and K_{Ic} under the assumption of leak-before-break[24] is represented by

$$K_c^2 = K_{Ic}\left[1 + \frac{1.4}{B^2}\left(\frac{K_{Ic}}{S_y}\right)^2\right] \qquad (6.16)$$

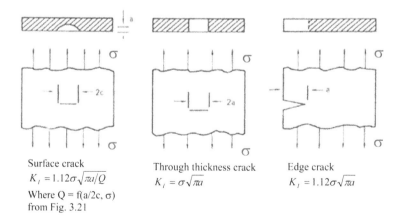

Surface crack
$K_I = 1.12\sigma\sqrt{\pi a/Q}$
Where Q = f(a/2c, σ)
from Fig. 3.21

Through thickness crack
$K_I = \sigma\sqrt{\pi a}$

Edge crack
$K_I = 1.12\sigma\sqrt{\pi a}$

FIGURE 6.33   Typical stress intensity factors.

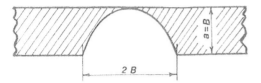

FIGURE 6.34 Through-thickness crack condition for Eq. (6.17).

Hence, combining Eqs. (6.15) and (6.16) gives, for $a = b$

$$\frac{\pi \sigma^2 B}{1 - \frac{1}{2}(\sigma/S_y)^2} = K_{Ic}^2 \left[1 + 1.4 \left(\frac{K_{Ic}^4}{B^2 S_y^4} \right) \right] \tag{6.17}$$

The formula given by Eq. (6.17) assumes the crack surface reaches the surface of the vessel wall as shown in Fig. 6.34 just before the breakthrough at the intersection of the two surfaces.

The work of several investigators[31] and other experience suggest that there are uncertainties in applying the leak-before-break criteria with respect to stress levels at intersections and the role of residual stresses, so that a conservative approach is warranted. This can be accomplished by increasing the nominal design stress, so that $\sigma \cong S_y$. Consequently, Eq. (6.17) can be reduced to

$$0.223\beta^2 + 0.159 = 1/\beta \tag{6.18}$$

where

$$\beta = \frac{1}{B} \left(\frac{K_{Ic}}{S_y} \right)^2 \tag{6.19}$$

For the stress ratio $\sigma/S_y = 0.5$, the leak-before-break criterion becomes

$$1.4\beta^2 + 1 = 1/\beta \tag{6.20}$$

Solving the cubic equation (6.18), the real root of the equation is $\beta = 1.505$, which gives fracture toughness as a function of S_y and B:

$$K_{Ic} = 1.227 S_y (B)^{1/2} \tag{6.21}$$

The relevant family of design curves for $\sigma = S_y$ under conditions of leak-before-break is given in Fig. 6.35. From the solution of the cubic equation (6.20) follows the parameter $\beta = 0.6375$, which yields the formula

$$K_{Ic} = 0.798 S_y (B)^{1/2} \tag{6.22}$$

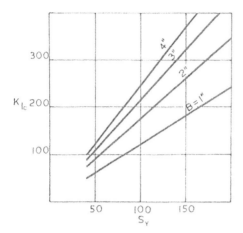

FIGURE 6.35 Design chart for leak-before-break at $\sigma = S_y$ (dimensions: K_{Ic} in ksi (in.)$^{1/2}$ and S_y in ksi).

The design chart for $\sigma = 0.5S_y$ (leak-before-break) is shown in Fig. 6.36. If required, the design charts for other σ/S_y ratios can be developed, utilizing the following calculation steps:

1. Substitute the required σ/S_y ratio in Eq. (6.17).
2. Transform Eq. (6.17) in terms of a dimensionless parameter β given by Eq. (6.19), so that only the numerical and β terms remain.

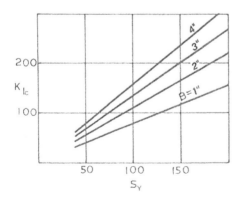

FIGURE 6.36 Design chart for leak-before-break at $\sigma = 0.5S_y$ (dimensions: K_{Ic} in ksi (in.)$^{1/2}$ and S_y in ksi).

3. Solve the resulting cubic equation for β using a conventional algebraic procedure, graphical method of quadratic and linear parts of the equation, similar in form to Eqs. (6.18) and (6.20), or by a cut-and-try approach.

4. Use Eq. (6.19) to establish the plane-strain fracture toughness K_{Ic} as a function of yield strength S_y and plate thickness B, as shown, for instance, by Eqs. (6.21) and (6.22).

The analysis shows in general that by increasing the wall thickness of the vessel, say, from 1 to 5 in., the design stress σ should be decreased by a factor of about 3.[31] In using the leak-before-break criteria the design engineer selects the nominal yield strength steel and wall thickness of the vessel according to conventional formulas of stress analysis so that the wall thickness is sufficient to withstand a given internal pressure. Then comes the choice of the minimum required toughness to meet the leak-before-break criterion followed by cost of material, fabrication, certification, and other technical and economic decisions. The charts given in Figs. 6.35 and 6.36 are certainly consistent with the general practice of leak-before-break criteria[24] and are rather conservative for a vessel thickness above 2 in.

The design technique described briefly in this section has a rational basis and practical justification. However, other criteria methods are also available based on the elastic–plastic approach and principles of fracture-safe design.[10,24]

PLANE STRESS AND PLASTIC BEHAVIOR

While the bulk of this book is reserved for the use of LEFM in design, there are a number of topics in fracture mechanics that are researched and used in special situations. The first topic in this category relates to the constraint conditions where the plane strain cannot be achieved. In such a case we deal with the plane-stress fracture toughness, mentioned previously in Chapter 2 (Fig. 2.10) and in Chapter 4 (Fig. 4.1) in connection with the concept of the limiting thickness expressed by Eq. (4.1), and the plastic zone radius defined by Eq. (4.3). Nominal stresses exceeding the yield strength of the material can exist when the material thickness complies with the criterion given by Eq. (4.4). In other words, we use the parameter K_{Ic} to deal with the relaxation of the constraint consistent with the plane-stress environment. This is, of course, a practical approach because the determination of the K_c values would call for the use of elastic–plastic fracture mechanics (EPFM), not at all the simplest alternative.

In dealing with low- and medium-strength steels found in such structures as bridges, ships, or pressure vessels, the LEFM approach involving K_{Ic} can be invalidated when confronted with a thickness insufficient to maintain plane-strain

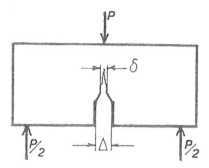

FIGURE 6.37 Basic CTOD model.

conditions. Out of several ways of treating this problem, the crack tip opening displacement (CTOD) method has distinct practical advantages.

The basic concept of CTOD was first proposed as the result of observing the opening of the notch faces, ahead of a sharp crack.[32] The typical test setup, shown in Fig. 6.37, permits measurements of crack opening displacements at elastic–plastic as well as fully plastic deformations. The force P can be measured and plotted as a function of Δ, which is the displacement indicated by the clip gauge. A typical load displacement characteristic is sketched in Fig. 6.38.

While this technique suggested that the CTOD values could be directly related to the critical crack extension force and the plane-strain fracture toughness,[24] the subsequent solution of the problem was accomplished by utilizing a "crack tip plasticity model" (also known as the "strip-yield model"), relating the CTOD to the applied nominal stress and the crack length.[33] The basic

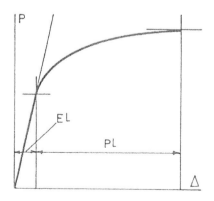

FIGURE 6.38 Load displacement curve (EL, elastic strain; PL, plastic strain).

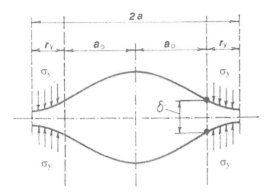

FIGURE 6.39 Strip-yield model.

dimensional quantities of the strip-yield model are indicated in Fig. 6.39. The displacement near the original crack tip is denoted by δ, as previously shown in Fig. 6.37, and it represents the actual value of the CTOD. This parameter increases with the increase in the real length of the crack and the higher external loading. It is also shown in Fig. 6.39 that the plastic zone radius at each end of the original crack contributes to the total crack length, so that $a = a_o + r_y$.

However, the equation derived using crack tip plasticity[33] has indicated a complex relationship between the CTOD (denoted by δ) and a number of stress analysis parameters in addition to the crack length:

$$\delta = \frac{8aS_y}{\pi E} \ln \sec\left(\frac{\pi\sigma}{2S_y}\right) \qquad (6.23)$$

Here a = half-length of real crack (in.), σ = nominal design stress (ksi), S_y = yield strength of the material (ksi), and E = elastic modulus of the material (ksi).

The main advantage of the CTOD concept is that δ values can be monitored throughout the elastic, elastic–plastic, and plastic conditions. The nature of the problem is quite different with the K_{Ic} analysis, because K_{Ic} requires the existence of the plane-strain region, or a suitable approximation of it, during the early stages of the elastic–plastic behavior.

Considerable effort was spent in the United Kingdom on evaluating the various methods of testing and measuring the critical values of CTOD. The principles of the LEFM approach were restricted to those cases where the stress intensity factors dominated the stress field in high-strength steels, aluminum alloys, and similarly behaving materials. The problem became even more demanding in dealing with the lower yield materials, and with welding of large vessels and other thick parts. The differences between the elastic solutions and

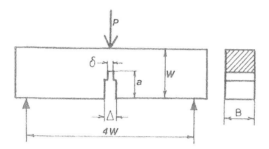

FIGURE 6.40 CTOD test piece notation and proportions.

the results requiring a "plastic correction" increased as the plastic zone became larger. All this pointed to the need to improve the testing methods for dealing with the new fracture toughness parameter, δ_c, related to the critical crack size. The result was a slow-bend test specimen (Fig. 6.40) and a new British standard.[34] The test setup illustrated in Fig. 6.40 shows a specimen similar to that for a K_{Ic} slow-bend test in the ASTM E-399 test method.[46] The CTOD formula in the British standard is based on converting the clip gauge displacement Δ into a crack tip value δ, as shown below:[36]

$$\delta = \frac{K^2(1 - v^2)}{2S_yE} + \frac{0.4(W - a)\Delta_p}{0.4W + 0.6a + Z} \tag{6.24}$$

The width of the beam cross-section is B, with $W/B = 2$. The specimen is precracked (in fatigue) to have a/W ratios between 0.45 and 0.55. The parameter Δ_p signifies the plastic component of the displacement, obtained as the offset shown in Fig. 6.41. Other terms are $a =$ total crack length (in), $B =$ specimen thickness (in.), $E =$ elastic modulus (ksi), $v =$ Poisson's ratio, $K =$ stress intensity factor at maximum loading (not at 5% offset as in Fig. 4.20), and $Z =$ dimensional correction (mm). In the British standard, Z is a dimensional correction (in mm), which is usually small and can be neglected as a relatively small quantity.

It should be noted that the formula for estimating the critical value of CTOD, Eq. (6.24), is based on an elastic and plastic component of CTOD and it consists of two parts. The first part contains the stress intensity factor and the elastic modulus while the second part derives from the geometry of specimen deformation and from the assumed location of the center of rotation of the two halves of the test piece under load. The distance to this center, measured from the crack tip, is $0.4(W - a)$. It is known, of course, that the position of the crack tip and the center of rotation will change with the crack growth. These fine points of yielding fracture mechanics are not accounted for in Eq. (6.24),

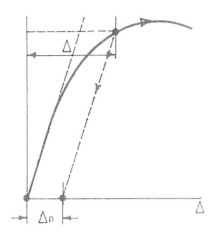

FıGURE 6.41   Clip gauge displacement.

which can only be approximate. Special techniques, however, are available to correct this problem.[36]

Since the parameter δ_c denotes a critical value of CTOD, it is also a material property, which (as expected) must depend on temperature, rate of loading, and specimen thickness. In general, the effects are similar to those found with the K_{Ic} or CVN test results. The δ_c parameter increases with increasing temperature. Also, an increased rate of loading or plate thickness decreases δ_c. Static and dynamic CTOD test results are available for a number of structural steels.[24] Although considerable scatter of the results was found in the temperature transition region, the lower-bound portions of the curves could be established. The general character of a typical transition curve is shown in Fig. 6.42.

One of the practical advantages of the δ_c parameter is its ability to correlate with the plane-stress fracture toughness K_c and plane-strain fracture toughness K_{Ic}, using the following relationships:

$$K_c = 0.89[E(S_y + S_u)\delta_c]^{1/2} \tag{6.25}$$

and

$$K_{Ic} = 0.77[E(S_y + S_u)\delta_c]^{1/2} \tag{6.26}$$

where E = elastic modulus (ksi), S_y = yield strength of the material (ksi), S_u = ultimate strength of the material (ksi), and δ_c = critical CTOD value (in.).

Another way of finding the K_{Ic} parameter involves a two-stage procedure,[24] provided the K_{Id} parameter can be obtained from a correlation

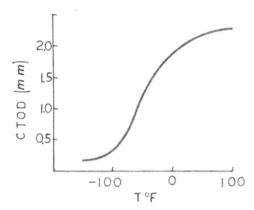

FIGURE 6.42 Approximate shape of transition curve, (A156 steel, according to Ref. 21).

formula, such as

$$K_{Id} = [5E(CVN)]^{1/2} \tag{6.27}$$

The temperature shift between K_{Ic} and K_{Id}, such as that sketched in Fig. 4.7, can be estimated from Eq. (4.5). For a discussion involving CTOD experimental and theoretical results, the concept of the J-integral, and the R-curve analysis of extending the LEFM into the plastic range, the interested reader is directed to the chapter on elastic–plastic fracture mechanics (EPFM) by Barsom and Rolfe.[24]

Before leaving this problem area dealing with the extension of LEFM theory to the plastic range, it is well to summarize the basic definitions of the three more important techniques, including practical formulas for correlation purposes.

CTOD

CTOD represents a measure of the opening displacement mode of the crack face. The crack tip opening displacement is regarded as a measure of the plastic strain at the crack tip. Several values of this parameter can be calculated from the load displacement records related to such characteristics as unstable cleavage fracture during and after ductile tearing, maximum load condition, first pop-in at brittle crack extension, and the onset of stable crack extension.

J-Integral

J-integral denotes a mathematical expression of a line- or surface-type integral that encloses the crack front from one crack surface to the other, used to characterize the local stress–strain field around the crack front. In the United States the *J*-integral approach to EPFM is rather popular. The *J*-integral defines the plastic stress and strain intensity in a manner similar to the way the *K* parameter in LEFM represents the stress intensity of the surrounding elastic field, in the crack vicinity. The *J*-integral depends on stress, strain, crack size, and the geometry of the crack and body.[37–39] It should be noted here that the basic concepts of LEFM and EPFM are rather similar, with the exception of the parameter describing the crack tip condition such as the size of the plastic zone. In the case of EPFM the plastic zone, characterized by the tensile stress field of yield strength level, is considerably larger than that for LEFM. If the parameter *J* is analyzed in the region of a smaller plastic zone, corresponding to the linear elastic stress intensity factor K_I, then there is a direct relationship between *J* and K_I of the type

$$J = \frac{K_I^2}{E} \tag{6.28}$$

It may be recalled that the strain energy release rate *G* correlates well with the stress intensity factor K_I for the plane-stress and plane-strain conditions, as shown by Eqs. (2.11) and (2.12). Also, the main portion of Chapter 3 is devoted to the K_I parameter for the most common crack and test specimen geometries. Equation (6.28) also indicates that the *J* dimension is lb/in. or N/mm as the case may be, if $J = G$. Recent investigations of the relationship between the *J* and δ parameters[40] show that

$$J = 0.5m(S_y + S_u)\delta \tag{6.29}$$

where the *m* factor varies between 1 and 2. It should be noted in general that the EPFM approach is not without some limitations. For instance, the fracture zone must be relatively small in comparison with the larger plastic region as defined by the *J*-integral approach. It also should be significantly smaller than the thickness *B* of the cracked body. The choice of *J*-integral over the CTOD approach has been driven by the academic circles dedicated to two- and three-dimensional finite-element modeling of *J*-integral functions, which has proven to be extremely complex and expensive.

R-Curve

The *R*-curve technique has been developed for the purpose of characterizing a material's fracture resistance as a function of crack growth. The *R*-curve can be expressed in terms of the stress intensity factor *K* or the crack opening displacement parameter δ as long as the plastic zone is relatively small. The use of this curve in conjunction with the plane-stress fracture toughness K_c represents an extension of LEFM into EPFM criteria. Assuming that the crack growth resistance is defined as K_R, the information can be presented as the variation of K_R as a function of crack length. However, this technique is normally applicable to linear behavior of high-strength steels and aluminum alloys. The elastic–plastic approach requires the use of the parameter δ.

The foregoing brief summary does not include any characterization of high-temperature and time-dependent fracture criteria under which steady-state creep affects the crack growth.

Although this book is essentially dedicated to the LEFM and transition temperature approaches, there may be some confusion with the use of the acronym FAD. For instance, Fig. 6.13 illustrates FAD, which stands for fracture analysis diagram, where nominal stresses are plotted as a function of temperature. All materials exhibit transition temperature characteristics, performing under plane-strain, elastic–plastic, or even fully plastic conditions. However, there is also another analysis technique having the same acronym, which represents an entirely different plot known as "failure analysis diagram." This technique was developed in Great Britain in the late 1970s and it provides a convenient approach to the analysis of fractures, covering the entire range of behavior between brittle and fully plastic conditions. In this case, FAD is obtained by plotting the stress intensity (say, K or K_c) as a function of the fracture stress, based on the residual strength in the presence of a crack. In other words, as the crack becomes longer, the residual strength and the factor of safety decline. The plot may represent directly stress intensity vs. the fracture stress, or K/K_c vs. stress/strength, in which case we have a normalized diagram. Essentially then, the stress is limited by collapse while the toughness is the natural limit of stress intensity. The British concept of the diagram is described in practical terms by Broek.[21]

ENVIRONMENTAL EFFECTS

It is sufficient to have only a passing glance at any paper or book dealing with the degradation of a material's integrity in any combination of mechanical, thermal, or other environment to realize the presence of countless effects that nature has in store to assure that a man-made structure of any kind will be broken and destroyed in the end. The only purpose of engineering is to postpone this occurrence for an acceptable time interval. Countless learned papers continue to be published that are intended to work against a stacked deck and to bring some order to totally disordered systems where statistics reigns supreme. It is fortunate that some materials such as metals and alloys may be less susceptible to certain failure mechanisms, such as stress corrosion, which is only one out of many. For this reason many tests have been developed to study stress-corrosion cracking in spite of the difficulties created by the complex chemical, mechanical, and metallurgical interactions. The reader should realize that after all the experimental effort in this area, we still have extensive data scatter and poor correlation with service experience.

To begin, most conventional metals corrode in service, although it appears that while modern mild steel rusts rather easily, old-type wrought or cast irons are in better shape in such an environment. It is also surprising that steel embedded in concrete does not rust, and that stainless steel in such an environment does not always perform well in a corrosive medium — and it is certainly expensive to fabricate. The use of aluminum alloy, as an alternate solution, may not be the best choice because of inadequate stiffness and the potential extra cost since it may be necessary to use more energy to produce aluminum than steel. The foregoing aspects of environment and the various technical decisions regarding the choice of the material for a particular application present formidable problems that often require analysis and tests under closely controlled conditions. The main difficulty there is that the analysis and tests must reflect structural performance of a particular component or a mechanical system under development. Metal fatigue here is the most frequent failure mechanism since stress-corrosion cracking, for instance, can be so insidious. This is usually the case in brittle fracture with the principal features of instability and sudden propagation of an existing crack under an applied stress much lower than the expected yield strength of the material. The insidious feature of this process is that a catastrophic structural failure can be triggered by a micromechanical event involving minute volumes of the material. Such an event can be started by a crack of millimeter dimensions.

The first approach to evaluating the stress-corrosion behavior of a material is based on the time required to induce a failure of a smooth or a mildly notched specimen. This is known as the "time-to-failure" technique, which represents the total time needed to initiate and propagate the crack to critical dimensions. The research in this area has shown the importance of separating the initiation and propagation phases of this process.[41] Furthermore, the fracture mechanics approach to the study of stress-corrosion cracking has, so far, involved precracked specimens and the established concept of the stress intensity factor K_I. The use of this factor is similar to that encountered in the study of plane-strain fracture toughness K_{Ic}, where the state of plane strain is enforced at the crack tip. This is also consistent with the assumption of a small plastic zone close to the tip of the crack, and with the limitation of the specimen size. Study of the environmental effects on precracked specimens has eventually led to the introduction of the threshold parameter K_{Iscc}, which, at a given temperature and environment, constitutes the stress intensity factor value below which subcritical crack extension does not take place.

The environmental effect in a liquid is established by comparing the test results in air and the liquid. In the case of an air environment, we may have moisture and hydrogen. Steel can be affected by hydrogen during a melting process, and ultrahigh-strength steel in particular is susceptible to stress-corrosion cracking in air due to the presence of water vapor. However, steels of intermediate strength do not appear to be affected by such an environment.

The experimental determination of stress-corrosion cracking involving precracked specimens falls into two categories. These are the "time-to-failure," mentioned above, and the "crack growth rate" categories. Unfortunately the latter tests are complex and require more sophisticated instrumentation. However, the crack growth rate technique is better for research purposes because it provides information on the kinetics of stress-corrosion cracking and the threshold behavior of K_{Iscc}. The most popular test specimens for studying stress-corrosion cracking include the cantilever beam under constant load and the wedge opening loading (WOL) specimen under constant displacement.[42] The geometry and dimensional details for the design of the two types of specimens are given in the literature.[24] The working equations intended for calculating the relevant stress intensity factors during laboratory experiments are quite similar to those discussed in Chapter 3. For instance, the theory given for the cantilever beam specimen, used in testing stress-corrosion cracking, corresponds to a class of similar configurations such as that shown in Fig. 3.14. However, Eq. (3.19) and Fig. 3.15 were based on a different source, and the crack length was assumed to be small compared to the depth of the beam. The formulas for the stress intensity factor[24] suggested for the WOL specimen in the study of stress-corrosion cracking are essentially identical, with the expressions for the wedge-opening mode given previously as Eqs. (3.31) and (3.33). The work of

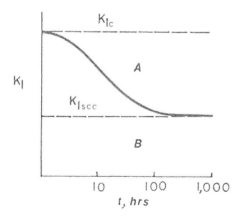

FIGURE 6.43 Typical curve for cantilever test (ASTM conditions observed).

Bueckner is linked with the mathematical developments of equations for stress intensity factors involved in edge cracking of round bars and flat strips.[28,43] The results derived from cantilever beam experiments, widely accepted, are characterized by the typical curve shown in Fig. 6.43.[24] The level of K_{Ic} indicates that the conditions of plane-strain constraint are fully satisfied, including the section size requirements. Also, by analogy to Eq. (4.1), the validity of the K_{Iscc} parameter is assured by compliance with the basic equation, as

$$B \geq 2.5 \left(\frac{K_{Iscc}}{S_y} \right)^2 \tag{6.30}$$

Careful experiment is required for the determination of the lowest possible stress intensity K_I. Pellini[10] shows the test procedure for the case of 4340 high-strength steel (215 ksi yield) free corrosion in synthetic seawater (3% NaCl), plotted in Fig. 6.44. The K_{Iscc} parameter is an asymptotic value of K_I below which no crack growth is expected. Region A, shown between K_{Ic} (dashed line) in Fig. 6.43 and the continuous curve, corresponds to fracture caused by subcritical crack extension. There is a no-failure area B below the full line. A similar area of no break B is indicated in Fig. 6.43. The minimum thickness of the specimen

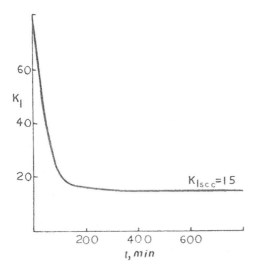

FIGURE 6.44 Test curve for minimum K_{Iscc} value for 4340 steel. Stress intensity is given in ksi (in.)$^{1/2}$.

follows from Eq. (6.30)

$$B = 2.5(15/215)^2$$
$$= 0.012 \text{ in. } (0.3 \text{ mm})$$

Experience with various specimen geometries and the methods of loading indicates the K_{Iscc} parameter is a material property. Also, for a given environment–material system K_{Iscc} is independent of specimen size above a specific geometry limit.[44] At the same time, the relevant nominal stress appears to be highly dependent on the specimen geometry and size. Other elements enter the equation of the effects, such as corrosion products and thermomechanical processing. The entire topic of corrosion-fatigue crack propagation, that is, fatigue in a hostile environment, is bound to be highly complex. While the extensive investigations in this field help with the process of material selection, little has been accomplished in establishing the essence of the mechanism of corrosion fatigue. The major problem here is that mechanical and chemical processes interact at the crack tip. The only way of approaching this complex situation is to compare the fatigue crack growth in hostile and benign environments.

Although standard methods for stress-corrosion cracking have still a way to go, the necessity of validating the K_{Iscc} data for the existence of plane-strain conditions at the crack tip has influenced numerous investigators determined to develop a classification routine for apparent K_{Iscc} behavior in geometrically proportionate specimens. The proposed classification[45] covers three types of behavior, corresponding to valid, partially valid, and invalid K_{Iscc}. This classification has led to the establishment of two boundaries. Boundary one is a conservative requirement of plane strain satisfying four equations, which according to the cantilever notation in Fig. 6.45, are

$$a_{\min} = 2.5\left(\frac{K_I}{S_y}\right)^2 \tag{6.31}$$

$$B_{\min} = 2.5\left(\frac{K_I}{S_y}\right)^2 \tag{6.32}$$

$$(W - a)_{\min} = 2.5\left(\frac{K_I}{S_y}\right)^2 \tag{6.33}$$

$$W_{\min} = 5\left(\frac{K_I}{S_y}\right)^2 \tag{6.34}$$

The second boundary was determined to correspond to the highest stress intensity factor K_I, signifying the initial appearance of a plastic hinge for an elastic–

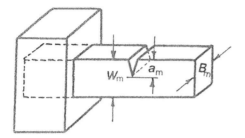

FIGURE 6.45 Key cantilever notation ($W_m = W_{min}$; $a_m = a_{min}$; $B_m = B_{min}$).

perfectly plastic material,[24] in bending, as shown in Fig. 6.45. The relevant equations for this boundary are

$$a_{min} = 1.0 \left(\frac{K_I}{S_y} \right)^2 \tag{6.35}$$

$$B_{min} = 1.0 \left(\frac{K_I}{S_y} \right)^2 \tag{6.36}$$

$$(W - a)_{min} = 1.0 \left(\frac{K_I}{S_y} \right)^2 \tag{6.37}$$

$$W_{min} = 2.0 \left(\frac{K_I}{S_y} \right)^2 \tag{6.38}$$

Equations (6.31) through (6.38) help to evaluate the effect of specimen size on K_{Iscc} behavior. The second important influence is concerned with the duration of the test, using a cantilever beam method and covering the range of test duration between 100 and 10,000 hours, for an apparent yield strength of a material equal to 180 ksi, which in this particular case represented high-alloy steel tested in synthetic seawater at room temperature. The test indicated the decrease of the apparent K_{Iscc} value with an increase in test duration. In general, the K_{Iscc} parameter decreases with the increase of yield strength. Barsom and Rolfe[24] provide K_{Iscc} data for several steels in aerated 3% NaCl solution of distilled water. These include structural steels such as A36, A588 Grade A, A588 Grade B, and A514 Grade E.

The basic design rule is not any different whether we use, say, K_{Ic} or K_{Iscc} parameters. For instance, if the stress intensity factor determined for the structural member at hand is not higher than K_{Ic} or K_{Iscc}, then the basic condition of structural integrity is complied with. The designer should always keep in mind the fundamental principle that the stress intensity factor K or K_I should be less than the critical stress intensity factor, in the same manner that the design stress is kept below the yield strength of the material in conventional stress analysis, that is, $\sigma < S_y$. The brief introduction given in this section to the problem of fracture toughness determination points to a number of constraints in selecting the parameter such as K_{Iscc}. K_{Iscc} tests, like other LEFM tests, involve limitations of thickness, remaining ligament, or the size of the plastically deformed region just ahead of the crack tip. The experience also indicates that the size of flaws and cracks in actual structures may be small compared to the fatigue crack in a test piece used in LEFM experiments. Small imperfections can be surrounded by the plastic zone, such as in the case of a welded region dominated by a high residual stress at a discontinuity.

Since K_{Iscc} is the plane-strain stress intensity factor for a given environment, and because structures can be subjected to load fluctuations in service, Barsom and Rolfe[24] suggest that even on the basis of limited information,[46] the K_{Iscc} parameter can be regarded as the corrosion fatigue limit value similar to the ΔK_{th} term referred to previously in Eq. (5.15) and cited in other sections of Chapter 5 as the threshold strain intensity range. Hence the (K_{Iscc}) parameter should be considered as a practical tool in judging service performance of fully immersed and cyclically loaded structural members. However, this statement is not intended to imply that our current knowledge of fundamental theory of stress-corrosion cracking is complete.[47]

As stated previously in this section, there is some justification for separating the initiation and propagation phases of corrosion-fatigue cracking, and the total useful life of a structure, submerged in an aggressive environment, is measured in a number of cycles. The relative magnitudes of initiation and propagation portions of corrosion fatigue depend on several factors such as material properties, geometry of the structure, applied stress levels, and environment. Also, for a given environment, the corrosion-fatigue process will be influenced by metallurgical, mechanical, and electrochemical factors. The total corrosion-fatigue damage happens more rapidly than the expected algebraic sum of individual effects of fatigue induced by corrosion, or stress-corrosion. This fact alone makes the understanding of the entire corrosion-fatigue process very difficult, particularly since the established behavior for one set of conditions at hand may not apply to the other. Details of corrosion-fatigue crack initiation behavior are well described by Barsom and Rolfe[24] with special regard to constructional steels tested in 3.5% NaCl solution. The text includes details of modified compact tension and single-edge notched specimens. The procedure involves, first, the establishment of the fatigue crack initiation threshold in the absence of aggressive environment, followed by the tests of corrosion fatigue crack initiation behavior for such steels as A36, A588 Grade A, A517 Grade F, and V150. The parameter of interest is $[\Delta K/(\Delta)^{1/2}]_{th}$, which defines the fatigue crack initiation threshold discussed previously in Chapter 5. This parameter has the dimension of ksi, if the notch tip radius Δ is given in inches. The specific values quoted[24] are featured in Table 6.1 and represent the fatigue testing threshold in air. The relevant expressions for the correlation between the threshold parameter and the number of cycles for crack initiation N_i, as quoted[24] for the steels, are stated below, including standard deviation (SD) values.

A36

$$\log(N_i) = 12.170 - 3.793 \log\left[\Delta K/(\rho)^{1/2}\right] \qquad SD = 0.093 \qquad (6.39)$$

A588 Grade A

$$\log(N_i) = 11.730 - 3.477 \log\left[\Delta K/(\rho)^{1/2}\right] \qquad SD = 0.187 \qquad (6.40)$$

A517 Grade F

$$\log(N_i) = 10.884 - 3.023 \log\left[\Delta K/(\rho)^{1/2}\right] \qquad SD = 0.180 \qquad (6.41)$$

V 150

$$\log(N_i) = 11.061 - 3.236 \log\left[\Delta K/(\rho)^{1/2}\right] \qquad SD=0.119 \qquad (6.42)$$

The foregoing formulas are the best-fit equations for corrosion-fatigue crack initiation. Since there are only small differences between Eqs. (6.39) through (6.42), it is possible that a superposition of all data for the four steels can be represented by one best-fit equation with a standard deviation of only 0.191.[21]

$$\log(N_i) = 11.444 - 3.357 \log\left[\Delta K/(\rho)^{1/2}\right] \qquad (6.43)$$

What is rather remarkable is that, even with significant differences in chemical composition, microstructure, and yield strength, the corrosion-fatigue crack initiation behavior was very similar. Additional effects such as frequency of load cycles at the ratio of minimum to maximum nominal stresses in the fatigue process appear to have only a secondary effect on the corrosion-fatigue crack initiation behavior of the steels investigated.[24] Hence for design purposes, the equation solved for the number of cycles for crack initiation N_i and recommended by Barsom and Rolfe,[24] is

$$N_i = 3.56 \times 10^{11}\left[\Delta K/(\rho)^{1/2}\right]^{-3.36} \qquad (6.44)$$

The prediction of corrosion-fatigue crack initiation life based on the lower bound of the data may be too conservative. The best-fit line given by Eq. (6.44) provides the number of cycles when ΔK is in ksi (in.)$^{1/2}$ and (Δ) is in inches. Caution should be exercised in using Eq. (6.44) when severe stress raisers or imperfections exist in the notch area.

While the crack initiation phase of corrosion fatigue is sufficiently defined for practical purposes, the crack propagation behavior depends strongly on many variables in addition to frequency, waveform, and stress ratio, making the entire problem unusually complex and very costly to tackle, especially in the case of the threshold phenomena. The research continues, of course, because of the existing scientific challenge, and significant accomplishments have been and will be made for the sake of better understanding of the fundamental problem of crack propagation in a hostile environment. However, this is still a long way from a proven design methodology and engineering standards, and it raises a pragmatic question: Under which set of conditions is crack growth analysis worthwhile? Should the stress-corrosion cracking be prevented or controlled? In mathematical terms of fracture mechanics the K_{Ic} parameter is simply replaced by K_{Iscc}, and as long as the appropriate stress intensity factor K, calculated using closed-form solutions or computer routines, is kept below the K_{Iscc} value, the stress-corrosion fatigue cracking should not occur. Unfortunately even a cursory glance at the number of tests and papers, and at the growing list of variables, clearly shows another picture.[24] The topics of prevention and controls are discussed in Chapter 12.

Since the subject of environmental effects is highly complex, this brief section can only provide a cursory view of the various topics. The following summary may help the designer to zero in on a practical slant of LEFM related to crack propagation in a corrosive environment.

LEFM

Linear elastic fracture mechanics (LEFM), which is highly developed, is applicable to various aspects of subcritical crack growth behavior during fatigue in benign and hostile environment, arrest of a fast crack, stress-corrosion, and the process of fracture.

Stress-Corrosion Cracking

Stress-corrosion cracking is a time-dependent process during which susceptible materials can fracture in tension in a corrosive environment.

Fatigue Crack Growth in a Benign Environment

Generalized fatigue crack growth as a function of ΔK is best represented on a log–log basis as $\mathrm{d}a/\mathrm{d}N$ vs. ΔK, in the following way:

$$\frac{\mathrm{d}a}{\mathrm{d}N} = C_{\mathrm{o}}(\Delta K)^n \tag{6.45}$$

where C_{o} is an empirical constant, n denotes the slope, and ΔK is the stress intensity factor fluctuation (or range).

There are three regions of this relationship, defining certain characteristics. Region I signifies low ΔK values at which existing cracks do not grow under continued load cycling. This ΔK level is known as ΔK_{th}, or the fatigue crack growth threshold level. Region II is fully represented by Eq. (6.45). Region III corresponds to unstable crack growth and final fracture, and it is of limited practical importance.

Applied (*K*) Level

The basic feature of dealing with the subcritical crack growth is to compute the appropriate K level for the selected geometry and loading. This allows the proper balance between the crack-driving force at the crack tip and the material resistance to fracture.

K_{Iscc}

This material property must be determined for the particular material and the environment by evaluating the K parameter as a function of time to failure (such as in Figs. 6.43 and 6.44) until the threshold of K_{Iscc} is reached, defined by K_{Iscc}.

K_{Iscc} Variables

These effects can be grouped together under the headings materials, environment, and loading. The composition and the strength level of the material are extremely important items in K_{Iscc} and da/dt considerations, where the latter represents the crack growth rate, say, in inches per unit of time, such as minutes. The directional effects are of interest when dealing with materials having directional properties, and residual stresses can either speed up or retard the mechanism of stress-corrosion cracking.

The environmental effects include chemistry, cathodic protection, temperature, and pressure. Cathodic protection and increasing pressure can increase stress-corrosion cracking susceptibility. The chemistry can alter K_{Iscc}, and the increasing temperature may lower the cracking susceptibility.

In the category of loading, exposure time is very important in testing and application while a deviation from section size may affect the plane-strain condition, a requirement for a valid K_{Iscc}.

While the appropriate research continues, it is of major importance to identify the mechanics of the K_{Iscc} and da/dt parameters, and to quantify the effects of numerous variables for practical use in design.

SYMBOLS

a	General symbol for crack size, in. (mm)
a_{min}	Minimum size of edge crack, in. (mm)
a_o	Original size of half crack, in. (mm)
da/dt	Change of crack length with time

$\mathrm{d}a/\mathrm{d}N$	Change of crack length with cycles
B	General symbol for thickness, in. (mm)
B_{min}	Minimum width of cantilever, in. (mm)
C_{o}	Empirical constant
c	Half-length of surface crack, in. (mm)
E	Elastic modulus, ksi (MPa)
$f(W)$	Correction factor
G	Elastic energy release rate, lb/in. (N/mm)
J	Symbol for J-integral, lb/in. (N/mm)
K	General symbol for stress intensity, ksi (in.)$^{1/2}$ [MPa (m)$^{1/2}$]
K_{I}	Stress intensity factor (Mode I), ksi (in.)$^{1/2}$ [MPa (m)$^{1/2}$]
K_{Ic}	Plane-strain fracture toughness, ksi (in.)$^{1/2}$ [MPa (m)$^{1/2}$]
K_{c}	Plane-stress fracture toughness, ksi (in.)$^{1/2}$ [MPa (m)$^{1/2}$]
K_{Id}	Dynamic fracture toughness, ksi (in.)$^{1/2}$ [MPa (m)$^{1/2}$]
K_{R}	Crack growth resistance, ksi (in.)$^{1/2}$ [MPa (m)$^{1/2}$]
K_{Iscc}	Threshold stress intensity factor for stress-corrosion cracking (plane strain), ksi (in.)$^{1/2}$ [MPa (m)$^{1/2}$]
m	Correlation factor for J-integral
n	Slope of linear plot
N	Number of cycles
N_{i}	Number of crack initiation cycles
P	Concentrated load, lb (N)
Q	Shape factor
r_{y}	Plastic zone size, in. (mm)
SD	Standard deviation
S_{u}	Ultimate strength, ksi (MPa)
S_{y} (also YS)	Yield strength, ksi (MPa)
T	Plate thickness, in. (mm)
t	Unit of time, min or s
W	Depth of cantilever specimen for crack tip opening displacement tests, in. (mm)
W_{min}	Minimum depth of cantilever, in. (mm)
Z	Thickness of knife edge, in. (mm)
w	Half-width of panel, in. (mm)
β	Dimensionless parameter
β	Crack opening displacement, in. (mm)
β_{c}	Critical crack tip opening displacement (CTOD) and when K becomes K_{Ic}, in. (mm)
Δ	Displacement of clip gauge, in. (mm)
Δ	Notch tip radius, in. (mm)

ΔK_{th}	Fatigue crack growth threshold, stress intensity factor fluctuation, ksi (in.)$^{1/2}$ [MPa (m)$^{1/2}$]
Δ_p	Plastic component of displacement, in. (mm)
$\Delta \sigma$	Stress range, ksi (MPa)
ΔT_w	Temperature increment, oF
ν	Poisson's ratio
σ	General symbol for stress, ksi (MPa)
σ_{CA}	Limiting design stress, ksi (MPa)
σ_w	Working stress, ksi (MPa)

REFERENCES

1. Griffith, A.A. The phenomena of rupture and flow in solids. Phil. Trans. Royal Society, **1920**, *221*, 163–198.
2. Irwin, G.R. The crack extension force for a part through crack in a plate. Trans. ASME J. Appl. Mech. **1962**, *29* (4), 651–654.
3. Irwin, G.R. Analysis of stresses and strains near the end of a crack traversing a plate. J. Appl. Mech. **1957**, *24*, 361–364.
4. Orowan, E. Fracture strength of solids. In *Report on Progress in Physics*; Physical Society of London: London, 1949; Vol. 12, 185–232.
5. Bluhm, J.I. A model for the effect of thickness on fracture toughness. Proc. ASTM **1961**, *61*.
6. Raring, R. Load deflection relationships in slow bend tests of Charpy V-notch specimens. Proc. ASTM **1952**, *52*.
7. Hertzberg, R.W., *Deformation and Fracture Mechanics of Engineering Materials*, 3rd Ed.; John Wiley: New York, 1989.
8. Witherell, C.E. *Mechanical Failure Avoidance*; McGraw-Hill: New York, 1994.
9. Irwin, G.R. Fracture Dynamics. In *Fracturing of Metals*; American Society of Metals: Cleveland, 1948.
10. Pellini, W.S. *Principles of Structural Integrity Technology*; Department of the Navy, Office of Naval Research: Arlington, VA, 1976.
11. Pellini, W.S.; Puzak, P.P. *Fracture Analysis Diagram Procedures for the Fracture-Safe Engineering Design of Steel Structures*, NRL Report 5920; U.S. Naval Research Laboratory: Washington, DC, 1963.
12. Puzak, P.P.; Pellini, W.S. *Standard Method for NRL Drop-Weight Test*, NRL Report 5831; U.S. Naval Research Laboratory: Washington, DC, 1962.
13. NavShips 250-634-3. *Method for Conducting Drop-Weight Test to Determine Nil-Ductility Transition Temperature of Ferritic Steels*; 1962.
14. Irwin, G.R. Relation of crack toughness measurements to practical applications. Weld. J. **1962**, *41* (11).
15. Robertson, T.S. Propagation of brittle fracture in steel. J. Iron Steel Inst. (London) **1953**.
16. Mosborg, R.J. An investigation of welded crack arresters. Weld. J. **1960**, *39* (1).
17. Pellini, W.S. *Criteria for Fracture Control Plans*, NRL Report 7406; U.S. Naval Research Laboratory: Washington, DC, 1972.

18. Blake, A. *Fracture Control and Materials Technology for Downhole Emplacement*, UCRL-53398; Lawrence Livermore National Laboratory: Livermore, CA, 1983.

19. Rybicki, E.F.; Stonesifer, R.B. Computation of residual stresses due to multi-pass welds in piping systems. J. Press. Vessel Technol. **1979**, *101*.

20. Rybicki, E.F.; Schmueser, D.W.; Stonesifer, R.B.; Groom, J.J.; Mishler, H.W. A finite element model for residual stresses and deflections in girth-butt welded pipes. J. Press. Vessel Technol. **1978**, *101*.

21. Broek, D. *The Practical Use of Fracture Mechanics*; Kluwer Academic Publishers: Dordrecht, 1988.

22. Watkinson, F. et al. The fatigue strength of welded joints in high strength steels and methods of its improvement. Proceedings: *Conference on Fatigue of Welded Structures*, The Welding Institute, Brighton, England, July 1970.

23. Fisher, J.W. et al. *Fatigue Strength of Steel Beams with Transverse Stiffness and Attachments*, NCHRP Report 147; Transportation Research Board: Washington, DC, 1974.

24. Barsom, J.M.; Rolfe, S.T. *Fracture and Fatigue Control in Structures*, 2nd Ed.; Prentice-Hall: Englewood Cliffs, NJ, 1987.

25. Barsom, J.M. Fatigue considerations for steel bridges. In *Fatigue Crack Growth Measurement and Data Analysis*, ASTM STP 738; American Society for Testing and Materials: Philadelphia, 1981.

26. Clark, Jr., W.C. *Some Problems in the Application of Fracture Mechanics*, Scientific Paper 79-1D3-SRIDS-PI; Westinghouse Research and Development Center: Pittsburgh, 1979.

27. American Association of State Highway and Transportation Officials. *Standard Specifications for Highway Bridges*; AASHTO: Washington, DC, 1977.

28. Paris, C.P.; Sih, G.S. Stress analysis of cracks. In *Fracture Toughness Testing and Its Applications*, ASTM STP 381; American Society for Testing and Materials: Philadelphia, 1965.

29. Brown, W.F.; Srawley, J.E. *Plane Strain Crack Toughness Testing of High Strength Metallic Materials*. ASTM Special Technical Publication, No. 410; American Society for Testing and Materials: Philadelphia, 1967.

30. Hahn, G.T.; Rosenfield, A.R. *Sources of Fracture Toughness: The Relation Between K_{Ic} and the Ordinary Tensile Properties of Metals*, ASTM STP 432; American Society for Testing and Materials: Philadelphia, 1968.

31. Irwin, G.R.; Krafft, J.M.; Paris, P.C.; Wells, A.A. *Basic Aspects of Crack Growth and Fracture*, NRL Report 6598, U.S. Naval Research Laboratory: Washington, DC, 1967.

32. Wells, A.A. Unstable crack propagation in metals — cleavage and fast fracture. Proceedings of *Cranfield Crack Propagation Symposium*, 1961.

33. Dugdale, D.S. Yielding of steel sheets containing slits. J. Mech. Phys. Solids, **1960**, *8*.

34. British Standards Institution. *Methods for Crack-Tip Opening Displacement (CTOD) Testing*; BS5762; BSI: London, 1979.

35. American Society for Testing and Materials. Standard method of test for plane-strain fracture toughness of metallic materials. In *ASTM Annual Standards*, ASTM Designation E-399-83, Vol. 03.01, 1987.

36. Knott, J.F.; Withey, P.A. *Fracture Mechanics — Worked Examples*; Institute of Materials, The Bourne Press: Bournemouth, England, 1993.

37. Rice, J.R. A path independent integral and the approximate analysis of strain concentration by notches and cracks. J. Appl. Mech. **1968**, *35*, 379–386.

38. Rice, J.R. *Mathematical Aspects of Fracture*; Academic Press: New York, 1968, Vol. 2.

39. Paris, P.C. Fracture mechanics in the elastic–plastic regime, in *Flaw Growth and Fracture*, ASTM STP 631, ASTM: West Conshohocken, PA, 1977.

40. Wellman, G.H.; Rolfe, S.T.; Dodds, R.H. Three dimensional elastic plastic finite element analysis of three point bend specimens. Weld. Res. Coun. Bull. **1984**, *299*.

41. Brown, B.F.; Beachem, C.D. A study of the stress factor in corrosion cracking by use of the precracked cantilever-beam specimen. Corrosion Sci. **1965**, *5*.

42. Novak, S.R.; Rolfe, S.T. Modified WOL specimen for K_{Iscc} environmental testing. J. Mater. **1969**, *4* (3).

43. Bueckner, H.F. Coefficients for computation of the stress intensity factor K_I for a notched round bar, In *Fracture Toughness Testing and Its Applications*, ASTM STP 381; American Society for Testing of Materials: Philadelphia, 1965.

44. Novak, S.R.; Rolfe, S.T. Comparison of fracture mechanics and nominal stress analysis in stress corrosion cracking. Corrosion **1970**, *26* (4).

45. Novak, S.R. Effect of prior uniform plastic strain on the K_{Iscc} of high strength steels in sea water. Eng. Fract. Mech. **1973**, *5* (3).

46. Fessler, R.R.; Barlo, T.J. The Effect of Cyclic Loading on the Threshold Stress for Stress Corrosion Cracking in Mild and HSLA Steels. *ASME 3rd National Congress on Pressure Vessel and Piping Technology*, San Francisco, 1979.

47. Staehle, R.W. *Evaluation of Current State of Stress Corrosion Cracking, Fundamental Aspects of Stress Corrosion Cracking*, NACE-1; National Association of Corrosion Engineers: Houston, 1969.

7

Experimental Determination of Stress Intensity Factor K_I

GENERAL COMMENT

There are many complex situations of structural geometry and loadings that make analytical solutions of Mode I stress intensity factors impossible and numerical simulations extremely difficult and expensive. Under these conditions the designer resorts to experimental determinations of fracture parameters. Also, many times it is advisable to conduct experiments to verify the results obtained through numerical analysis. Today, the experimentalist has many choices of using different methods for the determination of K. These experimental techniques include: the strain gage method, optical method of photoelasticity, shadow method of caustics, optical fiber sensors, Moiré interferometry, holographic interferometry, coherent gradient sensing technique, and so on. The first four experimental techniques, which are relatively simple to use, are discussed in this chapter.

STRAIN GAGE METHOD

Electrical resistance strain gages have been commercially available for about 60 years, and they are widely used in industry in different applications for strain measurements. Of all the experimental methods available for measuring strain (mechanical, electrical, optical, Moiré, etc.) the electrical resistance strain gage is the easiest and the least expensive to use. The strain gage also demonstrates all the optimum requirements of a strain measurement system, such as good stability with respect to temperature and time, accuracy of ± 1 μm/m over a range of $\pm 5\%$ strain, smaller gage length and width, minimal inertia, linear response and minimal skill requirement for installation and readout of the gage.[1] Although Irwin suggested the use of strain gages in fracture mechanics a long time ago, their use has only become common in the last 15 years. In this method, the strain at a point near the crack tip is measured and is then used to determine K. There are three basic electrical strain gages available; namely metal foil, semiconductor, and liquid metal gages. Although all three gage types have their own merits and demerits, metal foil gages are widely used for determining K. In this section, a simple method of determining K using a single strain gage is discussed.

Before going into the details of this method, it is essential to understand the nature of the area around the crack tip in order to position the strain gage at the right location. The whole area around the crack tip, can be divided into three regions as shown in Fig. 7.1. Region I, close to the crack tip is usually not a valid region for measuring the strain. This region is predominantly nonlinear with three-dimensional and plasticity effects present due to localized yielding near the crack tip. The intermediate region, region II, has a stress field that is dominated by the singular term, namely the stress intensity factor K. This is a valid area for determining strain field around the crack tip and thus K with required accuracy. The inner boundary of region II is approximately half the thickness of the specimen and the outer boundary about the plate thickness. The outer region, region III, is far away from the crack tip and needs complex analysis to obtain a stress intensity factor.

After identifying the region for locating the strain gage, the strain gage can be oriented at any angle to the crack tip as shown in Fig. 7.2. The notation $X-Y$ represents crack tip coordinates and $x-y$ represents strain gage coordinates. A

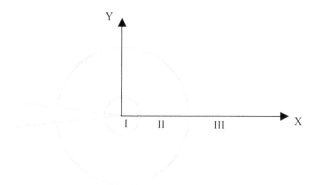

FIGURE 7.1   Schematic illustration of the three regions around the crack tip.

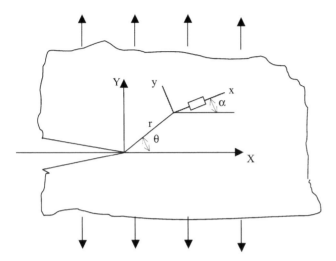

FIGURE 7.2 Strain gage orientation with respect to crack tip coordinates.

strain gage is located along the x-direction. The choice of the angles α and θ as shown in the figure depends only on the Poisson ratio of the material and not on the specifications of the geometry[2] (Table 7.1). Consequently, strain gages can be attached to any real structures, and a stress intensity factor can be easily determined. Dally and Sanford[2] showed that the angles α and θ can be represented as shown in Eq. (7.1):

$$\cos 2\alpha = -(1 - \nu)/(1 + \nu), \qquad \tan(\theta/2) = -\cot 2\alpha \qquad (7.1)$$

Now, a strain gage can be bonded at a distance slightly beyond half the thickness of the specimen and oriented along angles α and θ. A strain measurement system can be used to record the strain as a function of load. For

TABLE 7.1 Typical Values of Angles α and θ as a Function of Poisson's Ratio.

Poisson's ratio, ν	Angle θ	Angle α
0.20	83.64	65.91
0.30	65.16	61.29
0.33	60.00	60.00
0.40	50.76	57.69
0.50	38.97	54.74

homogeneous and isotropic material, the measured strain ε_{xx} along the gage is related to the opening-mode stress intensity factor K_I as shown in Eq. (7.2).

$$2\mu\varepsilon_{xx} = \frac{K_I}{\sqrt{2\pi r}}\left[\kappa\cos\frac{\theta}{2} - \frac{1}{2}\sin\theta\sin\frac{3\theta}{2}\cos 2\alpha\right.$$
$$\left. + \frac{1}{2}\sin\theta\cos\frac{3\theta}{2}\sin 2\alpha\right] \tag{7.2}$$

where, $\kappa = (1 - v)/(1 + v)$ and μ is the shear modulus of the material. It can be observed from Eq. (7.2) that the only variable in determining K_I is strain recorded by the strain gage and the term in the bracket is a constant, which depends on Poisson's ratio.

Design Problem 1

An aluminum edge crack specimen of thickness 10 mm is loaded under opening-mode loading. Design the location of the strain gage for this material and obtain a simple expression for opening-mode stress intensity factor for this plate.

Solution

For aluminum, $v = 1/3$, and using Eq. (7.1), $\cos 2\alpha = -0.5$ and thus $\alpha = 60°$. Now, $\tan(\theta/2) = 1/\sqrt{3}$ and $\theta = 60°$. For a thickness t of 10 mm, the radial distance r can be 6 mm away from the crack tip since $t/2 < r > t$. By substituting $\theta = \alpha = 60°$ in Eq. (7.2), a simple expression for K_I can be obtained:

$$K_I = 2\mu(1 + v)\sqrt{\frac{8}{3}\pi r\varepsilon_{xx}} \tag{7.3}$$

which can also be written as

$$K_I = E\sqrt{\frac{8}{3}\pi r\varepsilon_{xx}} \tag{7.4}$$

Design Problem 2

A strain gage mounted on a 2024 aluminum plate (Poisson's ratio $= 0.33$) with $\theta = 60°$ at a distance of 5 mm from the crack tip gives a strain reading of 1000 $\mu\varepsilon$ for a tensile load of 5000 N. Determine the stress intensity factor K_I for this load. Also, determine the strain reading at which this plate will fail if K_{IC} equals 10.42 MPa m$^{1/2}$.

Solution

For $\theta = 60°$, the stress intensity factor can be directly determined from Eq. (7.4). The elastic modulus E for 2024 aluminum is 70 GPa. Now substituting the above data, $r = 5$ mm or 0.0005 m and $\varepsilon_{xx} = 1000 \ \mu\varepsilon$ or 10^{-3} m/m into Eq. (7.4), the value for K_I can be obtained:

$$K_I = 4.53 \ \text{MPa m}^{1/2}$$

Now, if K_{IC} equals 10.42 MPa m$^{1/2}$, the corresponding strain value can also be obtained using Eq. (7.4):

$$\varepsilon_{xx} = 2300 \ \mu\varepsilon \qquad \text{for failure}$$

OPTICAL METHOD OF PHOTOELASTICITY

Photoelasticity is one of the oldest and most widely used full-field optical methods for determining fracture parameters. In a discussion to a paper by Wells and Post[3] in 1958, Irwin[4] observed that the photoelastic isochromatic fringes formed closed loops at the crack tip. This was also predicted by his recently developed stress field equations for crack tips. The experiments of Wells and Post also showed that these fringes were tilting forward, an effect of stress field parallel to the crack. Irwin was able to use these features and couple the photoelastic information with his field equations to determine both the value of K as well as the stress field parallel to the crack.

The photoelastic method (for more details see Dally and Riley[5] is based on the phenomenon that some transparent materials show temporary double refraction or birefringence under the influence of external loading. For stress-induced birefringence, the normally incident polarized light is split into two components along the principal stress directions in a plane perpendicular to the direction of light propagation, and are transmitted only along these planes through the material. Maxwell, who reported this phenomenon in 1853, noted that change in load-induced refractive index is linearly proportional to the principal stress components.[6] This fundamental relationship between stress and optical parameters is also known as the stress-optic law. For two-dimensional plane-stress bodies, the above theory can be reduced to a relative stress-optic law as given in Eq. (7.5):

$$\tau_{\max} = \frac{\sigma_1 - \sigma_2}{2} = \frac{N f_\sigma}{h} \tag{7.5}$$

where $\tau_{\max}$ is the maximum shear stress component, σ_1 and σ_2 are in-plane principal stress components ($\sigma_1 > \sigma_2$), N is the fringe order, f_σ a is material fringe value which is a property of the material for a given wavelength, and h is the thickness of the material.

It can be observed from Eq. (7.5) that the stress state and, eventually, the stress intensity factor around a crack tip can be obtained by determining the fringe order N from the fringes of constant maximum shear stress for a given load. Durelli and Shukla[7] have outlined procedures for the identification of these fringes. The fringes of constant maximum shear stress are also called iso-chromatics. The isochromatics of a loaded model can be obtained by placing the model in a circular polariscope as shown in Fig. 7.3. The circular polariscope consists of a light source, a polarizer, an analyzer and two quarter-wave plates. The cracked model that is to be studied is placed between the quarter-wave plates. A typical photograph showing isochromatic fringes in a single-edge notched ring specimen is shown in Fig. 7.4a.

Irwin[4] in his paper used the fact that $\partial \tau_m / \partial \theta = 0$ at the apogee (extreme position) on the fringe loop is zero to calculate both the value of K and the far field stress parallel to the crack σ_{ox}. The expressions for K and σ_{ox} are

$$K_I = \frac{2\tau_m \sqrt{2\pi r_m}}{\sin \theta_m} \left[1 + \left(\frac{2}{3 \tan \theta_m}\right)^2\right]^{-1/2} \left[1 + \left(\frac{2 \tan (3\theta_m/2)}{3 \tan \theta_m}\right)^2\right] \quad (7.6)$$

$$\sigma_{ox} = \frac{-K_I}{\sqrt{2\pi r_m}} \left[\frac{\sin \theta_m \cos \theta_m}{\cos \theta_m \sin (3\theta_m/2) + 3/2 \sin \theta_m \cos (3\theta_m/2)}\right] \quad (7.7)$$

Irwin's method of finding K from photoelastic data is easy to use and gives reliable estimates of K. For more elaborate and sophisticated methods of finding K from photoelastic data the reader is referred to a text by Dally and Riley.[5]

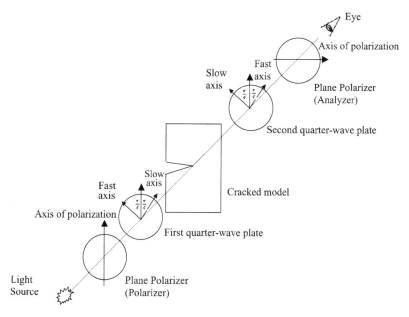

FIGURE 7.3 A cracked photoelastic model in a circular polariscope.

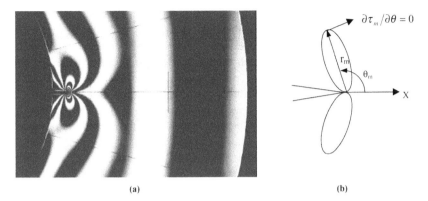

(a)   (b)

FIGURE 7.4 (a) Typical isochromatic fringe pattern around the crack tip in a ring specimen; (b) Characteristic geometry of an isochromatic fringe.

SHADOW METHOD OF CAUSTICS

The shadow method of caustics is another simple experimental technique for determining K_I for a cracked body. This method, first proposed by Mannog in 1964, is based on the concept that collimated light rays deflect due to stress gradients around the crack tip, creating a shadow spot.[8] Consider a plate with a single-edge crack subjected to tensile loading as shown in Fig. 7.5.

The far-field tensile loading introduces localized contraction in the z-direction around the crack tip due to the Poisson effect. In transparent materials the localized stress gradients at the crack tip also cause the refractive index of the material to change. These two effects cause the local area close to the crack tip

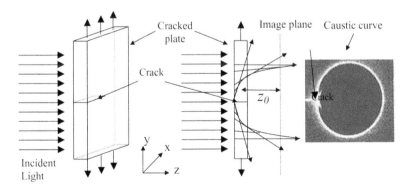

FIGURE 7.5 Deflections of light rays near the crack tip in a cracked plate under tensile loading.

to behave as a diverging lens and deflect the light rays away from the incident path as shown Fig. 7.5. However, the light rays far away from the crack tip pass through the plate with zero deflection. If a screen is placed at some distance from the plate, the light pattern with a shadow spot can be observed. The light concentration around the shadow spot is called the caustic curve. A typical shadow spot for a PMMA material under tensile loading is shown in Fig. 7.5. This method can also be used for opaque materials such as metals by reflecting the light from the polished metal surface. In this case the caustic is produced only because of the change in thickness at the crack tip.

The stress intensity factor can be determined from the caustic by measuring the transverse diameter of the caustic as shown Fig. 7.6. The relationship between the transverse diameter of the caustic and the stress intensity factor is given in the Eq. (7.8):

$$K_1 = 0.0934 \frac{D_0^{5/2}}{|z_0||C_1|h} \tag{7.8}$$

where z_0 is the distance between the specimen plane and the image plane as shown in Fig. 7.5, C_1 is the shadow optical constant, which is a material property, and h is the thickness of the specimen. Further details of this technique can be obtained from reference 1.

The caustic curve in the image plane is generated by a circular region on the specimen, and the radius r_0 of this circle can be obtained from the following relationship:

$$r_0 = \left[\frac{3K_1}{2\sqrt{2\pi}} |z_0||C_1|h \right]^{2/5} \tag{7.9}$$

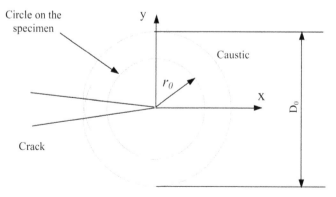

FIGURE 7.6 The caustic curve on the image plane and caustic generating circle on the specimen.

Design Problem 3

A caustic curve due to a stress singularity at a crack tip is formed on an image plane located 2 m from a 5 mm thick single-edge notch specimen of PMMA (plexiglass) under tensile loading. The diameter D_0 of the shadow spot is 10 mm. Determine K_I and then the radius r_0 of the region on the specimen that generates the caustic curve.

Solution

From the given data in the problem, $z_0 =$ distance between the specimen and image plane $= 2$ m, $h =$ thickness of the specimen $= 0.005$ m, $D_0 =$ diameter of the caustic $= 0.01$ m. For plexiglass, the shadow optical constant, $C_1 = -108$ $\mu m^2/$ N. Using Eq. (7.8), we can determine the stress intensity factor, K_I:

$$K_I = 0.865 \, \text{MPa} \, \text{m}^{1/2}$$

Now using Eq. (7.9) we can obtain the radius of the circle on the specimen for the given caustic diameter, D_0:

$$r_0 = 3.16 \, \text{mm}$$

Note: The region on the specimen r_0 from which the caustic is generated on the image plane should be at least half the plate thickness to ensure the stress field is outside the three-dimensional region and is K dominated. $r_0 = 3.16$ mm satisfies this condition.

Design Problem 4

For the same stress intensity factor, K_I obtained in Design Problem 3, if the image plane is moved to a new position so that $z_0 = 1$ m, determine the caustic diameter D_0 of the shadow spot and also find the new radius r_0 of the circle on the specimen.

Solution

From Design Problem 3, $K_I = 0.865 \, \text{MPa} \, \text{m}^{1/2}$, and from the given data $z_0 = 1$ m. Now writing Eq. (7.8) for D_0 and substituting the given values:

$$D_0 = \left[\frac{K_1 z_0 C_1 h}{0.0934} \right]^{2/5} = 7.6 \, \text{mm}$$

Using Eq. (7.9), we can obtain the value of r_0 for this caustic curve:

$$r_0 = 2.4 \, \text{mm}$$

This value of r_0 is less than half the plate thickness and is unacceptable. To obtain the correct value, the distance z_0 should be increased in the experiment.

FIBER-OPTIC SENSORS AND FRACTURE MECHANICS

The development of fiber-optic sensors first started in the mid-1970s and many researchers have since entered this field and accelerated its progress.[9] In general, fiber-optic sensors are based on the detection of modulated light intensity, frequency, polarization, or phase of the optical beam propagating through the fiber. Fiber-optic sensors have been used in applications for sensing various physical parameters such as strain, temperature, pressure, vibration, and so on. Their dielectric construction, geometric versatility, low weight, and small size make them an attractive choice for K measurements in both attached as well as embedded applications. Incorporating optical-fiber sensors into a material during manufacturing opens an avenue for real-time health monitoring throughout the lifetime of the material. This section presents a brief discussion on fiber-optic sensors and their application to fracture mechanics.

Fiber-Optic Sensors

Three types of fiber-optic interferometric sensors, namely Mach–Zehnder, Michelson, and Fabry–Perot, are discussed in this chapter. Interferometric sensors usually operate in two regimes. In the first case the relative phase difference is less than 2π and the shift of the interference fringes is less than a full fringe. In the second case the phase shift is much larger than 2π and many fringes pass by the observation point. The extent of the shift of the optical fringes is related to the parameter being measured. In all the experiments discussed in this chapter the interferometers were operating in the latter regime.

Mach–Zehnder Interferometer

An experimental setup of a Mach–Zehnder interferometer is shown in Fig. 7.7. Light from a He–Ne laser is split into two beams and each is then coupled into an optical fiber. One fiber is exposed to the applied strain while the other serves as a

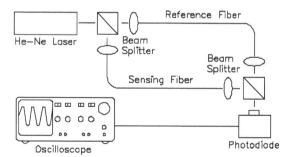

FIGURE 7.7   Experimental setup for the Mach–Zehnder interferometric sensor.

reference path. The two beams are recombined and made to interfere. A strain applied to the sensing fiber causes a phase shift and a corresponding displacement of the fringe pattern. Fringes are detected by a photodiode and the data is stored in a digital storage oscilloscope, from which it is transferred to a computer for further processing.

Butter and Hocker[10] have shown that for a surface-mounted sensor subjected to a uniaxial strain, ε_x, the relative optical phase change between the sensing and the reference arm of the Mach–Zehnder interferometer is given by

$$\delta\Phi = \left(\frac{2\pi n \varepsilon_x L}{\lambda}\right)(1 - c) \tag{7.10}$$

where

$$c = \left(\frac{n^2}{2}\right)[(1 - v)p_{12} - vp_{11}] \tag{7.11}$$

and where λ is the vacuum wavelength of the optical beam passing through an optical fiber of length L and refractive index n, p_{11} and p_{12} are the strain-optic coefficients, and v is the Poisson ratio of the fiber core. The phase shift required to move one interference fringe past a given point is 2π; therefore, the number of fringes passing a given point caused by the strain ε_x can be written as

$$N = \left(\frac{\delta\Phi}{2\pi}\right) = \left(\frac{n\varepsilon_x L}{\lambda}\right)(1 - c) \tag{7.12}$$

Equation (7.12) can be rearranged to give

$$\varepsilon_x = DN, \qquad D = \left(\frac{\lambda}{nL}\right)\left(\frac{1}{1 - c}\right) \tag{7.13}$$

Thus, the number of fringes passing a given point is proportional to the axial strain.

Sirkis and Haslach[11] extended Butter and Hocker's theory and derived a more complete phase–strain relationship for embedded interferometric optical-fiber sensors.

Michelson Interferometer

The experimental setup for this sensor remains the same as for the Mach–Zehnder interferometer except that instead of recombining the two beams at the end they are both reflected back through the fibers by mirroring the fiber ends. The two reflected beams are combined and made to interfere. The light intensity output from this sensor can be related to the applied axial strain in the same manner as that for the Mach–Zehnder sensor. In this case the effective gage length of the sensor is optically doubled due to the fact that the sensing beam traverses the sensing zone twice.

Extrinsic Fabry–Perot Interferometer (EFPI)

Figure 7.8a shows the experimental setup for a Fabry–Perot interferometric sensor. The actual sensor, shown in Fig. 7.8b, is a low-finesse Fabry–Perot cavity created by inserting two fibers with partially mirrored ends into a hollow-core fiber. Two separate end reflections take place, as shown in Fig. 7.8b. These two light beams combine to produce an interference pattern. A change in the "air gap," due to applied strain, causes the interference pattern to change and fringes are registered on the photodiode. The number of fringes N can be related to the applied strain, ε_x, as

$$N = \frac{\Phi}{2\pi} = \frac{2L}{\lambda}\varepsilon_x \tag{7.14}$$

where L is the gage length of the sensor and λ is the wavelength of light. The number of fringes passing a given point, N is proportional to the applied axial strain, ε_x.

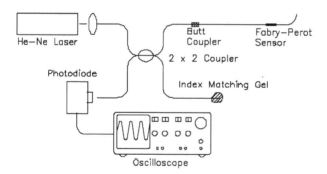

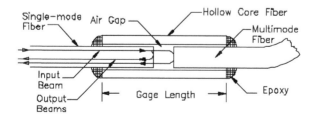

FIGURE 7.8 (a) Experimental setup for the Fabry–Perot interferometric sensor; (b) The extrinsic Fabry–Perot interferometric sensor.

Determination of K Using Fiber-Optic Sensors

In recent years fiber-optic sensors have become popular and their use in fracture mechanics has been clearly demonstrated.[12,13] This section is divided into two parts. The first part describes the experimental procedure and fracture mechanics results obtained by attaching the fiber-optic sensors to the surface of a fracture specimen and the second describes embedding the fiber-optic sensors in a fracture specimen to evaluate the stress intensity factor.

Surface-Attached Sensors

In order to evaluate the strain sensed by the optical fiber from the light intensity output of the interferometers, the proportionality constant D in Eq. (7.13) has to be determined. This constant can be calculated if the material parameters of the optical fiber, p_{11} and p_{12}, the strain-optic constants and the Poisson ratio ν are known. But these quantities are not generally known accurately for the fiber used; thus it is necessary to calibrate the Mach–Zehnder and Michelson sensors. The EFPI does not require calibration because the constant D is only a function of the wavelength λ and the gage length L and these are known accurately.

Calibration Procedure

A tension experiment, as shown in Fig. 7.9, is utilized to calibrate the Mach–Zehnder and Michelson interferometric sensors. The optical fiber is attached on one side of a tension specimen and a strain gage is bonded on the other side, at the same location. The specimen is monotonically loaded and the corresponding optical fringes are recorded. Figure 7.10 shows the variation of light intensity with time along with the corresponding strain gage output. The two outputs are used to obtain the calibration constant D as given in Eq. (7.13). The calibration procedure is repeated for various gage lengths to obtain D as a function of gage length L.

Surface-Attached Fracture Experiments

This section shows results from a simple experiment using an aluminum specimen. A single-edge notched (SEN) aluminum specimen is used with attached

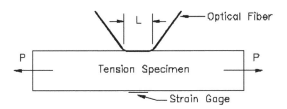

FIGURE 7.9 Tensile setup for calibrating attached fiber optic sensors.

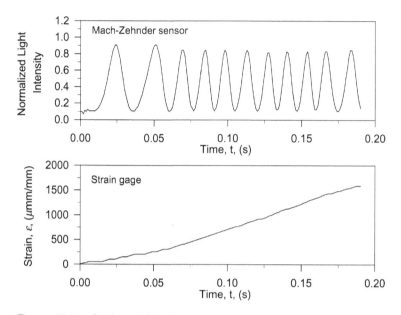

FIGURE 7.10 Strain and light intensity variation as function of time.

fiber-optic sensors to determine the stress intensity factor. An initial crack of length, a is saw-cut into the specimen as shown in Fig. 7.11. The optical fiber is attached to the specimen as shown in Fig. 7.11 at an angle $\alpha = \theta = 60°$ to the crack direction and at a distance r from the crack tip. To ensure perfect transfer of strain from the specimen to the fiber, the plastic coating covering the fiber is removed. The distance r is chosen so that no portion of the fiber is within half the plate thickness from the crack tip, where the three-dimensional effects are dominant. The attached fiber forms the sensing arm of the Mach–Zehnder interferometer. The specimen is monotonically loaded, and the load and the corresponding number of fringes are recorded simultaneously. Using the calibration constant D

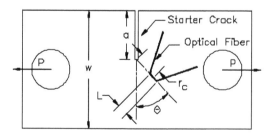

FIGURE 7.11 Single edge notched (SEN) specimen used in the static fracture experiment.

light intensity data is converted into axial strain ε_{xNxN}. This procedure is repeated for different crack lengths a. The plots of strain ε_{xNxN}, vs. load P are shown in Fig. 7.12. As expected, the strain varies linearly with load P. The strains obtained from the fiber-optic sensor are used to calculate the stress intensity factor K for the crack tip using Eq. (7.4):

$$K_I = E\sqrt{\frac{8\pi r}{3}}\varepsilon_{x'x'}$$

From the experimental data obtained, the value of K_I is evaluated, for various crack length-to-width ratios, as a function of increasing load. The results are shown in Fig. 7.13. The experimental results are compared with the theoretical handbook values of K. For a single-edge notched geometry

$$K_I = Y\sigma\sqrt{a} \tag{7.15}$$

where

$$Y = 1.99 - 0.41\left(\frac{a}{w}\right) + 18.7\left(\frac{a}{w}\right)^2 - 38.48\left(\frac{a}{w}\right)^3 + 53.85\left(\frac{a}{w}\right)^4 \tag{7.16}$$

and σ = far-field stress. As seen from Fig. 7.13, the experimental and theoretical values agree very well. The experiment is also conducted using the Michelson and the Fabry–Perot sensors. Figure 7.14 shows the stress intensity factor K_I as a function of applied load, for the three sensors.

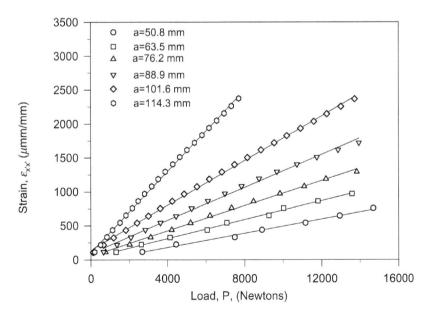

FIGURE 7.12 Plot of strain vs. load for various crack lengths.

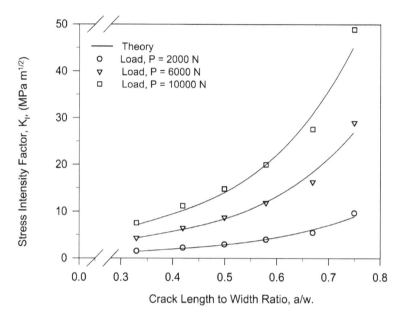

FIGURE 7.13   Stress intensity factor as a function of the crack length-to-width ratio.

Embedded Sensors

In this section results from a simple experiment conducted with embedded sensors on an acrylic material are shown. The material chosen to embed the optical-fiber sensors is plexiglass, which has a much lower Young's modulus than glass fiber.[12] Unlike the surface-attached case, the fibers need to be bent, in some cases, to be embedded. The optical fibers are very fragile without the buffer coating and it is necessary to leave this coating in place. Therefore, the calibration of the fiber has to be conducted with the buffer coating before fracture experiments.

Calibration Procedure

A tension experiment, as shown in Fig. 7.15, is used in the study of embedded fiber sensors. A small rectangular groove is milled into the specimen and an optical fiber is embedded as shown. The groove is filled with EnviroTex (polymer coating), the same material as plexiglass, and cured overnight. The embedded fiber forms the sensing arm of the Mach–Zehnder interferometer. The specimen is monotonically loaded and the corresponding optical fringes recorded. These are then used to obtain the calibration constant D using Eq. (7.13).

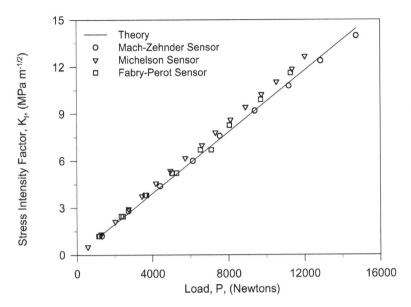

FIGURE 7.14   Plot of stress intensity factor vs. load for the various sensors used.

Embedded Fracture Experiments

Two fracture experiments using in-plane embedded optical-fiber sensors and transversely embedded optical-fiber sensors respectively to determine stress intensity factor K are described here.

In-Plane Embedded Optical Fibers.   The optical fibers are embedded at a depth d from the surface as shown in Fig. 7.16, and they form the sensing arm of a Mach–Zehnder interferometer. The specimen is monotonically loaded and the load and the corresponding optical fringes are recorded simultaneously. The

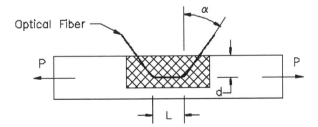

FIGURE 7.15   Tensile setup used for calibration of in-plane embedded fiber-optic sensors.

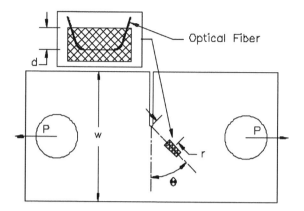

FIGURE 7.16 Single-edge notched specimen used in the in-plane fracture experiments.

light intensity data are converted to axial strain. The strain data obtained from the fiber-optic sensor are used to calculate the stress intensity factor using Eq. (7.4). Figure 7.17 shows the stress intensity factor K_I as a function of load for several depths d. The plots are linear but the slopes vary within 15% of the

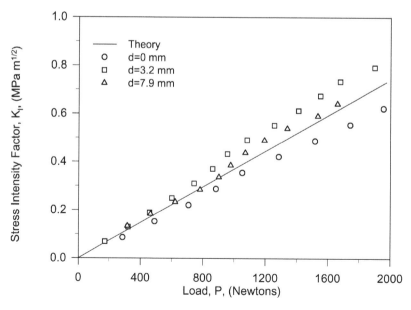

FIGURE 7.17 Plot of stress intensity factor vs. load for various embedded depths.

theoretical values. This is mostly due to the errors in the estimate of the effective gage length of the sensor.

Transversely Embedded Optical Fibers. In this case the fiber is embedded in the transverse direction, as shown in Fig. 7.18. The specimen is loaded and the corresponding transverse strains were measured using the fiber-optic sensor. These strains are used to calculate the stress intensity factor K_I associated with the crack tip. Using linear elastic fracture mechanics,

$$K_I = \left[\frac{E\sqrt{2\pi r}}{2\nu \cos(\theta/2)} \right] \varepsilon_{zz} \qquad (7.17)$$

where K_I is the stress intensity factor and ε_{zz} is the transverse strain.

The transverse strain data are used to determine the stress intensity factor K_I. The transverse embedded experiments are also conducted using the Michelson and Fabry–Perot sensors and the corresponding plots are shown in Fig. 7.19. The stress intensity factors determined by the Mach–Zehnder and the Michelson interferometers deviate more from the theoretical values than the Fabry–Perot sensor.

DISCUSSION

The experimental techniques available for K determination are many, ranging from very simple to quite complex. The simplest and the least expensive of all the techniques is the use of strain gages for measuring stress intensity factors. In recent years, strain gages have been used to evaluate K in many easy to difficult fracture problems. Shukla and colleagues[14,15] have used strain gages to successfully study dynamic fracture in metals and plastics. These[16,17] have also studied static and dynamic fracture in composite materials using strain gages. Very recently, Ricci et al.[18–20] have used strain gages to study both static and dynamic fracture at interfaces between dissimilar materials. Since the strain gage technique is robust, it can be used with ease in field applications also.

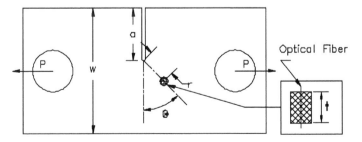

FIGURE 7.18 Single-edge notched specimen used in transverse fracture experiments.

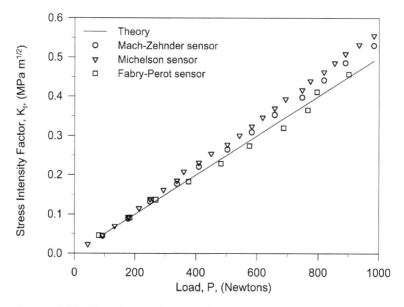

FIGURE 7.19 Plot of stress intensity factor vs. load for various sensors used.

The methods using photoelasticity and caustics for K determination are also well developed but are difficult to use in field applications. These require model materials or highly polished metal surfaces that are only possible in the laboratory environment.

The Mach–Zehnder and Michelson sensors are very cheap and easy to construct and use in a laboratory environment. However, they are not as useful in field applications. The entire sensing fiber is sensitive to environmental effects such as temperature, strain, vibration, and so on. This requires either a close coupling of the reference and sensing fibers, or a controlled environment. The Fabry–Perot sensor is not as easy to construct. However, its unique construction has a number of advantages. Only one fiber is required, which simplifies the setup. Both the reference and sensing beams propagate through the same fiber, which results in freedom from environmental effects. Also, the gage length can be controlled more accurately during sensor fabrication, as compared to the other two sensors. This sensor is most promising for field applications and in use with smart materials.

SYMBOLS

a	Crack length
C_1	Shadow optical constant

D	Diameter
E	Elastic modulus
$f\sigma$	Material fringe value
h	Thickness
K	Stress intensity factor
N	Fringe order
n	Refractive index
p	Strain-optic coefficient
r	Radial coordinate
w	Width
α	Gage orientation angle
$\delta\Phi$	Phase change
ε	Strain
θ	Angular coordinate
λ	Vacuum wavelength of light
μ	Shear modulus
ν	Poisson ratio
σ_1, σ_2	In-plane principal stresses
τ	Shear stress

REFERENCES

1. Kobayashi, A.S. *Handbook on Experimental Mechanics*; John Wiley and Sons, 1993
2. Dally, J.W.; Sanford, R.J. Strain-gage methods for measuring the opening mode stress-intensity factor, K_1, Experimental Mechanics, **1987**, *Dec*, 381–388.
3. Wells, A.A.; Post, D. The dynamic stress distribution surrounding a running crack — A photoelastic analysis. Proceedings of the SESA, **1958**, *16* (1), 69–92.
4. Irwin, G.R. Discussion of: The dynamic stress distribution surrounding a running crack — A photoelastic analysis, Proceedings of the SESA, **1958**, *16* (1), 93–96.
5. Dally, J.W.; Riley, W.F. *Experimental Stress Analysis*, 3rd Ed.;, College House Enterprises: Knoxville, TN.
6. Maxwell, J.C. On the equilibrium of elastic solids. Trans. R. Soc. Edinburgh **1853**, *XX* (1), 87–120.
7. Durelli, A.J.; Shukla, A. Identification of isochromatic fringes. Experimental Mechanics **1983**, *23* (1) 111–119.
8. Mannog, P. *Anwendug der Schattenoptik zur Untersuchung des Zerreissgangs von Platten*; Dissertation, Freiburg, Germany, 1964.
9. Giallorenzi, T.G.; Bucaro, J.A.; Dandridge, A.; Sigel, G.H.; Cole, J.H.; Rashleigh, S.C.; Priest, R.G. Optical fiber sensor technology. IEEE J. Quant. Elect. **1982**, *QE-18*, 626–665.
10. Butter, C.D.; Hocker, G.B. Fiber optics strain gauge. Appl. Opt. **1978**, *17*, 2867–2869.

11. Sirkis, J.S.; Haslach, Jr., H.W. Complete phase–strain model for structurally embedded interferometric optical fiber sensors, J. Intell. Mater. Sys. Struct., **1991**, *2* (1), 3–24.

12. Narendran, N.; Shukla, A.; Letcher, S. Application of fiber-optic sensor to a fracture mechanics problem. Eng. Fracture Mech. **1991**, *38*, 491–498.

13. Narendran, N.; Shukla, A.; Letcher, S. Determination of fracture parameters using embedded fiber optic sensors. Expt. Mech. **1991**, *31*, 360–365.

14. Shukla, A.; Agarwal, R.K.; Nigam, H. Dynamic fracture studies on 7075-T6 aluminum and 4340 steel using strain gages and photoelastic coatings. Eng. Fracture Mech. **1988**, *31*, 501–515.

15. Khanna, S.K.; Shukla, A. On the use of strain gages in dynamic fracture mechanics. Eng. Fract. Mech. **1995**, *51*, 933–948.

16. Shukla, A.; Agarwal, B.D.; Bhushan, B. Determination of stress intensity factor in orthotropic composite materials using strain gages. Eng. Fract. Mech. **1989**, *32* (3), 469–477.

17. Khanna, S.K.; Shukla, A. Development of stress field equations and determination of stress intensity factor during dynamic fracture of orthotropic composite materials. Eng. Fract. Mech, **1994**, *47*(3), 345–359.

18. Ricci, V., Singh, R.P.; Shukla, A. Evaluation of fracture mechanics parameters in bimaterial systems using strain gages. Eng. Fract. Mech. **1997**, *58* (4) 273–283.

19. Ricci, V.; Shukla, A.; Kavaturu, M. Using strain gages to investigate subsonic interfacial fracture in an isotropic-isotropic bimaterial. Eng. Fract. Mech. **2003**, *70*, 1303–1321.

20. Ricci, V.; Shukla, A.; Chalivendra, V.; Lee, K.H. Subsonic interfacial fracture using strain gages in an isotropic–orthotropic bimaterial. J. Theoret. Appl. Fracture Mech. **2003**, *39*, 143–161.

8

Dynamic Fracture

This chapter focuses on crack initiation under high rates of loading and rapid crack propagation and arrest of such cracks. The information about dynamic fracture in fracture mechanics literature is relatively scarce because dynamic fracture is analytically more difficult to analyze and experimentally very challenging to study. Nonetheless, there are many practical applications where the knowledge of dynamic fracture is needed and as such a brief discussion is included in this book.

The constitutive and fracture behavior of materials to a great extent is influenced by the rate at which the load is applied. Significant increase in yield strength with associated reduction in failure strain has been reported for many materials at high strain rate loading. The fracture characteristics of many materials, in particular polymers, are also influenced by the rate of loading. Dynamic fracture in general is composed of three different events in the sequence: crack initiation, crack propagation with or without branching, and crack arrest.

Crack initiation is characterized by the fracture toughness of the material, which itself is affected by the rate of loading. Once the crack initiates, subsequent crack propagation and crack arrest is dictated by the relative dominance of the energy available to drive the crack over the fracture resistance of the material. Inertia effects become significant whenever a stationary crack is subjected to rapid load or when the crack propagation speeds are as high as 10% of the shear wave speed in the material. The knowledge of the dynamic initiation toughness is particularly important in engineering design and analysis in order to establish the fracture tolerance of components experiencing rapid loading. The rapid loading could be because of an impact of a foreign object, an explosion, a thermal shock, and so on. The knowledge of crack arrest toughness is also of significance in situations where one needs to estimate the amount of crack growth before the crack arrests. Typical situations in this regard are crack propagation in long pipelines and explosive quarrying. In the former the interest is to restrict the crack propagation distance whereas in the latter the interest is to have maximum propagation of the explosively induced cracks. In the mid-1970s there was considerable interest in the nuclear industry in dynamic crack arrest in the pressure vessels of the reactors in case of loss of coolant accidents. During such an accident the vessel would get very hot and restarting the coolant flow would create a thermal shock that could cause dynamic crack propagation. In this chapter the test methods for obtaining dynamic initiation toughness are discussed, followed by a brief section on the behavior of propagating cracks. The standard test procedure employed for characterizing the arrest toughness of materials is also presented.

DETERMINATION OF DYNAMIC INITIATION TOUGHNESS

As mentioned in Chapter 2 the fracture toughness is a material property that characterizes the threshold for unstable propagation of an existing crack. Fracture toughness is defined as the stress intensity factor at the onset of unstable crack propagation provided certain conditions are met. Well-defined methods for obtaining fracture toughness under quasistatic conditions exist (ASTM-339). However, the value of this initiation stress intensity factor is observed to depend to a great extent on the rapidity of the applied load.

The major impediments in obtaining reliable and repeatable results in the measurement of dynamic initiation toughness are the transient nature of the phenomena and effects of inertia. When subjected to rapid loading, the rate of load increase is fast enough to be comparable to the stress wave travel times in the material. This factor brings in uncertainties in relating the load, usually measured at a location away from the crack tip, to the near-tip parameters such as the stress intensity factor. One could potentially use the full field optical methods described in Chapter 7, along with high-speed photography, to obtain the fracture parameters directly from the near-tip fields. However, this approach is not well suited as a regular testing procedure. Although there is no ASTM established standard testing procedure for measuring the dynamic initiation toughness, three different approaches with varying degrees of simplicity and limitations have been used to measure the dynamic initiation toughness.

Dynamic Initiation Toughness by Instrumented Impact Testing

Instrumented impact testing is one of the common methods used to determine the dynamic initiation toughness. In this method, a single-edge cracked specimen is subjected to three-point bending by impact. The specimen geometry along with the loading configuration is shown in Fig. 8.1. The span of the specimen S is four times the depth W, and the initial crack size a should be such that a/W is in the range 0.45–0.55. The specimen thickness B is chosen such as to maintain plane-strain conditions at the crack tip.

Typically, a drop weight arrangement is used to dynamically load the sample. The drop weight has an instrumented tup to monitor the load history, which is recorded using a data acquisition system. The dynamic initiation toughness K_{Id} is calculated from the load at crack initiation using the relation given in Eq. (8.1):

$$K_{Id} = \frac{F_i S}{BW^{3/2}} \frac{3\sqrt{x}\{1.99 - x(1 - x)[2.15 - 3.93x + 2.7x^2]\}}{2(1 + 2x)(1 - x)^{3/2}} \qquad (8.1)$$

where $x = a/W$ and F_i is the load recorded at crack initiation. One of the uncertainties in this method is the identification of the exact time of crack initiation, the load corresponding to that time, and the applicability of Eq. (8.1), which is based on quasistatic loading conditions. As a rule, the peak load registered coincides with crack initiation and Eq. (8.1) is applicable, if the crack initiation takes place sufficiently later in time so that the loading waves can travel several times in the sample before fracture initiation. Such delayed crack initiation can be expected only in cases where a notch instead of a sharp crack is used or when the fracture is elasto-plastic involving yielding at the crack tip before fracture. It should be realized that a J-integral type of characterization is more appropriate if the fracture is elasto-plastic. In general the recorded load history is not smooth but exhibits superimposed oscillations with decreasing amplitude. The stress intensity factor varies erratically during the initial stages due to the

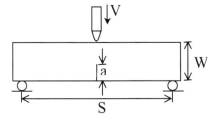

FIGURE 8.1 Schematic of dynamic initiation toughness measurement using instrumented impact.

constructive and destructive interference of the loading waves and boundary reflected waves at the crack tip.

Nakamura et al.[1] have investigated the inertia and transient effects in a dynamic loaded three-point bend specimen and have established conditions under which the use of Eq. (8.1) can be justified. Their analysis revealed that the short-time response of the specimen is dominated by discrete stress waves whereas the long-time response is essentially quasistatic. At intermediate times, the local oscillations at the crack tip are small even though global inertia effects are significant. Nakamura et al.[1] have established the notion of a transition time t_τ that defines the temporal boundary between dominance of inertia effects and deformation effects. After this transition time, quasistatic relations [Eq. (8.1)] can be used to calculate the stress intensity factor from the load. When crack initiation occurs at earlier times, alternate approaches like using response curves have been proposed by Kalthoff.[2] Another approach using the instrumented impact technique involves measuring the deflection of the beam itself through a laser-based sensor.[3] The initiation toughness is then calculated from the deflection of the beam at crack initiation. This technique, however, needs knowledge of specimen compliance to calculate the fracture load from the deflection. A discussion on the details of these approaches is not within the scope of this chapter.

Dynamic Initiation Toughness Measurement Using Hopkinson Bar Technique

In this method, the Hopkinson bar is used to load an edge crack specimen in three-point bending as shown in Fig. 8.2. The test setup consists of a cylindrical bar with a slightly rounded end. The bar is instrumented with strain gages at mid-length. The test sample is supported as shown against a heavy block acting as a momentum sink. A compressive stress pulse is generated in the bar by impacting it with a projectile made of the same material as that of the bar and the strain histories are recorded using the strain gages. Using a set of diametrically opposed gages and averaging the strain signals will eliminate any bending waves generated during the impact due to alignment inaccuracies. On reaching the bar–specimen interface, part of the compressive pulse gets transmitted into

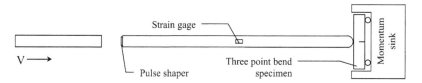

FIGURE 8.2 Schematic of dynamic initiation toughness using the Hopkinson bar.

the specimen and the rest gets reflected back into the bar due to mismatch of impedance.

If the incident and reflected waves do not interfere at the gage location, the load history at the end of the bar $F(t)$ and load point displacement history $u(t)$ can be calculated from the strain histories recorded by the strain gages using Eq. (8.2). The rate of loading can be controlled by inserting a pulse shaper, usually a thin disc made of a soft material such as solder lead, copper and so on, for metallic bars, and rubber or paper board for polymeric bars.

$$F(t) = AE\{\varepsilon_I(t) + \varepsilon_R(t)\}$$

$$u(t) = -c_b \int_0^t \{\varepsilon_I(t) - \varepsilon_R(t)\}dt$$

$$c_b = \sqrt{\frac{E}{\rho}}$$

(8.2)

where $\varepsilon_I(t)$ is the incident wave, $\varepsilon_R(t)$ is the reflected wave, A is the area of cross-section of the bar, E is the dynamic elastic modulus of the bar, c_b is the bar wave velocity, and ρ is the density of the bar material. It should be noticed that Eq. (8.2) is valid only as long as the bar remains elastic and the incident and reflected waves do not interfere at the gage location. The first condition can be satisfied by proper choice of the bar material for a given test and the second condition can be ensured by using a projectile of less than half the bar length. From the knowledge of the load at the time of crack initiation, the dynamic initiation toughness can be calculated using Eq. (8.1). However, as discussed earlier for instrumented impact the fracture should occur later in time in order to obtain a realistic estimate of the initiation toughness.

The advantage of using the Hopkinson bar instead of an instrumented impact is that the loading is well controlled. In the instrumented impact method, it has been observed that the tup loses contact momentarily with the sample and the sample loses contact momentarily with the supports during the initial stages of impact. These issues can be easily avoided in the case of Hopkinson bar loading by using incident pulses of sufficient length so that a positive loading is maintained until fracture. This, however, calls for a longer projectile and also a bar longer than twice the projectile length to avoid signal interference at the gage location. An alternate method to evaluate load and load displacement even when the strain waves interfere is to use the two-point strain monitoring method used by Bacon et al.[4]

When testing brittle polymers, which fail at very small values of load, use of metallic loading bars may not give sufficiently resolvable strain pulses and also results in higher loading rates causing fracture at earlier times. For testing such materials, use of polymeric loading bars has been pursued with success by

Martins and Prakash[5] and Evora and Shukla.[6]. Evora and Shukla[6] have also used high-speed photography in conjunction with the Hopkinson bar to evaluate the fracture initiation time and showed that it coincides with the peak load measured using Hopkinson bar. This is shown in Figs. 8.3 and 8.4. Figure 8.3 shows isochromatic fringe patterns associated with the crack tip up to the time of crack initiation. These optical patterns are analyzed to obtain K as a function of loading time. These values of K are compared with those obtained from the Hopkinson bar in Fig. 8.4. Since in the Hopkinson bar method K is calculated from measured loads, Fig. 8.4 shows that crack initiation does occur at peak load.

Dynamic Initiation Toughness Through Explosive Loading

In this method of obtaining the dynamic initiation toughness, proposed by Dally and Barker [7], a specially designed dog-bone specimen with an edge crack as shown in Fig. 8.5 is subjected to explosively generated tensile loading pulse. A strain gage (G in Fig. 8.5) of very small grid size (less than 1 mm) is mounted near the crack tip at a radial distance of $r_g > 0.5\,h$ (h is the specimen thickness) but close enough to be within the singularity dominant zone. The gage is placed with an orientation angle of $\alpha = 60°$, and positioned along the radial line making an angle of $\theta = 60°$ as shown in Fig. 8.5. The rationale for choosing these angles is that for a Poisson ratio of 0.3, these angles eliminate errors due to a single parameter representation (singular term only) of the strains at this location.[8]

The crack is formed by initially machining a notch of the required length. A sharp natural crack is subsequently extended from the notch tip by tapping a razor blade into the notch root in the case of brittle plastics or by fatigue loading as explained in the ASTM standard procedure for metals. Four symmetrically placed explosive charges are then ignited simultaneously and the strain histories are recorded through a high bandwidth strain conditioner and oscilloscope. The crack initiation time T_i and the corresponding strain are obtained from the strain

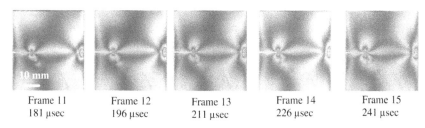

| Frame 11 | Frame 12 | Frame 13 | Frame 14 | Frame 15 |
| 181 μsec | 196 μsec | 211 μsec | 226 μsec | 241 μsec |

FIGURE 8.3 Photoelastic fringe pattern in the polyester specimen before crack initiation.

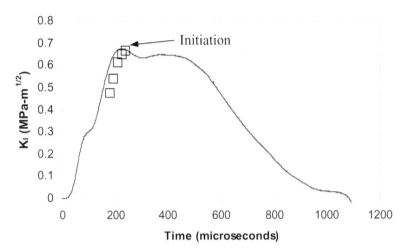

FIGURE 8.4 Dynamic stress intensity factor profile as obtained by the Split Hopkinson Bar and dynamic photoelasticity using high-speed photography for a polyester specimen.

history as explained in Eq. (8.3):

$$T_i = T_d - \frac{r_g}{c_s} \tag{8.3}$$

where T_d is the time at which the strain history starts deviating from linearity and c_s is the shear wave speed in the material. The stress intensity factor at the onset of crack propagation, which is in fact the dynamic initiation toughness K_{Id} can be

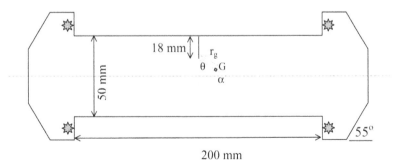

FIGURE 8.5 Geometry of dog-bone specimen and details of gage location for measuring dynamic initiation toughness (from Ref. [7]). Stars indicate the explosive location.

calculated from the strain value at initiation through the relation.[8]

$$K_{\text{Id}} = \left\{ \frac{8}{3} \pi r_{\text{g}} \right\}^{\frac{1}{2}} E \varepsilon_{\text{gi}} \tag{8.4}$$

where ε_{gi} is the strain at crack initiation and E is the elastic modulus of the material. Dally and Barker[7] validated the results of dynamic initiation toughness obtained from strain measurements with those obtained from dynamic photo-elasticity. This explosive loading falls in the higher end rapid loading and the typical rate of change of the stress intensity factor reported are of the order of 80 MPa-m$^{1/2}$/μs. Kavaturu and colleagues[9] proposed a simpler geometry as shown in Fig. 8.6 to study the complete history of dynamic fracture, including crack initiation, propagation, and arrest.

Rate Effects on Dynamic Initiation Toughness

The dynamic initiation toughness at very high rates of loading of materials is higher than the quasistatic fracture toughness. The dynamic initiation toughness has been observed to increase with the rate of buildup of K or, simply, with the loading rate. One plausible explanation of this behavior is the activity at micro length scales taking place inside the fracture process zone. With the load increasing, the strain at the crack tip increases leading to the development of a fracture process zone ahead of the crack tip within which formation of microcracks, voids, and their coalescence takes place in a stable manner before crack initiation could happen. This phenomenon has an inherent time frame

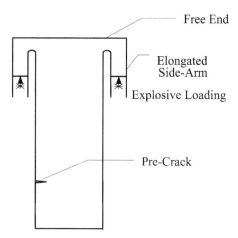

FIGURE 8.6 Geometry of simplified specimen to study complete dynamic crack propagation (from Ref. 9).

and when loaded dynamically the stress increases at a rate faster than that at which the fracture process zone activities develop. Therefore, the material withstands higher levels of stress before crack initiation, leading to the higher dynamic initiation toughness. Typical values of the dynamic initiation toughness available in the literature are listed in Table 8.1.

CRACK PROPAGATION

The study of fast propagating cracks is of significant interest due to its usefulness in estimating the integrity of structures subjected to extreme dynamic loadings such as impact and explosive blast. When materials are subjected to such extreme conditions, it is almost impossible to prevent fracture initiation and many times the fracture initiation and subsequent crack propagation and arrest occur at multiple sites. Typical examples of such situations are: integrity of vehicle or body armor after a shell impact, the propagation of geological faults during an earthquake, and so on. The knowledge of the already discussed dynamic initiation toughness helps only to the extent of predicting the circumstances under which fracture would initiate. In order to predict how far and how quickly the cracks would propagate, in what direction they would propagate, and whether they will branch or arrest, the fundamental phenomena associated with fast crack propagation need to be understood

Extensive research[13–15] has been carried out to investigate the stress fields around propagating cracks, terminal speeds attained by propagating cracks, and the conditions leading to arrest and branching of a propagating crack. It has been shown explicitly by several researchers that the leading term in the expansion of the stress field around propagating cracks retains the inverse square root singular behavior.[16,17] However, in addition to factors such as loading and geometry, the stress intensity factor for propagating cracks is also influenced by the crack speed. Many experimental techniques such as photoelasticity,[13] interferometry,[18] and strain gages[19] are used to study dynamic crack propagation. A typical

TABLE 8.1 Dynamic Initiation Toughness of Different Materials

Material	$\dot{K}_{Id}$ [(MPa m$^{1/2}$)/s]	K_{Id} (MPa m$^{1/2}$)	K_{IC} (MPa m$^{1/2}$)
Homalite-100[7]	0.76×10^5	0.655	0.445
ANSI steel[10]	10^5	61	55
Poly methyl methacrylate (PMMA)[11]	10^5	5.29	2.27
Graphite-Epoxy unidirectional composite[12]	10^5	7.5	2.3

photograph obtained using photoelasticity and high-speed photography of a running crack is shown Fig. 8.7. Such experimental data is analyzed to obtain the crack velocity, stress intensity factors, and energy release rates.

To maintain crack propagation, the crack driving force or the energy release rate available should be higher than the material resistance. Any surplus energy over the material resistance is converted into kinetic energy of the particles on either side of the crack path. For propagating cracks, the governing equation under elastodynamic conditions can be stated in a simple form as $K_I(t) = K_{ID}(v)$, in which $K_I(t)$ is the instantaneous dynamic stress intensity factor

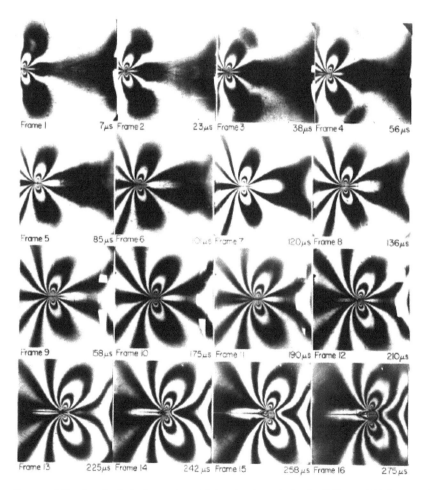

FIGURE 8.7   A set of photographs obtained using photoelasticity and high-speed photography of a running crack.

and $K_{ID}(v)$ is the material resistance which is a function of the crack speed, v. The instantaneous stress intensity factor, $K_I(t)$, is influenced by the loading and geometry. In the case of highly dynamic loading, $K_I(t)$ is influenced by arrival of loading waves and boundary reflected waves at the crack tip. The relationship between dynamic stress intensity factor and crack speed for a homogeneous material is well established. It has been shown theoretically that in homogeneous materials, the terminal speeds for propagating cracks in Mode I is the Rayleigh wave velocity, c_R, in the material. However, experimentally observed crack speeds are less compared to this limit. The dynamic stress intensity reduces to zero as the crack speed approaches the Rayleigh wave velocity irrespective of the far field loading, implying that it is impossible to drive a crack at the Rayleigh speed in a homogeneous material. This relationship can be written in the form of a simple expression as given in Eq. (8.5) and the trend is shown in Fig. 8.8:[20,21]

$$K_{ID} = \frac{K_{IA}}{1 - \{v/c_R\}^m} \qquad (8.5)$$

In Eq. (8.5), K_{IA} is the arrest toughness of the material and m is an experimentally determined constant. Such description of dynamic crack propagation is only partly correct. The vertical stem of crack velocity–stress intensity factor plot remains unique for a given material. This implies that the crack arrest toughness K_{IA} is a material property. However, the plateau region depends also on the nonsingular stress field components, for example the stress acting parallel to the crack. Thus, the velocity and stress intensity factor at which crack branching occurs can be influenced by the far field boundary conditions.[22]

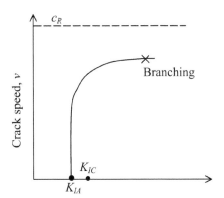

Dynamic fracture toughness, K_{ID}

FIGURE 8.8 K_{ID}–v relationship for a homogeneous material.

The dynamic propagation of cracks in nonhomogenous materials and at interfaces is very complex and has only recently received attention.[23,24] Recent investigation on cracks propagating along the interface between two dissimilar materials has shown that such systems can allow interfacial cracks to propagate faster than the shear wave speeds of the more compliant material.[23]

Propagating cracks leave behind a wealth of information in the form of fracture surface features. A detailed discussion of interpretation of these features is provided in Chapter 10 on fractography. These features provide information on the mechanisms involved in the crack propagation at microscopic levels. Additionally, these features can give qualitative information on the stress intensity factor and the speed of crack propagation. In general the fracture surface roughness depends both on stress intensity factor and crack speed. Increasing surface roughness has been observed with increasing stress intensity factor even under constant velocity crack growth. Three different zones, mirror, mist, and hackle, of increasing surface roughness are shown in Fig. 10.9.

Crack Branching

As mentioned in the previous section, when energy available for crack growth is in excess over the material resistance, this energy is converted into the kinetic energy of the material particles on either side of the crack and results in increasing the crack speed. Availability of energy considerably in excess leads to instability and the crack, instead of accelerating further, bifurcates into two or more cracks. Different instability criteria have been proposed to predict the onset of crack branching. Some of these are based on the concept of a critical velocity,[25,26] and others are based on the concept of either a critical stress intensity factor[27,28] or based on strain energy.[29–31] At a microscopic level, the crack propagation is not planar and the crack momentarily kinks out of the initial plane before kinking back into it, especially as the velocity increases. Under favorable conditions of velocity and energy, this phenomenon could lead to successful branching.

CRACK ARREST

Propagating cracks could arrest under several situations. A crack could propagate from a brittle weld into the tougher base metal and eventually arrest. When there exists a temperature gradient in a material exhibiting temperature-dependent fracture characteristics, cracks could initiate from a region of low temperature and eventually arrest when it reaches the region of normal temperature. In certain cases external elements such as a stiffener could inhibit crack propagation leading to crack arrest. In all cases, arrest takes place once the energy balance offsets in favor of the resistance to the crack propagation, that is, when the instantaneous stress intensity, $K_I(t)$, drops below the material resistance to crack growth $K_{ID}(v)$.

It was mentioned earlier that the material resistance to crack propagation, $K_{ID}(v)$, is a function of the crack speed and also that a stationary crack is stable as long as the stress intensity factor is less than the fracture toughness, K_{IC}, of the material. In this context, it is interesting to ponder whether or not a propagating crack will arrest once the stress intensity $K_I(t)$ drops below the fracture toughness K_{IC}. Experimental evidence suggest that the crack keeps propagating even after the stress intensity $K_I(t)$ has dropped below the quasi-static fracture toughness K_{IC}. The answer to this intriguing phenomenon lies in the arrest toughness of the material.

Similar to the fracture toughness, the arrest toughness is a material property, which can be defined as the minimum value of the stress intensity factor required keeping a crack propagating in the material. Once the stress intensity factor drops below this value, crack arrest follows. Crack arrest is more or less an abrupt phenomenon involving very high levels of deceleration.[32] It is an important property from a practical point of view, especially during situations where crack initiation and propagation cannot be eliminated altogether and crack arrest is the alternate chance of defense. To this extent the crack arrest toughness is the ability of the material to arrest a propagating crack.

The American Society of Testing Materials (ASTM) has outlined a standard procedure for determining the arrest toughness. This method will be discussed briefly in the following; however, a detailed description can be obtained from the ASTM standard E1221-88. As per the standard, the plane-strain crack arrest fracture toughness, K_{Ia}, is defined as the value of the stress intensity factor shortly after the crack arrests under conditions of crack front plane strain. The test primarily involves subjecting a compact crack arrest specimen to crack face pressure loading until crack propagation and rapid arrest occurs. The arrest toughness is then calculated from the crack mouth opening displacement history using quasi-static equations. The specimen geometry and dimensions are shown in Fig. 8.9.

The geometric proportions are

$$H = 0.6W \pm 0.005W$$
$$S = (B - B_N)/2 \pm 0.01B$$
$$N \leq W/10$$
$$0.15W \leq L \leq 0.25W$$
$$0.30W \leq a_0 \leq 0.40W$$
$$0.125W \pm 0.005W \leq D \leq 0.25W \pm 0.005W$$

The specimen thickness B is chosen based on plane-strain requirements discussed later. For low and intermediate strength steels, a brittle weld is deposited at the root of the machined groove of thickness N and subsequently a V-shaped notch is machined in the weld.

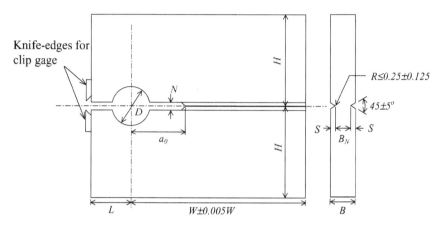

FIGURE 8.9 Specimen geometry of a compact crack arrest test specimen.

The specimen is loaded by pushing a wedge through a split-D pin inserted into the circular hole as shown in Fig. 8.10. Requirements on the wedge and split-D pin dimensions are discussed in ASTM E1221. The load applied by the wedge on the crack faces is not directly measured; instead the load is calculated from the crack mouth opening displacement CMOD, which is measured using a displacement transducer such as a clip gage. However, components of displacement that do not contribute to load can be present. These could be from the seating of the load platen and clip gage, local cracking in the brittle weld, or local yielding in the notch. A cyclic loading and unloading scheme is proposed in the ASTM standard to compensate for these factors.

In the test, the specimen is loaded to a predetermined displacement level, as prescribed in the standard. The wedge load and CMOD are recorded continuously. Neither the displacement transducer nor the load record is re-zeroed between cycles. This loading–unloading cycle is continued until rapid crack propagation and arrest occurs, which will be indicated by an abrupt load drop. Subsequently, the specimen is heat tinted at 250–350°C for about 10–90 minutes and split open to measure the arrested crack length, a_{a}. The provisional arrest toughness, K_{a} is calculated using Eq. (8.6).

$$K_{\mathrm{a}} = E\delta f(x)\frac{\sqrt{B/B_N}}{\sqrt{W}}$$

$$f(x) = \frac{2.24(1.72 - 0.9x + x^2)\sqrt{1 - x}}{(9.85 - 0.17x + 11x^2)}$$

(8.6)

where $x = a_{\mathrm{a}}/W$, E is the elastic modulus, and δ is the CMOD at 100 ms after the crack arrest adjusted for any noncontributing displacements as proposed in

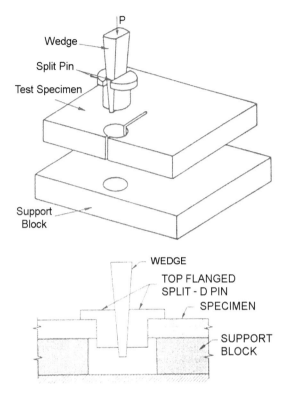

FIGURE 8.10 Schematic of loading arrangement for arrest toughness determination (ASTM E1221).

the standard. The provisional arrest toughness K_a, calculated above, represents the material arrest toughness K_{Ia}, only if the following conditions and all provisions in ASTM E1221 are satisfied.

$$W - a_a \geq 0.15W$$

$$W - a_a \geq 1.25 \left\{ \frac{K_a}{\sigma_{Yd}} \right\}^2$$

$$B \geq 1.0 \left\{ \frac{K_a}{\sigma_{Yd}} \right\}^2$$

$$a_a - a_0 \geq \frac{1}{2\pi} \left\{ \frac{K_a}{\sigma_{YS}} \right\}^2$$

TABLE 8.2 Crack Arrest Toughness of Some
Materials

Material	Arrest toughness, K_{IA} (MPa m$^{1/2}$)
Homalite-100	0.40
ANSI steel	52.00
Araldite B	0.67

where σ_{YD} and σ_{YS} are the dynamic and static yield strength, respectively. These conditions ensure that a state of plane strain exists around the crack tip and the crack jump is larger than the plane-stress plastic zone size. The measurement of the final crack length, a_a, could bring in uncertainties due to existence of uncracked ligaments, inclined crack fronts, crack tunneling, and so on. The appropriate guidelines for judging these situations are discussed in the standard. Table 8.2 shows some typical values of crack arrest toughness.

SYMBOLS

A	Area of cross-section
a	Crack length
B	Thickness
c_b	Bar wave velocity
c_R	Rayleigh wave speed
c_s	Shear wave speed
E	Elastic modulus
F	Force
F_i	Load at crack initiation
K	Stress intensity factor
K_{IA}	Opening mode arrest toughness
K_{IC}	Fracture toughness
K_{Id}	Dynamic initiation toughness
r	Radial coordinate
r_g	Gage radial position
S	Span
t	Time
T_i	Crack initiation time
u	Load point displacement
v	Crack speed
W	Depth
x	Normalized crack length (a/W)

α Gage orientation angle
ε Strain
δ Crack mouth opening displacement
θ Angular coordinate
ρ Mass density

REFERENCES

1. Nakamura, T.; Shih, C.F.; Freund, L.B. Analysis of a dynamically loaded three-point-bend ductile fracture specimen. Engg. Fracture Mech. **1986**, *25* (3), 323–339.
2. Kalthoff, J.F. On the measurement of dynamic fracture toughness — a review of recent work. Int. J. Fracture **1985**, *27*, 277–298.
3. Lorriot, T.; Martin, E.; Quenisset, J.M.; Sahraoui, S.; Lataillade, J.L. A methodological improvement of dynamic fracture toughness evaluations using an instrumented charpy impact tester. J. Physique, Colloque C8, Supplément au Journal de Physique III **1994**, *4*, 125–130.
4. Bacon, C.; Färm, J.; Lataillade, J.L. Dynamic fracture toughness determined from load point displacement. Exper. Mech. **1993**, *34* (3), 217–223.
5. Martins, C.F.; Prakash, V. Dynamic fracture of linear density polyethylene. In *The Minerals, Metals & Materials Science*; Srivatsan, T.S.; Lesuer, D.R.; Taleff, E.M. Eds., pp. 105–122, 2002.
6. Evora, V.; Shukla, A. Fabrication, characterization and dynamic behavior of polyester/TiO_2 nanocomposites. Mater. Sci. Engng **2003**, *A361*, 358–366.
7. Dally, J.W.; Barker, D.B. Dynamic measurements of initiation toughness at high loading rates. Exper. Mech. **1988**, *29* (3), 298–303.
8. Dally, J.W.; Sanford, R.L. Strain-gage methods for measuring the opening mode stress intensity factor, K_I. Exper. Mech. **1987**, *27* (4), 381–388.
9. Kavaturu, M.; Shukla, A.; Singh, R. Initiation, propagation and arrest of a bimaterial interface crack subjected to controlled stress wave loading. Int. J. Fracture **1997**, *83*, 291–304.
10. Kobayashi, T.; Dally, J.W. Dynamic photoelastic determination of the $\dot{a} - K$ relation for 4340 alloy steel. *Crack Arrest Methodology and Applications, ASTM STP 711*, ASTM: Philadelphia, 1980; 189–210.
11. Rittel, D. An investigation of dynamic crack initiation in PMMA. Mech. Mater. **1996**, *23*, 229–239.
12. Coker, D.; Rosakis, A.J. Dynamic fracture of unidirectional composite materials: mode I crack initiation and propagation. To be submitted to Int. J. Fracture **2004**.
13. Dally, J.W.; Shukla, A.; Kobayashi, T. A dynamic photoelastic study of crack propagation in a ring specimen. In *Crack Arrest Methodology and Applications, ASTM STP 711*, ASTM: Philadelphia, 1980; 109–127.
14. Kalthoff, J.F.; Beinert, J.; Winkler, S.; Klemm, W. Experimental analysis of dynamic effects in different crack arrest test specimens. *ASTM STP 711*, ASTM: Philadelphia, 1980; 161–177.

15. Kobayashi, A.S.; Seo, K.K.; Jou, J.Y.; Lirabe, Y. A dynamic analysis of modified compact tension specimens using Homalite 100 and polycarbonate plates. Exper. Mech. **1980**, *40*, 73–79.

16. Irwin, G.R. Series representation of the stress field around constant speed cracks. In *University of Maryland Lecture Notes*, 1980.

17. Freund, L.B.; Rosakis, A.J. The structure of the near-tip filed during transient elastodynamic crack growth. J. Mech. Phys. Solids **1992**, *40* (3), 699–719.

18. Tippur, H.; Rosakis, A.J. Quasi-static and dynamic crack growth along bimaterial interfaces: A note on crack tip field measurements using coherent gradient sensing. Exper. Mech. **1991**, *31* (3), 243–251.

19. Khanna, S.K.; Shukla, A. On the use of strain gages in dynamic fracture mechanics. Engng Fracture Mech. **1995**, *51*, 933–948.

20. Kobayashi, T.; Dally, J.W. The relation between the crack velocity and the stress intensity factor in birefringent polymers. *ASTM STP 627*; ASTM: Philadelphia, 1977; 257–273.

21. Rosakis, A.J.; Freund, L.B. Optical measurement of the plane strain concentration at a crack tip in a ductile steel plate. J. Engng Mater. Technol. **1982**, *104*, 115–120.

22. Shukla, A.; Anand, S. Dynamic crack propagation and branching under biaxial loading. In *ASTM STP 905*, ASTM: Philadelphia, 1986; *17*, 697–714.

23. Singh, R.P.; Lambros, J.; Shukla, A.; Rosakis, A.J. Investigation of the mechanics of intersonic crack propagation along a bimaterial interface using coherent gradient sensing and photoelasticity. Proceedings of The Royal Society (London) **1997**, *453*, 2649–2667.

24. Parameswaran, V.; Shukla, A. Crack tip stress fields for dynamic fracture in functionally graded materials. Mech. Mater., **1999**, *31*, 579–596.

25. Yoffe, E.H. The moving Griffith crack. Philosophical Magazine **1951**, *42*, 739–750.

26. Craggs, J.W. On the propagation of a crack in an elastic–brittle material. J. Mech. Phys. Solids **1960**, *8*, 66–75.

27. Clark, A.B.J.; Irwin, G.R. Crack propagation behaviors. Exper. Mech. **1966**, *6*, 321–330.

28. Congleton, J. Dynamic crack propagation; Sih, G.C. Ed. Noordhoff Publishing: Groningen, Netherlands, 1973; 427–483.

29. Ramulu, M.; Kobayashi, A.S. Mechanics of curving and branching — a dynamic fracture analysis. Int. J. Fracture **1985**, *27*, 187–201.

30. Theocaris, P.S.; Georgiadis, H.G. Bifurcation predictions for moving cracks by the T-criterion. Int. J. Fracture **1985**, *29*, 181–190.

31. Shukla, A.; Nigam, H.; Zervas, H. Effect of stress field parameters on dynamic crack branching. Engng Fracture Mech. **1990**, *36* (3), 429–438.

32. Dally, J.W. Dynamic photoelastic studies of fracture. Exper. Mech. **1979**, *19* (10), 349–361.

9

Fracture Toughness of Fiber Reinforced Composites

INTRODUCTION

A composite material is a combination of two or more materials at a macroscopic scale created to produce a more useful material. This resulting composite material has some properties that are superior to its constituents. Fibers of various kinds have been used in the past as reinforcements in a variety of matrix materials. The most popular of these are glass fiber reinforcements in a polymeric matrix. Various physical properties of these fiber reinforced plastics (FRP) such as the tensile strength, compressive strength, stiffness, and so on, have been well characterized; however, relatively little work has been carried out on the fracture properties of these materials. This chapter summarizes the important fracture mechanics issues such as the fracture mechanics concepts, fracture toughness testing, and fracture toughness results of the FRPs.

MEASURES OF FRACTURE TOUGHNESS FOR COMPOSITES

The property of the material that controls the onset of crack growth is called the fracture toughness of the material. The fracture toughness of composite materials is generally measured in terms of a stress-controlled parameter called the stress intensity factor or the energy based parameter, namely, the energy release rate G or the J-integral. A brief description of these fracture toughness measures is given below.

Stress Intensity Factor, K

The stress intensity factor is a parameter that denotes the intensity of the crack-tip stress field for a particular mode of loading in a linear–elastic material. Solution of the elastic stress fields in the vicinity of the crack tip for isotropic[1] as well as anisotropic[2] materials show that the stress fields are singular with $r^{-1/2}$ type of singularities. Westergaard,[1] Irwin,[3] Sneddon[4] and Williams[5] showed that the Mode I stress field in a linear–elastic cracked body, as shown in Fig. 9.1, is given by

$$\sigma_{ij} = \frac{K}{\sqrt{r}}f_{ij}(\theta) + \sum_0^\infty A_m r^{m/2} g_{ij}^{(m)}(\theta) \tag{9.1}$$

where σ_{ij} are the stress components, K is the stress intensity factor, A_m, where $m = 0, 1, \ldots, \infty$, are higher order coefficients, and $f_{ij}(\theta)$ and $g_{ij}(\theta)$ are dimensionless functions of θ.

For the most general case there could be three different modes of loading, namely, symmetric loading, in-plane shear, and out-of-plane shear as shown in Fig. 2.8 of Chapter 2. Associated with these loadings, there are three stress intensity factors (SIF), Mode I SIF or K_I, mode II SIF or K_{II} and Mode III SIF or K_{III}. For example, for an anisotropic plate containing a crack of half length a, oriented at an angle α to the leading direction (Fig. 9.2), the stress intensity factors K_I and

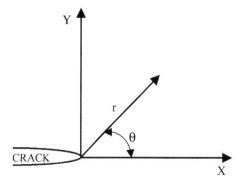

Figure 9.1 Crack tip coordinate system. The z-direction is normal to the page.

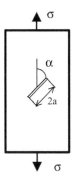

FIGURE 9.2 Anisotropic plate containing a crack oriented at an angle α to the loading direction.

K_{II} are given by

$$
\begin{aligned}
K_{\mathrm{I}} &= \sigma_\infty \sqrt{a}\sin^2 \alpha \\
K_{\mathrm{II}} &= \sigma_\infty \sqrt{a}\sin \alpha \cos \alpha
\end{aligned}
\tag{9.2}
$$

where σ_∞ is the far field stress.

The concept of stress intensity factor works well for translaminar fracture of composites. However, for interlaminar fracture the concept may be more difficult to apply,[6] particularly when different anisotropic properties are present above and below the plane of delamination. For this and other reasons it is customary to characterize interlaminar fracture with energy concepts.

The Energy Release Rate, G

The energy release rate G is defined as the amount of energy released per unit of new separational area formed due to cracking. The energy release rate is also defined as the crack extension force. In order to understand progressive fracturing quantitatively, it is necessary to define the force tending to cause crack extension. A simple procedure using energy concepts is utilized to develop an analytical description of the crack extension force.

In this procedure we assume our system is a plate with a crack in it as shown in Fig. 9.3. We also assume that our system has only one energy reservoir, that is, the elastic stress field energy, and that the increments of forward motion of the crack tip can be taken as increment dA of new separational area. From the principles of theoretical mechanics, it then follows that the crack extension

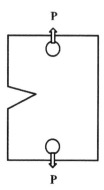

FIGURE 9.3 Loading of plate with edge crack.

force G is given by

$$G = -\frac{dU_T}{dA} \qquad \text{(system-isolated)} \tag{9.3}$$

where U_T is the total stress field energy.

The system-isolated requirement simply means that no energy enters or leaves during crack extension. The system-isolated requirement can be removed if the energy entering the stress field through motion of the loading forces is taken into account.

Let the separation of the loading points due to application of load P be Δ. If the displacement changes by an amount $d\Delta$ during the increment of crack extension dA, then dU_T is augmented by an amount $Pd\Delta$, thus the equation for G becomes

$$GdA = -(dU_T - Pd\Delta) \tag{9.4}$$

It has been assumed that the stress field is elastic. This means that the energy loss only occurs at the location of the leading edge of the crack. Actually, the energy devoted to fracturing is mainly dissipated into heat within a region of nonelastic strains adjacent to the leading edge of the crack.

Of course, the term $Pd\Delta$ in Eq. (9.4) must be generalized if there are more than one loading point.

For linear–elastic assumptions

$$\Delta = CP \tag{9.5}$$

where C is known as compliance. Furthermore

$$U_T = \frac{1}{2}P\Delta \tag{9.6}$$

Using Eqs. (9.4), (9.5), and (9.6) we obtain

$$G = \frac{P^2}{2} \frac{dC}{dA} \tag{9.7}$$

since $dA = B\, da$

$$G = \frac{P^2}{2B} \frac{dC}{da} \tag{9.8}$$

C and dC/da can be obtained experimentally, numerically, or analytically.

Relationship Between G and K

For linear elastic orthotropic composite materials, a simple relation relates G and K:

$$G = cK^2$$

where

$$c = \sqrt{\frac{a_{22}a_{11}}{2}} \left[\sqrt{\frac{a_{22}}{a_{11}}} + \frac{a_{66} + 2a_{12}}{2a_{11}} \right]^{1/2} \qquad \text{for Mode I} \tag{9.9}$$

and

$$c = \frac{a_{11}}{\sqrt{2}} \left[\sqrt{\frac{a_{22}}{a_{11}}} + \frac{a_{66} + 2a_{12}}{2a_{11}} \right]^{1/2} \qquad \text{for Mode II}$$

where a_{ij} are material properties given by

$$a_{11} = \frac{1}{E_{\mathrm{L}}}, \qquad a_{22} = \frac{1}{E_{\mathrm{T}}}, \qquad a_{66} = \frac{1}{\mu_{\mathrm{LT}}}, \qquad a_{12} = \frac{\nu_{\mathrm{LT}}}{E_{\mathrm{L}}} \tag{9.10}$$

J-Integral

The definition of the *J*-integral for elastic reversible material is the same as energy release rate G. Customarily, the crack is regarded as two-dimensional and the expression for J is a path-independent line integral given by[7]

$$J = \oint_{\Gamma} \left[U\mathrm{d}y - \vec{t}\, \frac{\partial \vec{u}}{\partial x}\, \mathrm{d}s \right] \tag{9.11}$$

where $\vec{u}$ is the displacement vector and $\vec{t}$ is the stress vector acting outward across the element $\mathrm{d}s$ of the line Γ, and U is the stress field energy density.

Begley and Landes[8–10] suggested an experimental method for obtaining a value of J. They measured it by using the energy rate interpretation of the *J*-integral given by Rice.[7] Rice showed that the *J*-integral might be interpreted as the potential energy difference between the two identically loaded bodies having neighboring crack sizes. This is stated mathematically as

$$J = -\frac{\partial U}{\partial a} \qquad (9.12)$$

where U is the potential energy and a is the crack length.

For experimental determination of the J-integral, several notched specimens with neighboring crack lengths are tested and load–displacement (at the point of load application) curves obtained. Areas under the load–displacement curves are obtained for different displacements and plotted against crack length. These curves of strain energy vs. crack length for constant displacement are usually referred to as the energy curves. The slope of an energy curve gives $\partial U/\partial a$ for a constant displacement, and thus the J-integral ($-\partial U/\partial a$) is obtained. In many cases energy curves are approximated for simplicity, by straight lines. A plot of the J-integral against displacement is a J-curve. The critical value of the J-integral, J_c, is obtained corresponding to the critical displacement beyond which the load decreases monotonically.

EXPERIMENTAL MEASUREMENT OF FRACTURE TOUGHNESS

Translaminar Fracture Toughness

There are many situations in which through-thickness translaminar fracture is of concern with laminates. Battlefield damage to composite structures can be through-thickness, as can inadvertent projectile impact with commercial aircraft structures. The use of laminates with too few cross plies is another situation in which translaminar fracture is important.

The ASTM Committee E-8 on Fatigue and Fracture has standardized the translaminar fracture toughness test and its designation is E1922-97. This test method covers the determination of translaminar fracture toughness, K_{TL}, for laminated polymer matrix composite materials of various ply orientations using test results from monotonically loaded notched specimens. During the testing load vs. displacement across the notch at the specimen edge, V_n, is recorded. The load corresponding to a prescribed increase in normalized notch length is determined from the load–displacement record. The translaminar fracture toughness, K_{TL}, is calculated from this load using standard fracture mechanics equations.

A testing machine that can record the load applied to the specimen and the resulting notch-mouth displacement simultaneously is utilized. A typical arrangement is shown in Fig. 9.5. Pin-loading clevises of the type used in Test Method E399 are used to apply the load to the specimen. A displacement gage is used to measure the displacement at the notch mouth during loading.

The required test and specimen configurations are shown in Figs. 9.4 and 9.5. Notch length, a_n, is between 0.5 and 0.6 times the specimen width, W. The notch width is $0.015W$ or thinner.

The specimen thickness B is the full thickness of the composite material to be tested. A value of W between 25 and 50 mm works well. The specimen dimensions are shown in Fig. 9.5. The notch can be prepared using any process that produces the required narrow slit. Prior tests[11,12] show that a notch width less than $0.015W$ gives consistent results regardless of notch tip profile.

During testing, the specimen is loaded at a slow rate such that the time from zero to peak load is between 30 and 100 s and load and the output of the displacement gage is recorded.

The value of K_{TL} is calculated from the following expression:[14]

$$K = \left[\frac{P}{BW^{1/2}} \right]$$

$$\times \frac{\alpha^{1/2}(1.4 + \alpha)[3.97 - 10.88\alpha + 26.21\alpha^2 - 38.9\alpha^3 + 30.1\alpha^4 - 9.27\alpha^5]}{(1 - \alpha)^{1/2}}$$

(9.13)

where K = applied stress intensity factor (MPa $m^{1/2}$), P = applied load (MN), $\alpha = a/W$ (dimensionless), a_n = notch length (m), B = specimen thickness (m),

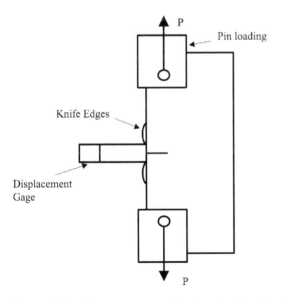

FIGURE 9.4 Test arrangement for translaminar fracture toughness tests.

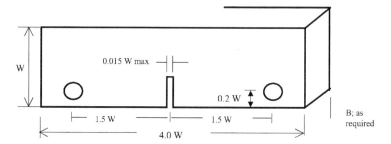

FIGURE 9.5 Translaminar fracture toughness test specimen.

and W = specimen width (m). The expression is valid for $0 \leq \alpha \leq 1$, for a wide range of laminates.[11]

This method gives the fracture toughness under Mode I loading conditions. The translaminar fracture toughness (K_{TL}) data can establish the effects of fiber and matrix variables and stacking sequence of the laminate on the translaminar fracture resistance of composite laminates. Thus, for quality control specifications, K_{TL} data can be used to establish criteria for material processing and component inspection. The translaminar fracture toughness K_{TL} determined by this method may be a function of the testing speed and temperature. Application of K_{TL} in design of service components should be made knowing that the test parameters specified by this test may differ from service conditions, possibly resulting in a different material response than that seen in service.

Measurement of Stress Intensity Factor K_I Using Strain Gage

The determination of stress intensity factor K from crack tip stress, strain, or displacement fields is much more complex for composite materials than isotropic materials. All the experimental techniques discussed in Chapter 7 with some modifications can be adopted to evaluate K in composites. The simplest and least expensive of all these techniques is the method of strain gages. A brief description on how to successfully use this method is given here. Shukla et al.[30] has used advanced elasticity concepts to develop a theoretical solution for strain along the strain gage direction as shown in Fig. 9.6. This strain is given by

$$
\varepsilon_{x'x'} = \frac{A_0}{2\alpha} \frac{1}{E_T} \left\{ \frac{1 - \nu_{LT}\nu_{TL}}{1 - \nu_{LT}} \right\} \left\{ \frac{\cos(\theta_1/2)}{\sqrt{r_1}} (\alpha - \beta) + \frac{\cos(\theta_2/2)}{\sqrt{r_2}} (\alpha + \beta) \right\}
$$
$$
+ \frac{A_0}{2\alpha} \frac{1}{G_{LT}} \left\{ \frac{\nu_{LT}}{(1 + \nu_{LT})\sqrt{\nu_{TL}}} \right\} \left\{ \frac{\sin(\theta_1/2)}{\sqrt{r_1}} - \frac{\sin(\theta_2/2)}{\sqrt{r_2}} \right\} \tag{9.14}
$$

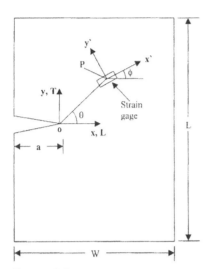

FIGURE 9.6

where

$$A_0 = \frac{K}{\sqrt{2\pi}}$$

$$\alpha^2 = \frac{a_{66} + 2a_{12}}{4a_{11}} - \frac{1}{2}\sqrt{\frac{a_{22}}{a_{11}}} \qquad (9.15)$$

$$\beta^2 = \frac{a_{66} + 2a_{12}}{4a_{11}} + \frac{1}{2}\sqrt{\frac{a_{22}}{a_{11}}}$$

and a_{ij} are material properties defined in Eq. (9.10).

$$r_1 = [x^2 + (\alpha + \beta)^2 y^2]^{1/2}, \qquad r_2 = [x^2 + (\beta - \alpha)^2 y^2]^{1/2} \qquad (9.16)$$

$$\theta_1 = \tan^{-1}\frac{(\alpha + \beta)y}{x}, \qquad \theta_2 = \tan^{-1}\frac{(\beta - \alpha)y}{x}$$

$$\tan^2 \phi = \frac{1}{\nu_{LT}} \qquad (9.17)$$

If the strain $\varepsilon_{x'x'}$ for a given geometry and load is measured by a strain gage then the only unknown in Eq. (9.14) is A_0 or the stress intensity factor K and can be easily determined. Every other quantity in Eq. (9.14) depends on the location of the strain gage or material properties.

Design Problem 9.1

A single-edge notched glass epoxy composite specimen is loaded in Mode I as shown in Fig. 9.3. A strain gage is to be used to measure the crack tip stress intensity factor K. Find the optimum location and position of the strain gage if the material properties of the composite are as follows:

> Longitudinal modulus: $E_L = 33.3$ GPa
> Transverse modulus: $E_T = 24.6$ GPa
> Poisson's ratio: $\nu_{LT} = 0.163$
> Shear modulus: $G_{LT} = 5.2$ GPa
> Thickness: $t = 2.9$ mm

Solution

Selection of Strain Gage Orientation (Φ). Using Eq. (9.17) and the value of $\nu_{LT} = 0.163$, we get the orientation angle $\Phi = 68°$. Thus the strain gage should be oriented at $68°$.

Selection of Strain Gage Angular Position (θ). The next step is to decide the angular position θ of the strain gage. The gage should be located in the region of high strains for better read out and low strain gradients to reduce error caused by mispositioning of the gage during mounting and averaging error resulting from the finite size of the gage. To ensure proper positioning of the gage, we first evaluate axial strain, $\varepsilon_{x'x'}$ from Eq. (9.14) as a function of angle θ for a fixed value of $K = 1$ MPa m$^{1/2}$ and $\Phi = 68°$. This strain is plotted, in Fig. 9.7, against θ for different values of radial position, r. It is observed that the peak value of strain $\varepsilon_{x'x'}$ occurs at $\theta = 38°$ independent of the radial position. Further, the strain plot is reasonably flat around this angle indicating very low strain gradients. Thus $\theta = 38°$ appears a good choice for angular position of the strain gage.

Selection of Strain Gage Radial Position (r). A plot of the strain $\varepsilon_{x'x'}$ is shown as a function of r in Fig. 9.8 for a fixed value of $\theta = 38°$, $\Phi = 68°$, and $K = 1$ MPa m$^{1/2}$. It can be seen that the strains drop rapidly as we move away from the crack tip. Strain gage should not be placed very close to the crack tip to avoid three-dimensional effects and other factors that make the two-dimensional plane-stress solution invalid. Also, the large strain gradients close to the crack tip would cause large averaging errors due to the finite gage size. On the other hand the strain gage cannot be placed too far from the crack tip because the K-dominated singular solution may not be valid in that region. Thus, looking at the strain gradient in Fig. 9.8, radial positions between $r = 5$ mm and $r = 9$ mm for mounting strain gages appear reasonable. The strain values at these positions are large enough to be accurately read on a recording device. Thus, the optimum location for positioning strain gage for this

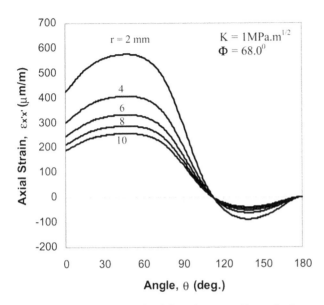

FIGURE 9.7 Variation of axial strain $\varepsilon_{x'x'}$ with angle θ.

material is

$$r = 5\text{–}9 \text{ mm}$$
$$\Phi = 68°$$
$$\theta = 38°$$

An experiment was conducted using the material from the above design problem and with the determined location and position of the strain gage. The specimen was 150 mm long, 50 mm wide, and 2.9 mm thick. The strain gage used was 3.18 mm long and 1.38 mm wide. The results are shown in Fig. 9.9. The stress intensity factor obtained using the strain gage matches very well with the theoretical solution.

Interlaminar Fracture Toughness

Since most composite materials are laminates, any delamination between layers or sufficiently high interlaminar stresses can cause interlaminar fracture. Thus, in recent years[15] many new test procedures have been devised to determine interlaminar fracture toughness of fiber composite materials. Some of these tests are discussed in brief here. These tests are mostly limited to unidirectional laminates where the crack propagates between the plies along the fiber direction.

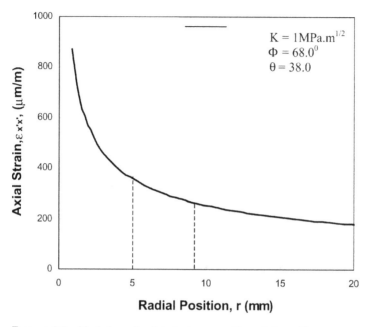

Figure 9.8 Variation of axial strain $\varepsilon_{x'x'}$ with radial position r.

Double Cantilever Beam Test

A schematic of the double cantilever beam (DCB) specimen and loading is shown in Fig. 9.10. The specimen is made of an even number of layers. A starter crack is introduced at one end of the specimen by introducing a piece of Teflon tape, 0.025 mm thick, during the fabrication process. The specimen is then mounted and tested in a properly aligned test frame. For load application and to prevent any bending moments, two door hinges are mounted at the crack end of the specimen. The specimens are generally 3 mm thick, 38 mm wide, and 229 mm long. The specimens are loaded at a crosshead rate of 1 to 2 mm/min. The specimen is unloaded after about 10 mm of crack extension.

The load–displacement data is recorded during loading and unloading. The procedure is repeated for 10 mm crack extensions each time until the crack is approximately 150 mm long. A typical load–displacement plot is shown in Fig. 9.11. From this plot one can calculate compliance C for each crack length a, from which the slope dC/da of the curve can be obtained for any crack length. Finally, the value of the critical energy release rate G_C is obtained from Eq. (9.8):

$$G_C = \frac{P_C}{2B}\frac{dC}{da}$$

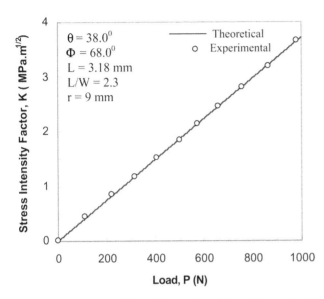

FIGURE 9.9 Comparison of experimental and theoretical stress intensity factor.

Since the critical load P_C is known for different crack length, several values of G_C are obtained. These are averaged to give the final value of G_C.

Another method to determine the fracture toughness using the data in Fig. 9.12 has been proposed by Whitney and Nuismer.[16] This method, called the area method, is a direct way to evaluate G_{IC}.

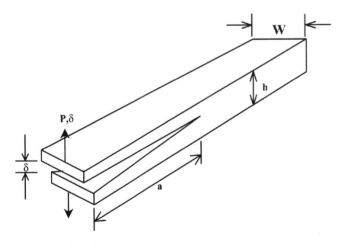

FIGURE 9.10 DCB specimen.

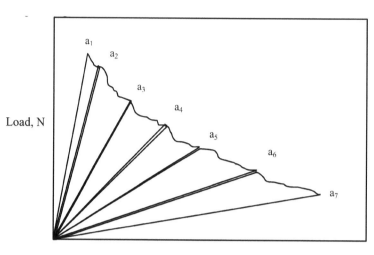

FIGURE 9.11 Load–displacement behavior for a DCB specimen at various crack lengths.

The critical strain energy release rate may be determined from a loading–unloading sequence as shown in Fig. 9.12. From the definition of the strain energy release rate, G_{IC} may be obtained as

$$G_{IC} = \frac{\Delta A}{W(a_2 - a_1)} \tag{9.18}$$

where ΔA is the area indicated in Fig. 9.12, $a_2 - a_1$ is the increment in crack length, and w is the width.

Some Results from Translaminar Fracture Testing

The definitive research on the development of translaminar fracture toughness test methods for carbon/epoxy laminates is the work of Harris and Morris.[11,18] More recently, the ASTM Committee E8 on Fracture and Fatigue, with the help of a number of universities, governmental and industry laboratories, has evaluated the fracture properties of some materials. Some of the details of this work are given in Underwood et al.[14] The materials tested by this group were carbon/polymer laminates of two types of symmetrical lay-ups (quasi-isotropic) [0/45/90] and [0/90]; two carbon fiber/epoxy materials — a relatively brittle T300 fiber/976 epoxy and a tougher AS4 fiber/977-2 epoxy; two laminate thickness – 2 mm and 4 mm; and three specimen configurations: (1) The standard three-point bend; (2) the compact configuration used for many types of fracture

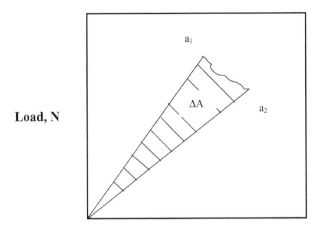

Load, N

Displacement, mm

FIGURE 9.12   Area method to evaluate G_{IC}.

tests, and (3) an extended compact specimen with arm-height to specimen width ratio of 1.9, compared to 0.6 for the standard compact specimen. The specimen geometries are shown in Fig. 9.13.

The quasi-isotropic [0/45/90] lay-ups were selected because of their frequent use in composite structures. The [0/90] lay-ups were used to investigate fracture testing of materials with considerable orthotropy. The specimens were loaded until fracture. During loading the load–displacement as well as the

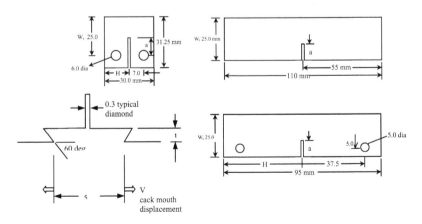

FIGURE 9.13   Specimen configurations.

crack growth measurements were made. Two calculations of the fracture toughness K at the maximum load were made using the length of the machined notch (K_{max}) and using the length of the machined notch plus any crack growth K_{max-0}. Typical results for T300/976 [90/-45/0/45] lay-up specimens are shown in Table 9.1 below.

Fracture Toughness of Short Fiber Composites

Short fiber composites are used in producing components by injection or compression molding. Some of the examples of short fiber composite products are safety helmets, household furniture and appliances, and so on. The fracture toughness of such composites has been extensively studied by Gaggar and Broutman[19-22] and Agarwal and colleague.[23-27]

For glass fiber mat reinforced epoxy the fracture toughness as a function of notch root radius is shown in Table 9.2. The notch root radius does not influence the fracture toughness of the material. Gaggar and Broutman[22] also showed that the effect of initial crack length on fracture toughness is minimal.

Fracture Toughness of Composite Laminates

Several studies have appeared over the past 30 years on the investigation of fracture toughness of composite laminates. Many different fracture models have been developed. Awerbuch and Madhukar[28] have published a detailed review of several of the fracture models. They have also compared some experimental results with fracture models. There is a good correlation between these fracture models and existing experimental notch-strength data. The simplest of all these models for obtaining the fracture toughness or notch strength of composite laminates is by Whitney and Nuismer.[16,29] Table 9.3 shows typical values of fracture toughness of some composite laminates.

Dynamic Fracture

Studies on dynamic fracture of fiber composites are almost nonexistent. Even for isotropic materials, very few studies have been conducted to determine the effect

TABLE 9.1   Translaminar Fracture Toughness for T300/976 (from Ref. 14)

	K_{max} (MPa m$^{1/2}$)	K_{max-0} (MPa m$^{1/2}$)
Three-point bend	33.4	39.3
Standard compact	29.3	32.9
Extended compact	32.2	44.1

TABLE 9.2 Effect of Notch Root Radius on Fracture Results of Epoxy Composites (from Ref. 22)

Notch root radius (mm)	Brittle resin composites		Flexible resin composites	
	K_Q (MPa m$^{-1/2}$)	Average	K_Q (MPa m$^{-1/2}$)	Average
0.178	8.14	10.45	10.01	11.00
	12.1		11.77	
	11.22		11.00	
0.254	11.0	11.11	9.57	11.22
	11.88		12.43	
	10.56		11.55	
0.305	10.45	11.22	12.10	12.10
	12.1		12.21	
	11.22		11.88	
0.508	11.33	10.78	11.77	11.77
	9.68		11.22	
	11.55		12.32	

of very high strain rate on fracture toughness (K_{IC}). Shukla and Kavaturu[30] have investigated dynamic fracture of bimaterial interfaces using dynamic photoelasticity and have reported that the dynamic initiation toughness of interfacial cracks is higher than the static value. Recently, Rosakis et al.[31] have studied the initiation fracture toughness for graphite/epoxy unidirectional composites as a function of the strain rate. The results shown in Fig. 9.14 indicate that the

TABLE 9.3 Typical Values of Fracture Toughness of Some Composite Laminates

Material and lay-up	K_{IC} (MPa m$^{-1/2}$)
Graphite–Epoxy	
[0 ± 45]$_S$	2.1–36.9
Quasi-isotropic	2.4–55.7
Cross-ply	2.8–54.1
Boron–Aluminum	
[O]$_8$ unidirectional	5.7–107.0
Quasi-isotropic	8.2–34.9
Cross-ply	4.4–41.5
Graphite–Polyamide	
Quasi-isotropic	7.6–57.6
Glass Epoxy	
Random short fibers	8.9–28.5

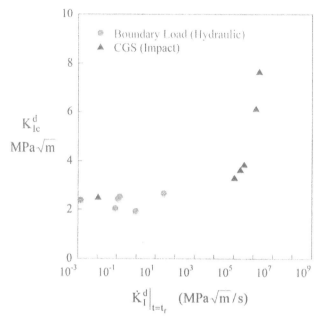

FIGURE 9.14 Dynamic initiation toughness of graphite/epoxy unidirectional composites as a function of the strain rate (from Ref. 31).

dynamic initiation toughness is sensitive to the strain rate and is about four times higher than the quasi-static value.

Closure

The fracture toughness of composites is a function of a wide variety of parameters. These parameters not only include material properties of the fiber and the matrix but also the geometry of the fiber lay-up, interfacial conditions, and the volume fractions of the fiber and the matrix. Furthermore, unlike metals where fracture toughness depends only on the energy dissipation mechanisms at the crack tip, in composites the fracture process could comprise energy losses in fiber breakage, matrix cracking, fiber pullout, fiber slipping, and fiber debonding. The fracture could also be affected by fiber bridging, in which the debonded fibers apply closure forces behind the crack tip. These mechanisms associated with fracture are schematically shown in Fig. 9.15. All these mechanisms affecting fracture toughness in composites could be gainfully tailored to improve the fracture toughness. With proper combination of fiber and matrix materials, composites with very high fracture toughness could be produced.

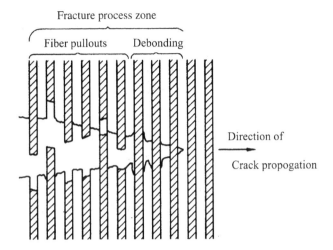

Fracture process zone

Fiber pullouts   Debonding

Direction of

Crack propogation

FIGURE 9.15 Physical mechanisms involved with fracture in composites (from Ref. 32).

Design Problem 9.2

A lumberjack uses a wedge force to split a piece of spruce wood. Spruce wood is orthotropic in nature with much stronger properties along the grain. Calculate the force required for crack extension assuming an initial central crack of length 3 in. and thickness 12 in. The crack is parallel to the grain of wood and critical energy release rate is 0.1 lb/in.

Solution

For spruce wood

$$E_L = \frac{1}{a_{11}} = 1.43 \times 10^6 \text{ psi}$$

$$E_T = \frac{1}{a_{22}} = 51.4 \times 10^6 \text{ psi}$$

$$\mu_{LT} = \frac{1}{a_{66}} = 0.8 \times 10^6 \text{ psi}$$

$$\nu_{LT} = a_{12}E_L = 0.4$$

For Mode 1 cracking

$$G = \sqrt{\frac{a_{22}a_{11}}{2}}\left[\sqrt{\frac{a_{22}}{a_{11}}} + \frac{a_{66} + 2a_{12}}{2a_{12}}\right]^{1/2} K^2$$

Substituting for the elastic constants

$$G = 1.1 \times 10^{-5} K^2$$

Now, for a central crack with splitting forces

$$K = \frac{P}{\sqrt{\pi a}}$$

where a is half the crack length and P is the load per unit thickness

$$0.1 = 1.1 \times 10^{-5} \frac{P^2}{\pi a}$$

Therefore

$$P^2 = \frac{0.1 \times \pi a}{1.1 \times 10^{-5}}$$

and

$$P = 207 \text{ lbs/in.}$$

Design Problem 9.3

A graphite–epoxy composite (AS4-3501-6, $[O_{24}]$) double cantilever beam specimen was tested for interlaminar fracture toughness. From the load–displacement
plot similar to Fig. 9.12, the area ΔA under the curve for a crack extension of
10 mm was found to be 20 N mm. Calculate the critical energy release rate if
specimen thickness was 10 mm. Also, if the material has the following elastic
properties $(E_{LT} = 42 \text{ GPa}, \ E_T = 9 \text{ GPa} \ \nu_{LT} = 0.27, \ \mu_{LT} = 6 \text{ GPa})$, determine
the stress at which a large plate containing a 30 mm central crack will fail in
tension.

Solution

Using the information given from the load–displacement data the critical energy
release rate can be defined using Eq. (12) as

$$G_C = \frac{\Delta A}{W(a_2 - a_1)}$$

$$= \frac{20 \text{ N mm}}{10 \text{ mm} (10 \text{ mm})}$$

$$= 200 \text{ N/m}$$

Now, $G = cK^2$, where c is given by Eq. (9) and

$$a_{11} = \frac{1}{E_L} = \frac{1}{42} = 2.38 \times 10^{-2}$$

$$a_{22} = \frac{1}{E_T} = \frac{1}{9} = 11.11 \times 10^{-2}$$

$$a_{66} = \frac{1}{\mu_{LT}} = \frac{1}{6} = 16.67 \times 10^{-2}$$

$$a_{12} = \frac{\nu_{LT}}{E_L} = \frac{0.27}{42} = 6.43 \times 10^{-3}$$

Therefore

$$c = 8.86 \times 10^{-2} / \text{GPa}$$

Now, for a large plate with central crack with remote tension loading

$$K = \sigma\sqrt{\pi a}$$

so

$$G = 200 = 8.86 \times 10^2 \times 10^{-9}\sigma^2(\pi \times 0.015)$$

and

$$\sigma = 6.92 \, \text{MPa}$$

SYMBOLS

a	Crack length
B	Width
C	Compliance
E_L	Young's modulus in lateral direction
E_T	Young's modulus in transverse direction
G	Energy release rate
J	J-integral
K	Stress intensity factor
K_{TL}	Translaminar fracture toughness
P	Load
U_T	Total stress field energy
Δ	Displacement
μ_{LT}	Shear modulus
ν_{LT}	Poisson's ratio

REFERENCES

1. Westergaard, H.M. Bearing pressure and cracks. J. Appl. Mech. **1939**, *6*, A49–53.
2. Sih, G.C.; Paris, P.C.; Irwin, G.R. On cracks in rectilinearly anisotropic bodies. Int. J. Fracture Mech. **1965**, *1* (3), 189–203.
3. Irwin, G. Analysis of stresses and strains near the end of a crack traversing a plate. J. Appl. Mech. **1957**, *24*, 361–364.
4. Sneddon, I.N. The distribution of stress in the neighbourhood of a crack in an elastic solid. Proceedings, Royal Society of London **1946**, *A-187*, 229–260.
5. Williams, M.L. On the stress distribution at the base of a stationary crack. J. Appl. Mech. **1957**, *24*, 109–114.
6. Wang, S.S. Fracture mechanics for delamination problems in composite materials. J. Comp. Mat. **1983**, *17*, 210.
7. Rice, J.R. A path-independent integral and the approximate analysis of strain concentration by notches and cracks. Appl. Mech. **1968**, *35*, 379–386.
8. Begley, J.A.; Landes, J.D. *Fracture Toughness, ASTM STP 514*; American Society for Testing and Materials; Philadelphia, 1972; 1–20.
9. Begley, J.A.; Landes, J.D. *Fracture Toughness, ASTM STP 514*; American Society for Testing and Materials: Philadelphia, **1972**, 24–39.
10. Begley, J.A.; Landes, J.D. *Developments in Fracture Mechanics Test Methods Standardization, ASTM STP 632*; American Society for Testing and Materials: Philadelphia, 1977; 57.
11. Harris, C.E.; Morris, D.H. A comparison of the fracture behavior of thick laminated composites utilizing compact tension three-point bend and center-cracked tension specimens. In *Fracture Mechanics: Seventeenth Volume, ASTM STP 9.5*; American Society for Testing and Materials: Philadelphia, 1986; 124–135.
12. Underwood, J.H.; Burch, I.A.; Bandyopadhyay, S. Effects of notch geometry and moisture on fracture strength of carbon/epoxy and carbon/bismaleimide laminates. In *Composite Materials: Fatigue and Fracture (Third Volume), ASTM STP 1110*; American Society for Testing and Materials: Philadelphia, 1991; 667.
13. Underwood, J.H.; Kortschot, M.T. Notch-tip damage and trans-laminar fracture toughness measurements from carbon/epoxy laminates. *Proceedings of 2nd International Conference on Deformation and Fracture of Composites*; The Institute of Materials: London, 1991.
14. Underwood, J.H.; Kortschot, M.T.; Lloyd, W.R.; Eidinoff, H.L.; Wilson, D.A.; Ashbaugh, N. Trans-laminar fracture toughness test methods and results from interlaboratory tests of carbon/epoxy laminates. *Fracture Mechanics: 26th Volume, ASTM STP 1256*; American Society for Testing and Materials: Philadelphia, 1995; 486.
15. Carlsson, L.A.; Pipes, R.B. *Experimental Characterization of Advanced Composite Materials*; Prentice-Hall, Inc.: New Jersey, 1997.
16. Whitney, J.M.; Nuismer, R.J. Stress fracture criteria for laminated composites containing stress concentrations. J. Compos. Mater. **1974**, *8*, 253–265.
17. Piascik, R.S.; Newman, J.C.; Underwood, J.H. The extended compact tension specimen. J. Fatigue Fracture Engng Mater. Structures **1997**, *20* (4), 559–563.

18. Harris, C.E.; Morris, D.H. *Fracture behavior of thick laminated graphite/epoxy composites*, NASA CR-3784; National Aeronautics and Space Administration: Washington, 1984.

19. Gaggar, S.K.; Broutman, L.J. Effect of crack tip damage on fracture of random fiber composites. Mater. Sci. Engng **1975**, *21*, 177–183.

20. Gaggar, S.K.; Broutman, L.J. Crack growth resistance of random fiber composites. J. Compos. Mater. **1975**, *9* (2), 216–227.

21. Gaggar, S.K.; Broutman, L.J. Strength and fracture properties of random fiber polyester composites. Fibre Sci. Technol. **1976**, *9* (3), 205–224.

22. Gaggar, S.K.; Broutman, L.J. *Fracture Toughness of Random Glass Fiber Epoxy: An Experimental Investigation Composites. ASTM STP 631.* In *Flaw Growth and Fracture*; ASTM: Philadelphia, 1977; 310–330.

23. Agarwal, B.D.; Giare, G.S. Crack growth resistance of short fiber composites: I—Influence of fiber concentration, specimen thickness and width. Fiber Sci. Technol. **1981**, *15*, 283–298.

24. Agarwal, B.D.; Giare, G.S. Effect of matrix properties on fracture toughness of short fiber composites. Materi. Sci. Engng **1982**, *52*, 139–145.

25. Agarwal, B.D.; Patro, B.S.; Kumar, P. Crack length estimation procedure for short fiber composite: an experimental evaluation. Polym. Compos. **1985**, *6*, 185–190.

26. Agarwal, B.D.; Patro, B.S.; Kumar, P. Prediction of instability point during fracture of composite materials. Compos. Technol. Rev. **1984**, *6*, 173–176.

27. Agarwal, B.D.; Patro, B.S.; Kumar, P. J integral as fracture criterion for short fiber composites: An experimental approach. Engng Fracture Mech. **1984**, *19*, 675–684.

28. Awerbuch, J.; Madhukar, M.S. Notched strength of composite laminates: predictions and experiments — A review. J. Reinforced Plas. Compos. **1985**, *4*, 3–159.

29. Nuismer R.J.; Whitney, J.M. Uniaxial failure of composites containing stress concentrations. In *Fracture Mecanics of Composites, ASTM STP 593*; American Society for Testing and Materials: Philadelphia, 1975; 117–142.

30. Shukla, A.; Kavaturu, M. Opening-mode dominated crack growth along inclined interface. Int. J. Solids Structures **1998**, *65*, 293–299.

31. Coker, D.; Rosakis, A.J. Dynamic fracture of unidirectional composite materials: mode I crack initiation and propagation. To be submitted to Int. J. Fracture, 2004.

32. Agarwal, B.D. *Analysis and Performance of Fiber Composites*; Wiley Interscience: Hoboken, NJ, 1990.

10

Post Mortem Failure Analysis

FRACTOGRAPHY

The occurrence of fracture in any type of material leaves behind a wealth of information in the form of fracture surface features. Fractography is the science of studying the fracture surface in order to determine the source of fracture and relationship between the mode of crack propagation and the microstructure of the material. Fracture surface appearances and features were used to assess the quality of iron and steel as early as the 17th century. With the advent of microscopes, observation of the fracture surface is possible in greater detail, and this has enabled better understanding of the microlevel mechanisms involved in the fracture process. Apart from this, fractography is extensively used in failure analysis, wherein the features of the fracture surface are effectively used to identify the reasons that led to the failure. Thus, design engineers can learn a lot from fractured components and can use this information in understanding why fracture occurred and how to design the component so that such failures do not occur in the future.

The techniques generally used in fracture surface analysis consist of visual examination, optical microscopy, scanning electron microscopy, and transmission electron microscopy, depending upon the extent of details required. The fracture surface features that can be identified depend to a great extent on the technique employed. These techniques will be reviewed very briefly in this chapter. However, the main focus will be on the various fracture surface features commonly observed, in different materials such as metals, polymers, and composites, and their interpretation at macroscopic and microscopic levels.

OPTICAL MICROSCOPY

In this technique the fracture surface is viewed through an optical microscope. For opaque materials light is shined onto the fracture surface and the reflections are viewed through the microscope. For transparent materials like polymers, the fracture surface can be cut in the form of a thin slice from the parent material, illuminated from beneath and viewed through the microscope. These microscopes have limited resolution and low depth of field. The maximum magnification possible using this technique is 1300 at a resolution of 1 μm. In spite of these limitations optical microscopy is still the most widely adopted technique during the initial stages of fractographic analysis, to obtain a general overview of the fracture surface and to identify different regions, which should be studied in detail. The salient features revealed by this technique are crack origination sites, crack propagation direction, and location of arrest marks.

SCANNING ELECTRON MICROSCOPY

In a scanning electron microscope (SEM), a focused beam of electrons scans the specimen (the fracture surface itself or a replica of it), which is placed in the SEM chamber. These electrons on impinging the specimen generate several signals, such as secondary electrons, back scattered electrons, auger electrons, cathodoluminescence, x-rays, and transmitted electrons. Of these, the secondary emitted electrons (emissive mode) and the back-scattered electrons (reflective mode) are of interest in fractographic analysis. These signals are collected and electronically processed to generate an image of the specimen surface. The principal contrast mechanism is due to the variation in the angle of incidence of the electron beam to the surface of the specimen, which is surface topography.

This technique allows magnifications varying from 5 to 240,000, but the useful upper limit for fractography is about 30,000. It has a resolution limit of 100 Å and depth of field varying from 1 mm at a magnification of 100 to 10 μm at magnification of 10,000. The SEM process frequently needs sectioning of the fracture surface or preparation of a replica of the fracture surface. The specimen or the replica has to be electrically conductive. Hence, nonconducting materials and plastic replica have to be provided with an electrically conductive coating. The SEM permits direct examination of the fracture surface without any surface preparation for conductive materials. The unique feature of this technique is that there is no magnification or image formation produced due to optical or magnetic lenses.

TRANSMISSION ELECTRON MICROSCOPY (TEM)

In transmission electron microscopy (TEM), a beam of high-energy electrons is focused by a set of magnetic lenses and this beam is made to pass through the specimen. The electrons, which are transmitted through the specimen, are collected and magnified by passing them through another set of magnetic lenses. Magnifications varying from 210 to 300,000 can be obtained in this type of microscope. The specimens in TEM must be reasonably transparent to electrons, and must have sufficient local variations in thickness and/or density, in order to provide adequate image contrast, and must be small enough to fit within the specimen chamber. This would many times call for preparation of a carbon or plastic replica of the fracture surface. These microscopes have better resolution of the order of 25–50 Å, when compared to SEM. Also, TEM photographs do not provide three-dimensional effects as well as an SEM photograph.

There are many factors that cause various types of artifacts in TEM and SEM fractographs, which can mislead the analysis. Experience, along with careful deliberation of multiple fractographs of the same region, is often required before reaching valid conclusions.

INTERPRETATION OF FRACTURE SURFACE FEATURES

Brittle Fracture in Metals

Brittle fracture is characterized by very little plastic deformation ahead of the crack tip. Cleavage fracture is the most brittle form of fracture, which occurs in metallic materials. Cleavage fracture in metals occurs by direct separation along certain preferred crystallographic planes due to simple breaking of atomic bonds. The salient feature of cleavage fracture, which is obvious even to visual examination, is the bright shiny appearance of the fracture surface. This is due to the high reflectivity of the flat cleavage facets. At higher magnifications under an optical or electron microscope, the cleavage facets appear to contain some irregularities.

The fracture surface at higher magnifications will reveal cleavage steps, river patterns, tongues, and fanlike markings, each of which is associated with a particular fracture mechanism at the microscopic level. Within a grain the crack may grow simultaneously on two parallel planes. These two cracks may join along the line where they overlap, through secondary cleavage or shear, forming a step in the fracture surface. A cleavage step may also form when a crack crosses a screw dislocation. Cleavage fracture takes place on certain preferred crystallographic planes called cleavage planes. When the cleavage planes of the adjacent grains have different orientation as in the case of a twist boundary, the crack must reinitiate on the differently oriented cleavage plane. It may do so at a number of places and spread out into the new grain [1] as explained in Fig. 10.1. This results in the formation of a number of cleavage steps. Merging

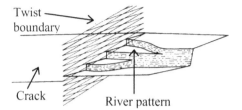

FIGURE 10.1   Formation of river pattern (from Ref. 1).

of cleavage steps results in the formation of river patterns on the fracture surface as shown in Fig. 10.2a. These river patterns coalesce downstream of crack propagation and thus give the possibility of determining the direction of local crack propagation from a fractograph. If the cleavage crack initiates at only one point along the grain boundary, it must fan out into the whole crystal from this one point, forming cleavage surfaces which resembles a fan, as shown in Fig. 10.2b.

Cleavage tongues are formed when a cleavage crack intersects a twin interface and propagates along the interface, before joining the main cleavage, which continues around the twin. Such tongues are shown in Fig. 10.2c. Another feature seen in cleavage fracture, called Wallner lines, is shown in Fig. 10.3. These marks are the result of the propagating crack front interacting with the elastic stress waves in the material. The Wallner lines appear as parallel cleavage steps, creating a rippled pattern. The significant difference between Wallner lines and fatigue striations is that Wallner lines cross each other, where as striations do not.

Ductile Fracture in Metals

The mechanism of initiation, growth, and coalescence of microvoids give rise to the characteristic fractographic features in a ductile fracture. Fractographs of ductile fracture obtained from SEM will show small dimples, which represent coalesced voids. Microvoids generally initiate from internal free surfaces created during casting or metalworking and also from the boundary between a second phase brittle particle and the matrix. Dimples appear in two different shapes, equiaxed or parabolic, as shown in Fig. 10.4. Equiaxed dimples indicate that loading was predominantly tensile, whereas elongated (parabolic) dimples occur under shear or tear mode.

Fatigue Fracture in Metals

Fatigue fracture surfaces usually exhibit features such as beach marks, striations, and ratchet marks. The beach marks, also referred to as clamshell marks, are shown in Fig. 10.5. They are visible at very low magnifications, in some cases

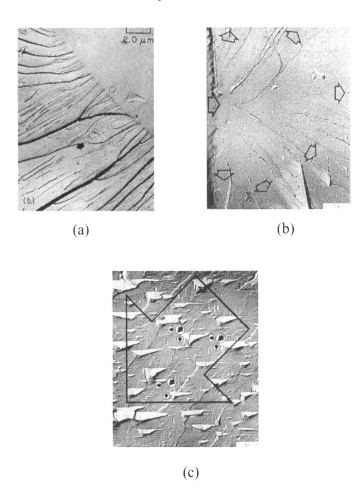

(a)   (b)

(c)

FIGURE 10.2 (a) River pattern, arrow shows the direction of crack propagation (from Ref. 2); (b) Fanlike markings, arrows indicate crack propagation direction (from Ref. 3); (c) Cleavage tongues, big arrow shows the overall direction of crack propagation, small arrow gives local fracture direction (from Ref. 3).

even to naked eye observation. The individual marks represent successive positions of the crack front where there is either a variation in the cyclic load or a difference in rate of oxidation or corrosion of the fracture surface. Their spacing gives an idea of the nonuniformity of the cyclic load. These marks are concentric about the crack origination site and grow outwards in the direction of crack growth. Although most of the time beach marks are associated with fatigue cracks, their presence does not always indicate a fatigue fracture.

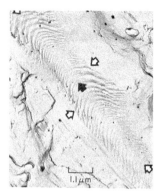

FIGURE 10.3   Wallner lines (black arrow indicates fracture direction (from Ref. 2).

Fatigue striations, shown in Fig. 10.6, are often seen under high magnifi-cations and the striation spacing indicate the progress of a crack during each load cycle. These are most often found within the fine structure of individual beach marks. The presence of these striations is definite evidence of fatigue-crack propagation, but their absence does not positively conclude the absence of fatigue-crack propagation. The formation of striations calls for continuity of crack front through adjacent grains and more than one crystallographic plane for crack growth. If these requirements are not met, well-developed striations cannot be seen on the fracture surface. The striations (1) are parallel to each other and at right angles to the local crack growth direction, (2) vary in spacing with cycle amplitude, (3) are equal in number to the number of load cycles, and (4) are generally grouped into patches within which all markings are continuous.

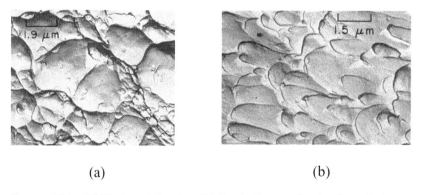

(a)   (b)

FIGURE 10.4   (a) Equiaxed dimples; (b) Parabolic tear dimples (from Ref. 2).

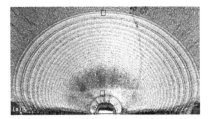

FIGURE 10.5   Beach marks (2×) (from Ref. 2).

Ratchet marks are macroscopic features due to multiple crack origin, each producing a separate fatigue zone.

Environment-Assisted Cracking in Metals

In both cleavage and fatigue, a major proportion of the advancing crack tip is exposed to the environment. Unless the crack propagates faster, the environment can flow to fill the crack extension. Environmental effect is more often recognized in stress corrosion cracking (SCC).

Both gaseous and liquid environments are known to cause cracking in various stressed materials. Stress corrosion cracking in steels and aluminum is intergranular, as shown in Fig. 10.7, in which the grain facets can be clearly seen. The dark patches shown by arrows are secondary cracks, which have grown into the grain boundaries. In stress corrosion cracking the surfaces exposed will normally be coated with a layer of corrosion products along with extensive secondary cracking. Figure 10.8 shows beach marks in steel, which has undergone stress corrosion cracking. These beach marks are the result of differences in the rate of penetration of corrosion on the surface.

FIGURE 10.6   Fatigue striations (from Ref. 2).

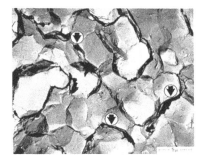

FIGURE 10.7 Intergranular stress corrosion cracking surface in 7075-T6 aluminum (from Ref. 3).

Brittle Fast Fracture in Polymers

Similar to metals, polymers also exhibit features on the fracture surface, which can be used to understand the fracture history. The fracture surface generated by a running crack at high velocity is shown in Fig. 10.9. This surface shows three distinctive regions. The "mirror" is the smooth fracture surface, which reflects light specularly. This is followed by a matte region called the "mist," and finally the relatively coarse "hackle" pattern. In this particular fracture the crack propagated under increasing stress intensity factor and velocity. The mirror zone under high magnification reveals steps (lines), running parallel to the crack propagation direction. This indicates simultaneous crack propagation on different parallel planes and happens when the running crack interacts with voids. In the mirror zone the crack tends to propagate along a single plane, but interaction with voids causes propagation of crack in different planes.

The mist zone reveals parabolic markings under high magnifications as shown in Fig. 10.10. These markings are formed when the crack front interacts

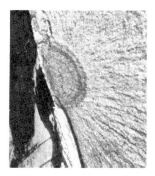

FIGURE 10.8 Beach marks due to stress corrosion cracking (4×) (from Ref. 3).

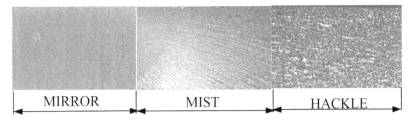

FIGURE 10.9 Different regions in brittle fast fracture.

with a flaw that has been growing ahead of the crack front. For this to happen the stresses ahead of the crack front should be high enough to activate these inherently present flaws. The openings of these markings indicate the crack propagation direction. The crack begins to propagate along several planes, resulting in the formation of lines (steps) parallel to the direction of propagation.

In the hackle region the parabolic markings are deeper than those seen in the mist zone, indicating that the stresses ahead of the crack front are high enough to activate flaws, which are further away from the main crack. The crack also propagates in several planes leading to the formation of lines in the direction of crack propagation. In addition to these features, river markings as shown in Fig. 10.11 are also seen in the hackle zone. These markings are caused by secondary crack propagation perpendicular to the direction of the main crack front.

Design Problem 10.1

The fractograph of a failed component, shown in Fig. 10.12, exhibits striations. Assuming that the loading is purely reverse ($R = -1$) and uniform, determine

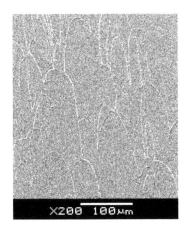

FIGURE 10.10 Parabolic markings in mist.

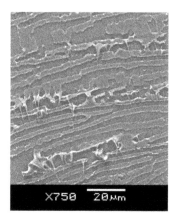

FIGURE 10.11　Hackle region showing river markings.

the applied stress intensity factor range using the Paris equation (Eq. 5.14). For this material, $A = 5 \times 10^{-31}$, $m = 3.2$.

Solution

$\mathrm{d}a/\mathrm{d}N \sim (1 \times 10^{-6})/10 = 10^{-7}$ m/cycle (about 10 striations can be seen in the space of 1 μm).

$$\frac{\mathrm{d}a}{\mathrm{d}N} = A(\Delta K)^m$$

$$\Delta K = \left\{ \frac{1}{A} \frac{\mathrm{d}a}{\mathrm{d}N} \right\}^{1/m} = 19.1\,\mathrm{MPa\,m}^{1/2}$$

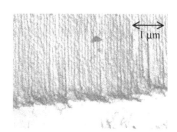

FIGURE 10.12　Fractograph of a failed component (from Ref. 2).

FIGURE 10.13 Fractograph of a failed component (from Ref. 2).

Design Problem 10.2

For the same component in Design Problem 10.1, the fractograph at another location is shown in Fig. 10.13. Determine ΔK. What could be the reasons for this higher ΔK compared to that in Design Problem 10.1?

Solution

$da/dN \sim (5 \times 10^{-6})/7 = 7 \times 10^{-7}$ m/cycle (about seven striations can be seen in the space of 5 μm).

$$\frac{da}{dN} = A(\Delta K)^m$$

$$\Delta K = \left\{\frac{1}{A}\frac{da}{dN}\right\}^{1/m} = 35.4 \text{ MPa m}^{1/2}$$

Why is ΔK larger than that in Design Problem 10.1?

1. If this region is further downstream (along the direction of crack growth) of the region in Design Problem 10.1, then crack length has increased. Thus ΔK is larger either due to the larger crack length or due to the increased crack length and an increase in load.
2. If the region is closer to that in Design Problem 10.1, then one could assume that the crack length was the same and therefore larger ΔK is entirely due to increase in load.

SYMBOLS

a	Crack length
A	Paris constant
N	Cycle number
m	Paris exponent
ΔK	Stress intensity factor fange

REFERENCES

1. Broek, D. *Elementary Engineering Fracture Mechanics*, 4th Ed.; Kluwer Academic Publishers: Boston, 1986; 33–71.
2. Fellows, J.A. Fractography and atlas of fractographs. In *Metals Hand Book*, 8th Ed.; American Society for Metals: Materials Park, OH, 1974; Vol. 9.
3. Beachem, C.D. Microscopic fracture process. In *Fracture — An Advanced Treatise*; Liebowitz, H., Ed.; Academic Press: New York, 1968; Vol. 1, Chapter 4, 244–347.

11

Case Studies

TYPICAL STRUCTURAL FAILURES IN SERVICE

The basic principles, diagrams, and definitions of the transition temperature approach to fracture-safe design are outlined in Chapter 6. That information is concerned with the flaw size, stress, and service temperature connected with the initiation and propagation of brittle fracture. The practical use of this methodology, including knowledge of the nil-ductility transition (NDT) temperature of the steel, has been applied in a review of a number of service failures, well described by Pellini and Puzak.[1] A large body of these test data has also provided a critical validity check of the fracture analysis diagram (FAD). Especially noteworthy were test results obtained from deliberately flawed vessels.

This section is based on the transition temperature methodology and on selected cases of fracture in service.[1] The discussion covers the sequence of steps normally involved in establishing the NDT temperature for a particular material, the applied stress level, and the flaw size parameters consistent with the particular examination of the fractured components. To assist the reader in following the various areas of the stress–temperature field, Fig. 11.1 provides a definition of fracture analysis zones, corresponding to the established categories. The various categories are defined as indicated below.

Category A. Small cracks (or flaws) are loaded above the yield strength of the material at a temperature lower than the NDT.

Category B. Here we can have small cracks just below the yield or within a high-level residual stress, still below the NDT.

Category C. Elastic stress loading occurs on increasingly higher dimensions of flaws (or cracks) below the NDT.

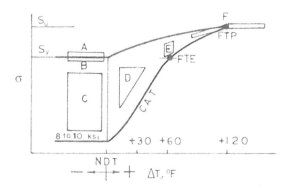

FIGURE 11.1 Location of fracture zones A to F (from Ref. 1).

Category D. Found between the (vertical) line of the NDT and the crack arrest temperature (CAT) curve are large flaws loaded by elastic stresses.

Category E. Just above "fracture transition elastic" (FTE), we find plastic strain loading of moderately large flaws.

Category F. In the region above FTE and "fracture transition plastic" (FTP), very large flaws are loaded near the ultimate strength of the material.

A large, thick-walled vessel constructed as a one-piece forging made of ASTM A293 steel (carbon 0.28, $S_y = 96$ ksi, $S_u = 116$ ksi, and elongation $\varepsilon = 18\%$) failed at the shell stress level of 30 ksi, or 0.31 S_y. The vessel had operated for several months (and 1550 cycles) when a small crack only 0.125 in. deep developed at the root of the first thread of a threaded closure, like that sketched in Fig. 11.2. The fractured surface had a "flat break" appearance, and there was a secondary fracture propagating longitudinally as indicated by the vertical arrow in Fig. 11.2. The shell stress level of 30 ksi drove the crack (from Category A to the C direction, Fig. 11.1) until it reached the CAT curve at about 8–10 ksi. At this stress level the crack was arrested as expected. The crack initiation at the root of the first thread and the subsequent behavior below the NDT (130°F for a typical Ni-Cr-Mo-V steel) were in conformance with the theoretical and actual behavior. The significance of the area of 8–10 ksi, below the NDT and CAT lines, is consistent with the "lower-bound stress" design practice described in Chapter 6.

What happens when the material selected represents poor chemical composition and heat treatment practice in spite of the fact that the configurational design of the structural part may be acceptable? A case in point is a heavy section cast propeller made of AISI Type 410 stainless steel (having 12% Cr), which was used in icebreaker service and failed five times in a rapid sequence of events. The problem was that the propellers were expected to endure plastic deformation

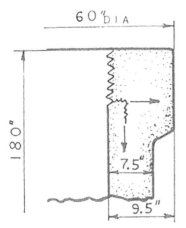

60" DIA

180"

7.5"

9.5"

F<small>IGURE</small> **11.2** Threaded closure detail.

overloads, while the casting contained numerous fabrication defects and deep-chipping, hammer marks. Although the range of the strength properties of the material had to be regarded as acceptable, the elongation at the lower strength level was only 5%. Plastic overload, small casting flaws, and brittle fracture below the NDT were consistent with Category A of failure (Fig. 11.1). The goal here was to find another solution for the ice environment, that is, for a service temperature of 30°F. This solution was ultimately found by modifying the Type 410 stainless steel, featuring low silicon with the addition of some molybdenum and nickel. This corrective action resulted in upgrading the strength properties and in shifting the NDT to −30°F, so that the sum NDT +60°F would be no higher than the temperature of the ice water, and consequently of the service temperature of the propellers. After the material upgrade, the propellers endured the plastic overloads since only the very large flaws would be able to initiate failure. For the record, the improved steel featured $S_y = 65$ ksi, $S_u = 90$ ksi, and an elongation of 18%.

It is always quite appropriate to look for a weakness in a pressure vessel design and construction when the conventional geometry is disturbed by a local welding pattern, such as in the case of a large vessel containing a manhole cover as the access point. This design in question involved a 3 in. thick, stress-relieved pressure vessel that failed during the hydrostatic test. The rough sketch in Fig. 11.3 indicates the location of the crack. Point (1) in the drawing refers to the contour of the failed wall with the manhole reinforcement pad (not shown). Point (2) gives the approximate original location of the pad, which was welded to the vessel wall. Point (3) shows the position of the initiating crack at the toe of the fillet weld and the corresponding location in the vessel wall. The starting

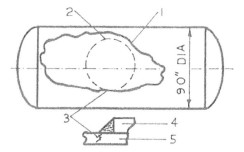

FIGURE 11.3 Location of original crack in welded vessel.

size of the crack was $1\frac{1}{2} \times \frac{3}{4}$ in. Points (4) and (5) refer to pad and vessel walls, respectively. The reinforcement pad (4) was welded to the inside of the vessel. The vessel and the pad were made from A302 B steel characterized by $S_y = 63$ ksi, $S_u = 85$ ksi, and an elongation of 24%. The root of the problem was reported to be the procedure of stress relief. The plate was annealed and furnace cooled at 1650°F, and stress relieved at 1150°F, resulting in a yield strength of about 40 ksi instead of the expected range of 60–65 ksi. The proper technique would have been normalizing heat treatment with air cooling from 1650°F.[1] The failure temperature was established to be 15°F below the NDT point. The fact that the actual yield strength was rather low explains the failure in accordance with the CAT principle because we have plastic strain loading in the presence of small flaws, consistent with Category A (Fig. 11.1).

Additional quoted experience[1] involved the same material and a vessel of a similar design that developed a brittle fracture and failed catastrophically at a temperature 40°F below the NDT. The crack initiation was traced to the heat-affected zone (HAZ) at the toe of a fillet weld for a manhole-port reinforcement plate. Although the crack propagated for a distance of 8 ft and was arrested, such weldments and the residual stresses often combine to create this and similar hazards for engineering design. It is particularly important to be aware of the various complications arising from structural discontinuities, pads, attachments, and patch plates, which are often difficult to design and fabricate without encountering surprises.

Another typical area of concern is the weld repair practiced in many branches of industry. The problem is particularly difficult when we have to deal with existing equipment and structures, which may have some cracks or evidence of mechanical or chemical abuse. Is there any right or wrong approach, say, to the task of salvaging repair?

A case in point, for instance, may be a pressure vessel that had been used in chemical processing for 20 years and had suffered local wall thinning due to corrosion. To save the vessel, it was decided to use four patch plates, welded to the interior surface. The salvage repair involved fillet and butt welds with the vessel shell acting as the "backup plate." The repair was accomplished "in place" and the vessel was not stress relieved. Local, small weld cracks migrated into the critical areas of the central welding and created sufficient flaws for the initiation of a fracture during the routine hydrotest. The vessel had $1\frac{1}{4}$ in. wall thickness and 5 ft diameter, as shown in Fig. 11.4. In this configuration, (1) denotes patch plates attached to the inside surface of the shell, (2) is a line tracing of the crack, (3) is the location of welding of patch plates, (4) indicates the cross-section of the patch plate, (5) identifies small weld cracks, (6) is the vessel wall, and (7) shows patch plate weld seams. The material was identified as ASTM A212 Grade B, which had $S_y = 40$ ksi, $S_u = 77$ ksi, and elongation of 26%. The expected NDT temperature for this material was 80°F, as shown in Fig. 11.5. The symbols (R) and (H) in this sketch refer to residual and hydrotest stresses, respectively. The symbol (FT) denotes here failure temperature, which was about 15°F below the NDT. The hydrostatic stress in the shell was close to 20 ksi. The problem developed because the vessel was not stress relieved after the completion of the weld repairs. Also, a combination of small cracks with residual stress just below the yield point left of the NDT line resulted in a vessel fracture consistent with the Category B zone of the summary of cases given in Fig. 11.1.

Another example of a failure due to small defects and high weld residual stresses is concerned with a large storage tank pressure vessel having the overall dimensions of 70 and 45 ft, as shown in Fig. 11.6. Details of a critical joint are given in Fig. 11.7. The fracture originated in the knuckle plate (B) caused by small weld cracks where the skirt (A) and the knuckle plate (B) of the bottom head were connected. The small cracks (D) then evolved into a branch of cracks (C) (Fig. 11.6), which propagated through the bottom head and lower parts of the vessel with catastrophic results, involving a process of fragmentation into many

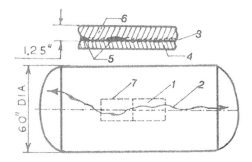

FIGURE 11.4 Repair weld concept.

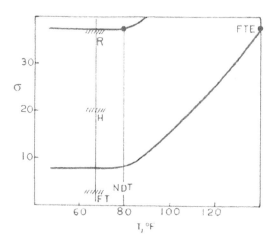

FIGURE 11.5   Crack arrest temperature (CAT) curve for salvage repair.

pieces. The material was ASTM A285 Grade C steel, with $S_y = 38$ ksi, $S_u = 61$ ksi, and elongation of 31%. The expected NDT for this type of material was 60°F. The FAD for the case at hand is shown in Fig. 11.8. The vessel was not stress relieved. It appears that the failure took place at a temperature around 15°F below the NDT. The symbols (R) and (H) define residual stress (yield strength level) and the hydrostatic shell stress (H) in the cylindrical portion, respectively. The failure temperature is denoted by (FT), as before. The stress in the cylindrical part was 18 ksi, while the knuckle portion stress was calculated as about 25 ksi.

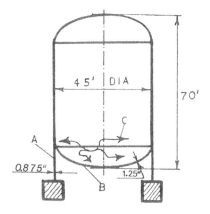

FIGURE 11.6   Configuration of storage tank.

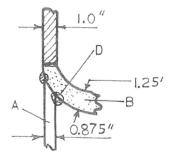

FIGURE 11.7   Detail of critical joint in storage tank.

The few cases discussed so far do not place the failure temperature above 70°F, which coincides with the review and findings related to ship experience during World War II. The majority of observed fractures occurred at temperatures of 30–50°F, and all of the cases involving conventional ship steels indicated failures below the crack arrest temperatures for the service stresses involved. The common denominators controlling the ship failures (irrespective of welding parameters) were NDT properties of the steels. Also, the most common failures resulted from a combined effect of arc strikes, weld cracks, and residual stresses caused by a welding procedure.

One of the remarkable failure cases quoted in the literature[2] is concerned with the anomalous brittle behavior of heavy steel forgings. The heavy retaining ring for an aluminum extrusion container had the dimensions shown in Fig. 11.9.

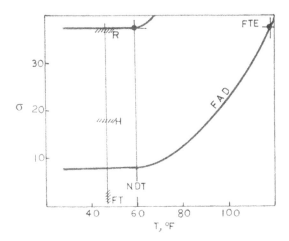

FIGURE 11.8   Fracture analysis diagram for failed storage tank.

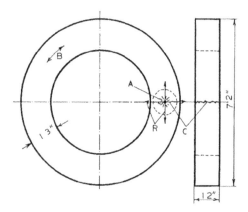

FIGURE 11.9 Dimensions, stresses, and crack orientation for retaining ring.

The ring material, ASTM A293 steel (0.30 carbon), had yield strength $S_y = 142$ ksi, ultimate strength $S_u = 157$ ksi, and an elongation of 12%. In a forged condition this material conformed to ASTM A293-55T Class 8 steel, representing a Cr-Mo-V forging grade for service conditions in the temperature range of 700–900°F. The intention was for the material to develop reliable creep resistance properties, which required low tempering temperature, and the resulting NDT temperature was 400°F, which must be regarded as extremely high. The sequence of stress conditions that followed the forging and tempering procedures was logical as well as startling. The function of the retaining ring was to provide a shrink-fit containment for the extrusion chamber, operating normally in the range of 700–900°F, required for the aluminum extrusion. Hence, control of the entire process involved the assurance of the appropriate prestress and information on the temperature conditions between the retaining ring and the internally heated extrusion chamber liner. The prestress action resulted in the tensile field of shrink-fit stresses in the ring shown in Fig. 11.9 by (B). To assure proper control of the temperature, thermocouples were silver brazed to the face of the ring, using an oxyacetylene torch. The localized torch heating to red heat became a source of a cluster of radial microcracks, indicated by (A) in Fig. 11.9. At the same time, a system of high residual stresses (symbol R in Fig. 11.9) was caused by the localized torch action, providing the deadly combination of small cracks and yield point level stresses that initiated the fracture. The propagation of the fracture was well accomplished by the shrink-fit level of stresses for the section of the order of 50 ksi. The resulting brittle fracture across the 12×13 in. section is shown at arrow (C) in Fig. 11.9. This was certainly a case of a spontaneous fracture of a large, forged portion of a material, defined as a high-strength steel. A summary of the failure conditions is given in Fig. 11.10. The complete fracture

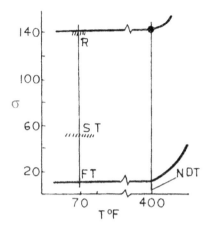

FIGURE 11.10   Failure diagram for thick retaining ring.

occurred at (FT), failure temperature of 70°F, which is very low in comparison
with the indicated NDT. The local residual stress (R) was of the order of the
yield strength of the material, while the shrink-fit stress was only 50 ksi, symbol
(ST) in Fig. 11.10. It is of course true that the ratio

$$S_y/S_u = 142/157$$
$$= 0.90$$

and the 12% elongation suggest certain natural tendencies of the material to be of
low ductility. The major culprit, however, was the combined effect of microcracks from torch brazing and a highly localized residual stress. This remarkable
effect has been proven to occur even in unusually thick sections of steel, provided
the stresses driving the fracture are higher than about 5–8 ksi. It may be recalled
that this limitation is essentially the lower-bound design stress discussed in
Chapter 6.

The foregoing case was based on service experience with the so-called
thick-ring geometry, for which the ratio of the average diameter to the radial
thickness amounted to

$$\frac{0.5(46 + 72)}{36 - 23} = 4.5$$

This number certainly indicates the appropriate ratio for the thick-ring criterion,
and a question may be raised regarding the behavior of a thin ring under similar
circumstances. This problem can be examined on the basis of a rather early
experience with a forged retaining ring in a turbine generator, having the proportions of a thinner ring, as illustrated in Fig. 11.11. The sketch includes a partial

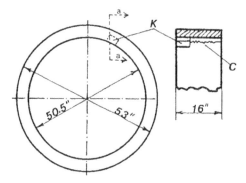

FIGURE 11.11   Thin retaining ring.

view of the ring cross-section in the direction of arrows a–a, featuring a keyway (K) for an electrical cable and the trace of a propagating crack (C). The ring material was SAE 3335 steel with $S_y = 99$ ksi, $S_u = 122$ ksi, and an elongation of 18%. The unusual feature of this case is the service record indicating that the fracture initiation condition had developed over a period of 40 years. The ring burst during an overspeed run and caused extensive damage to the turbine generator. The investigation of this accident at the time[3] raised the issue of how to distinguish between the brittle and tough response of steel on the basis of the then acceptable practice of heat treatment. The thin-ring material had a strength ratio of

$$S_y/S_u = 99/122$$
$$= 0.81$$

which, together with the improved elongation of 18%, compared with that for the thick-ring material, offered a higher level of ductility and perhaps a better resistance to fracture. Unfortunately, as the science of fracture mechanics progresses, the question of improved resistance based on the conventional mechanical properties still does not have a very clear-cut answer.

The initiation of fracture for the thin ring, defined in Fig. 11.11, was traced to the keyway and heavy electrical arcing caused by the breakdown of the electrical insulation. The process of arcing created local cracks. The field of high residual stresses, in combination with these cracks (also observed in the case of a thick ring), resulted in the initiation of the fracture. Since the material of the thin ring had been temper embrittled during the improper heat treatment procedure, the NDT temperature was established as 180°F, which proved to be very high. The principal elements of this failure analysis are given in Fig. 11.12. The service temperature at the time of the unfortunate combination of the various

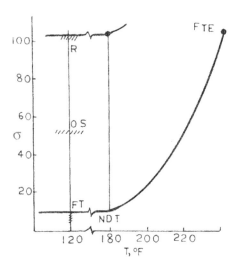

FIGURE 11.12 Elements of failure analysis for generator retaining ring.

effects was found to be about 60°F below the NDT. The hoop stress in the retaining ring for the turbine generator was about 50 ksi and represented a substantial crack-driving force for the propagation of the fracture from the corner of the keyway, across the entire cross-section, ($2\frac{1}{2} \times 16$ in.) of the forged ring. Evidently such extreme conditions had not existed during the prior 40 years of service life, until the failure of the electrical insulation.

The summary given in Fig. 11.12 includes three symbols that should be defined: The symbol (R), as in all diagrams presented in this chapter, represents the level of the residual stresses, which, as a rule, are considered to be quite close to the strength at yield for a particular material. The (OS) point of the diagram defines, in this case, the stress at the overspeed run, which drives the crack propagation. Without such a driving force, the crack may arrest. Finally, (FT) is the failure temperature, which completes the list of the conditions of a complete fracture as long as the point (FT) falls below the NDT.

The next topic of interest is concerned with the elastic loading of a large internal flaw discovered in a cast member such as, for instance, the extrusion press (hydraulic) cylinder shown as a half-view (not to scale) in Fig. 11.13. This sketch relates to an 80 ton extrusion press that failed by splitting along one side of the hydraulic cylinder because of a fracture that started from a large fatigue crack ($16\frac{1}{2}$ in. long) and propagated through the entire wall thickness of $13\frac{1}{2}$ in. The origin of the fatigue crack was traced to a casting shrinkage defect close to the inside wall surface, as shown in Fig. 11.13. From there, the fracture propagated in both directions, indicated by the arrows. The material

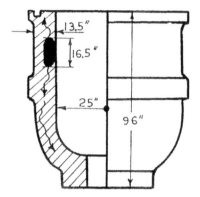

FIGURE 11.13   Part view of hydraulic cylinder (from Ref. 1).

was ASTM A27, Grade 70-36 cast steel with $S_y = 31$ ksi, $S_u = 69$ ksi, and an elongation of 28%. In terms of the mechanical properties ratio,

$$S_y/S_u = 31/69$$
$$= 0.45$$

which, together with the 28% elongation, gives full confidence that the material is totally ductile. Unfortunately the apparent ductile condition did not arrest the propagation of fracture, as shown by the fracture analysis diagram in Fig. 11.14; the cylinder failed, with the critical length of the crack at $16\frac{1}{2}$ in., at a shell stress of 13 ksi. (FT) denotes a temperature of 70°F at failure. It is clear from the diagram that a large-crack failure here corresponds to a relatively

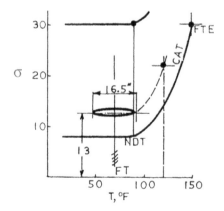

FIGURE 11.14   Fracture analysis for cylinder of 80 ton press.

small shell stress of 13 ksi. Pellini and Puzak[1] note that this level of stress is conventional for large extrusion presses. For instance, at 4440 ton ram pressure for the internal cylinder radius of 25 in. (Fig. 11.13), the operating pressure becomes

$$P = \frac{4400 \times 2240}{\pi \times 25^2}$$
$$= 5020 \, \text{psi}$$

and, using the average hoop formula, the corresponding shell stress is

$$\sigma = \frac{Pr}{t}$$

where

$$r = 25 + (0.5 \times 13.5)$$
$$= 31.75 \, \text{in.}$$

and

$$\sigma = 5020 \times 31.75/13.5$$
$$= 11,806 \, \text{psi}$$

For a 2750 ton ram pressure and a wall thickness of $t = 8.5$ in.[1] the approximate numbers become

$$P = \frac{2750 \times 2240}{\pi \times 625}$$
$$= 3138 \, \text{psi}$$

$$r = 25 + (0.5 \times 8.5)$$
$$= 29.25 \, \text{in.}$$

and

$$\sigma = 3138 \times 29.25/8.50$$
$$= 10,800 \, \text{psi}$$

Using Lame's formula, the shell stresses are 12,340 and 11,046 psi instead of 11,806 and 10,800 psi. This simple calculation shows the convenience and reasonable accuracy of the average hoop formula.

Assuming the same NDT temperature of 90°F for both presses, and an operating temperature of 120°F for the smaller press, the 13 ksi stress at the approximate flaw size of 16 in. would put the cylinder condition for the smaller unit to the right of the NDT line and below the CAT curve. Hence the service temperature of 120°F and the stress of about 13 ksi for the smaller press would not attain the conditions for fracture initiation, assuming the same material for both presses. The stress required for the fracture initiation in a cylinder of the smaller press under these assumptions would have to be significantly higher than 13 ksi. The new stress point could then be located at the intersection of the two dashed lines, roughly indicated above the CAT curve in Fig. 11.14.

As a follow-up on the example of the failure temperature location between the NDT and CAT curve, the case of hydrotest fracture in a high-pressure gas flask[1] may well be of interest. The material was ASTM A302 Grade B steel with carbon content $C = 0.24$, yield strength $S_y = 81$ ksi, ultimate strength $S_u = 104$ ksi, and an elongation of 22%. The basic configuration, dimensions, and the size of the initiating flaw are shown in Fig. 11.15. The flaw was classified as a large lamination that was not discovered during the high-quality fabrication procedure and x-ray inspection after the final stress relief heat treatment.

The sketch in Fig. 11.15 shows that a 30 in. outer diameter (OD) cylindrical forging (CF) is attached by welds (W) to a necked-down portion of the flask, and that it contains a radial lamination (L), 12 in. long, which constitutes the initiating flaw. The cracks (C) are shown emanating up and down from the lamination. The nondestructive techniques could not identify a tightly bound and totally contaminated lamination defect in the cylindrical forging. Modern techniques of detection and monitoring are focused on crack propagation rather than on the process of discovering potential areas of crack initiation, such as laminations. Still, even the fatigue monitoring technologies are confronted with many unknowns. These techniques include acoustic, electrical, thermal, neutron,

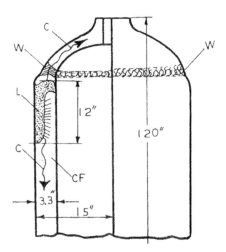

FIGURE 11.15 Part view of gas flask (from Ref. 1).

magnetic, and optical methods in addition to x-rays. And, if a careful search of crack initiators, such as laminations, using the x-ray technique, does not produce the desired results, the question is: Which other technique (from the foregoing listing) could be selected to assure that a crack initiator will be discovered? The only other approach to this problem is to acknowledge that the existence of some of the initiators is inevitable and to design a structure that could tolerate a certain amount of damage. This is, indeed, a tall order under the best of circumstances.

Since the selected material for the job indicated an NDT temperature of 30°F,[1] the fracture analysis diagram for the hydrotest of the gas flask can be illustrated as in Fig. 11.16, assuming a temperature at failure 15°F above the NDT and 57 ksi stress, developing in the wall of the cylindrical forging. This fracture stress should be consistent with the crack size of 12 in. The approximate location of the dashed curve in Fig. 11.16 was developed from a previous case described by Pellini and Puzak.[1] It will suffice here to make a general statement of practical significance. Assuming that a flaw size is fixed, a much larger stress would be required to cause the initiation of fracture above the NDT as compared to that below the NDT temperature.

The appearance of a fractured surface should correlate with the conditions below and above NDT. Structural failure fragmented into a large number of small pieces signifies a failure below the NDT. On the other hand, fracture resulting in a

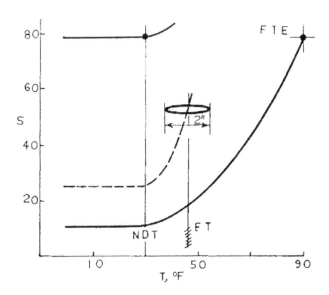

FIGURE 11.16 Approximate diagram for failure analysis of gas flask.

few large pieces, with shear lips, which are plainly visible, is consistent with failure conditions above the NDT.

The case of a gas flask failure represented by Fig. 11.16 is consistent with fracture Category D summarized in Fig. 11.1. The relevant failure in Fig. 11.16 is located between the NDT line and the CAT curve, where large flaws are loaded by elastic stresses. The next case to be examined is defined by Category E (above the FTE), where plastic strains act on moderately large flaws. Several tests conducted for the Pressure Vessel Research Council (PVRC) and the Atomic Energy Commission (AEC)[4,5] provided confirmation of Category E of structural failure.

The case in point was observed during a low-cycle fatigue test of a PVRC vessel, where deep cracking took place at a connection between the vessel shell and the nozzle. A part view of this junction, where the fatigue crack initiated the fracture, is given in Fig. 11.17. The shell (S) and nozzle (N) are joined by means of a weld (W). Continued testing at a high stress level developed a fatigue crack (FC) up to a length of 6 in., at which point a brittle fracture was initiated in the nozzle part and propagated in both directions, as shown by arrows (C). The resulting fracture of the shell indicated a heavy shear lip (SL), of the order of $\frac{1}{8}$ in. The strain gage record showed that the nozzle was loaded to unusually high stress levels reflecting plastic-strain conditions. The shell stress was of the order of 48 ksi. The tests were conducted at about $60-70°$F, with an average NDT temperature of $10°$F, corresponding to the nozzle material of the ASTM A105 forging. This steel had 0.25% carbon, $S_y = 36$ ksi, and $S_u = 67$ ksi, with a fine elongation of 34%. The S_y/S_u ratio of 0.53 suggests favorable ductility. In spite of these characteristics, however, a large fatigue crack developed to cause the initiation of fracture.

The fracture analysis diagram for both nozzle and the shell, using Category E criteria (Fig. 11.1), is now approximated by Fig. 11.18. According to the diagram, at the service temperature equivalent to the FTE for the steel, a large flaw and plastic stresses would be required to assure fracture initiation. The necessary flaw size of 6 in. appears to be consistent with the dashed line in Fig. 11.18. Also the size of the shear lip noted in the shell is typical for the fractures that propagate

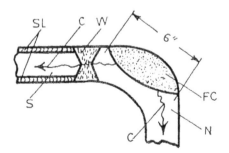

Figure 11.17   Shell-and-nozzle junction in PVRC vessel.

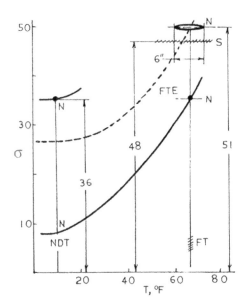

FIGURE 11.18 Approximate fracture diagram for PVRC vessel (from Refs. 4 and 5).

at the FTE temperature through the stress field at the level of yield strength. All locations in Fig. 11.18 denoted by (N) refer to nozzle characteristics. The stress corresponding to a 6 in. crack was taken at 51 ksi, consistent with the reported level[4,5] of a plastic strain of 0.47%. The failure temperature (FT) is shown to be less than 70°F. The symbol σ indicates the stress level of the shell material. The diagram is keyed to the NDT value of the nozzle material. Fracture initiation is shown to be about 60°F above the NDT.

The most likely situations in failure analysis boil down to the determination of the conditions for fracture initiation, presence of the nominal stress necessary to assure crack propagation, and the ultimate level of fracture development that renders the machine, equipment, or structure unfit for use. This pattern of behavior is particularly applicable to the general area of pressure vessel technology unless we encounter the so-called leak-before-break criterion reviewed previously in Chapter 6. The example that follows is intended to highlight the procedure for dealing with this problem in terms of transition temperature methodology and practical solutions.

This case is concerned with a PVRC vessel, using ASTM A302-B shell material and a modified version of ASTM A182-FI material for the nozzle part of the vessel. The strength and elongation properties of these two materials are rather close, with the resulting identical NDT temperature of about 10°F. The overall configuration of the pressure vessels involved in the entire series of full-scale pressure vessel tests conducted by the Southwest Research Institute is available.[4,5] However, this does not affect the nature of the results presented here related to the leak-before-break criterion or the sketch in Fig. 11.19.

Figure 11.19 indicates that two fatigue cracks developed simultaneously, one in the corner of the nozzle (port opening) and the other at the nozzle-to-shell junction. The larger crack broke through eventually (after 40,000 cycles), after a 2 in. travel through the wall, creating the classical condition of leak-before-break. The experiment for the PVRC indicated that the shell stress reached 32 ksi away from the junction, while the area close to the nozzle-to-shell transition was in the plastic regime, recording 52 ksi. However, neither of the crack and stress combinations noted was able to generate total fracture at the FTE temperature. The approximate diagram summarizing the flaw size and stress conditions is given in Fig. 11.20.

In this diagram the symbol (N) refers to the nozzle crack size and the mechanical properties of 62 and 86 ksi. The test stress for the shell part, σ, and the appropriate strength values were noted as 32, 58, and 80 ksi, respectively. The crack size of 5 in. and the plastic stress number of 52 ksi placed by the NTS symbol refer to the nozzle-to-shell junction. The fatigue test temperature (FTT) indicates a leakage condition rather than the conventional failure mode of the vessel. The levels of plastic stresses in Figs. 11.18 and 11.20 were originally inferred from the experiments.[1] Any estimates based on the plastic-strain values are likely to be approximate, at best.

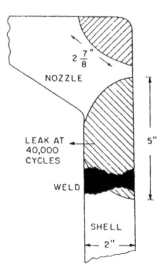

FIGURE 11.19 Part view of nozzle-to-shell connection (from Ref. 1).

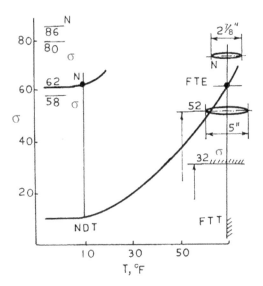

FIGURE 11.20 Flaw size and stress conditions for nozzle-to-shell junction.

The last category, denoted by (F) in Fig. 11.1, is concerned with the region above the FTE and "fracture transition plastic" (FTP), where very large flaws are loaded near the ultimate strength of the material.

The original information on near-ultimate tensile loading of very large flaws was obtained from an extensive series of deliberately flawed vessels in hydrostatic and pneumatic burst tests.[1] The test vessels were 15 ft long, with 22 in. outside diameter and 1.5 in. wall thickness. The materials included a large variety of steel chemistry and heat treatment, which, in turn, assured a large number of NDT temperatures, ranging from 15 to 200°F. The test temperatures varied between 40 and 70°F. The yield strength S_y and the ultimate strength values S_u had the ranges of 80–100 and 110–130 ksi, respectively.

A summary of the hydrostatic burst tests is illustrated in Fig. 11.21. All fractures can be arranged into three major groups depending on the extent of shear lips and the brittle appearance of the microstructure. The fracture surfaces characterized by lack of shear lips and fragmentation into several pieces are likely to represent the fractures below the NDT. Also, pressurization at higher NDT temperatures leads to increasing nominal burst stresses and distinctly different fracture appearance, signaling changes in fracture toughness of the material.

The definitions of the various symbols in Fig. 11.21 are as follows:

B   Brittle fracture
PT   Extent of prefracture tear

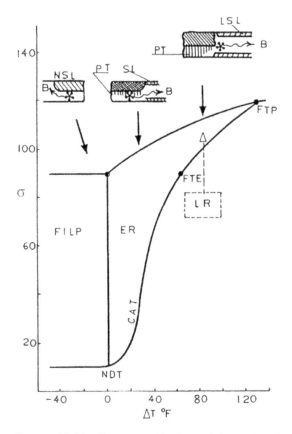

FIGURE 11.21 Summary of hydrostatic burst data (from Ref. 1).

NSL	No shear lips
LSL	Large shear lips, $\frac{1}{4}$ to $\frac{1}{2}$ in.
SL	Small shear lips, $\frac{1}{16}$ in.
LR	Limited ruptures, $\sim$2 ft each side
ER	Extensive ruptures
FILP	Fragmentation into large pieces

Small shear lips ($\frac{1}{16}$ in.) are expected from failures at $20-50°$F above the NDT temperature. Also a limited bulge can develop prior to fracture when toughness increases. The large shear lips ($\frac{1}{4}$ to $\frac{1}{2}$ in.) are characteristic of temperatures equal to NDT $+90°$F, at which time the fracture propagation velocity decreases. Under these conditions, the actual stresses at each end of the slit (introduced for test purposes) can be close to the ultimate tensile strength of the material. This is shown symbolically by the (LR) item in Fig. 11.21. The slit used in the hydrotest was about 10 in.

In going from pressurization by liquids to gases, the burst tests are conducted with 20 in., deliberately introduced slits. A brief summary of the pneumatic experiments is shown in Fig. 11.22. It should be added that a high degree of fragmentation is expected from fractures developed below the NDT, under pneumatic loading. In the hydro experiment we have fewer but larger pieces, and they are projected outward at lower velocities. Tremendous amounts of energy are stored in the compressed gases.

The definitions of the symbols in Fig. 11.22 are as follows:

FISP   Fragmentation into small pieces
FIP   Fragmentation into two or three pieces
SL   Shear lips
FSF   Full-shear fractures
CRST   Complete rupture "single tear"

It appears that all of the pneumatic-loaded vessels failed above the NDT at relatively low burst stresses. In contrast with the hydro test, the pneumatic press-

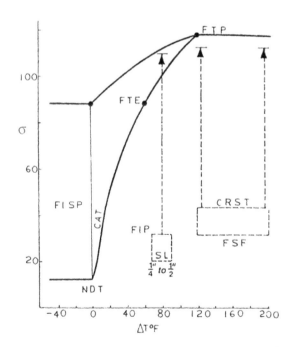

FIGURE 11.22 Summary of pneumatic burst data (from Ref. 1).

ure was not relieved by the escape of a small amount of the pressurizing medium. The compressed gas provided a "soft spring" across the open slit and allowed a continuous process of tearing and shearing. The burst stress levels are shown in Fig. 11.22 by the dashed arrows.

Several practical points should be noted from the pneumatic burst experiments:

1. The fracture-safe regions of the diagram in Fig. 11.22 represent safety, because the classical brittle behavior is avoided.
2. Any tendency to form a local bulge does not interfere with the process of the tearing action and fracture propagation.
3. This last item deals with fracture safety in the region to the right of the CAT curve. Pellini and Puzak[1] caution that the flaw must not be so large as to cause local plastic instability, leading to vessel destruction.

In closing this section on typical structural failures, based on the information collected over the years by Pellini and Puzak,[1] it is well to note that the transition temperature approach strongly depends on knowledge of the NDT properties of steel. This knowledge, in turn, is important in the development of new and improved steels.

It should be emphasized, in line with the purpose and scope of this book, that at the time of Pellini's original work, many papers and reports outside the Naval Research Laboratory (NRL) activities covered too narrow a span of interest for practitioners, who were always saddled with the task of solving industrial problems. These problems were on the "firing line" for the purpose of improving fracture-safe design. Thirty years later the general situation regarding the injection of pragmatism into the entire field of fracture mechanics and structural integrity technology is not much better. One can easily flounder on the shear impossibility of extracting practical data from a mass of theoretical papers and numerical solutions.

CRANE RETROFIT AND MATERIALS CONTROL

The topic of crane retrofit and materials control is included as an example of technical concern and to provide a brief historical record related to practical experience with the heavy lifting equipment supplied by the crane industry for all phases of underground nuclear testing. Because of national importance and the sensitive nature of the test program, it was necessary to evolve a set of conservative rules for retrofit and materials control activities consistent with the principles of fracture mechanics. It soon became clear that only a special blend of stress analysis, fracture mechanics, and materials technology could provide the basis for quality and safety assurance for downhole emplacement hardware and equipment. In addition to such areas as pressure vessel technology, aerospace, bridge structures, transportation on land, shipping, and processing equipment, it is necessary to recognize the importance of handling and moving equipment at all levels of construction. The material presented in this section is based on a critical survey[6] of the relevant experience and state-of-the-art knowledge pertinent to mechanical and structural engineering, supporting the test program. The nature of fracture control was regarded as the application of fracture mechanics and stress analysis to the load-carrying members with potential sensitivity to crack initiation and fracture.

It was recognized that, in the case of classical behavior of engineering materials, the traditional mechanical properties such as yield strength, ultimate strength, and elongation must be supplemented with appropriate knowledge of plane-strain fracture toughness and the transition temperature parameters. The general principles of fracture-safe design were expressed in terms of one of the following practical rules:

1. Selected materials should be inherently insensitive to brittle fracture in a given working environment.
2. The structural member is designed to a low stress level to assure insufficient supply of elastic energy to propagate any existing or "pop-in" type crack.
3. The part is designed with a redundant load path in order not to compromise nuclear or industrial safety.

The two approaches used in the preliminary design included transition temperature evaluation and LEFM. The key fracture mechanics parameters of interest were plane-strain fracture toughness K_{Ic} and NDT temperature. The two sources of information for the various technical decisions included the NRL[7] and the original publication of the Iron and Steel Institute.[8] These sources were particularly helpful in the evolution of fracture-safe design criteria and stress–temperature curves for crack arrest behavior of medium-strength steels below the yield level of about 120 ksi. Although the foregoing criteria were generally accepted, history shows that engineering solutions to fracture control were slow in coming because of theoretical and economic constraints.[9,10] Although under ideal conditions the parameter K_{Ic}, which is a unique property of the material, should be determined by test, a first approximation could be found from the correlation techniques that were available in the form of elementary equations for estimating purposes, as follows:

$$K_{\text{Ic}} = \left(5CVNS_y - 0.25S_y^2\right)^{1/2} \tag{11.1}$$

and

$$K_{\text{Ic}} = [(0.6DT + 75)S_y - 0.25S_y^2]^{1/2} \tag{11.2}$$

where K_{Ic} = plane-strain fracture toughness at a slow rate of loading (ksi in.$^{1/2}$), CVN = standard Charpy V-notch (CVN) test value at upper shelf (ft-lb), DT = dynamic tear value at upper shelf (ft-lb), and $S_y = 0.2\%$ offset yield strength at upper shelf temperature (ksi).

It was important to realize that a brittle fracture in general was of a catastrophic nature and occurred at speeds of crack propagation of up to 7000 fps, or even higher.[11] However, the experience of the test program with LEFM techniques indicated that LEFM was not directly applicable to the majority of structural problems. Consequently, the design engineer wanted the material to behave in a generally yielding fashion in order to avoid a confrontation with brittle behavior. In addition, the LEFM technique had a measure of constraint in the form of the actual crack (or flaw) size, which became an indispensable design parameter.

At this point of the program (that is, the late 1960s and 1970s), various techniques were available for evaluation of the fracture toughness transition curves as a function of temperature. However, because of economic considerations and plain, deep-rooted resistance to change in industry, the CVN technique was the desirable test method for the determination of NDT temperature. In a number of steels this point corresponds to the temperature at which the fracture toughness begins to increase significantly from plane-strain to fully ductile behavior. Also, during this process of transition, the material goes through the elastic–plastic change, indicating an increase in the size of the plastic zone at the crack tip. However, it soon became obvious that the CVN test should not be regarded as an invariant source for defining NDT values for broad families of steels. The relatively shallow notch and the limited constraint provided by a small CVN test piece causes a shift of the transition curve toward the lower temperature. This effect does not apply to the conventional dynamic tear (DT) data.[12] In considering the importance of NDT it was concluded that the use of the CVN method required the development of specific correlations for each grade of steel, so that the practical solution was the DT method, which gave direct indexing of the NDT. Unfortunately, the majority of procurement specifications were still written in terms of the CVN criteria. Clearly, the economic rather than the technical considerations prevailed, and the crane industry was no exception. It was difficult to extract all the information because certification of structural performance needed a quantitative measure of fracture toughness as well as NDT. The steel-producing industry had a very difficult time practicing material characterization using DT tests, which were recommended by the American Society for Testing and Materials. Detailed descriptions of the DT and other specimens are given in Chapter 4.

Almost 30 years ago and very early into the test program, certain problems started to surface that ended with several incidents of structural failure involving handling and lifting equipment. This development became the precursor of the crane retrofit and materials control effort, which lasted for many years and resulted in a unique, conservative philosophy of design, procurement, and certification of industrial cranes intended for critical service in support of underground nuclear testing as well as related activities.

The first case deals with the fracture of the shank from a crane hook assembly in a catastrophic manner at the ambient temperature of 28°F and at a very small fraction of the rated capacity.[13] According to the original records, the failure load was 9.3% of the rated value, 4.6% of the yield load, and only 2.6% of the maximum theoretical capacity. The sketch in Fig. 11.23 shows the location of the fractured area of the shank. The fracture took place across the minimum area of cross-section, as could be expected in a typical eyebar configuration where the effects of tension and bending are combined. A significant amount of galling was noted on the inner surface of the eye, where the shank pin exerts the maximum contact pressure. The fractured area had chevron markings, implying high velocity of crack propagation, compatible with brittle conditions of the material. The crack initiated the fracture in the weld repair area, and subsequent magnetic particle inspection revealed a $\frac{3}{4}$ in. length of the original flaw. It was determined that the flaw was in the HAZ of the repair weld. The broken parts did not fit well together, indicating a significant amount of permanent deformation, which had developed during the process of fracture. Although the fine chevron pattern was identified, essentially, with the brittle behavior, there were also certain coarser markings suggesting the presence of some shear deformation that developed during the final stages of the fracture when the velocity of crack propagation had ultimately decreased.

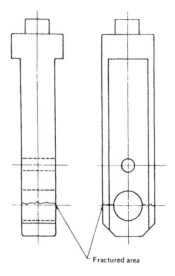

Fractured area

FIGURE 11.23 View of shank from crane hook assembly.

Closer scrutiny of the entire process indicates clearly the sequence and great complexity of crack initiation and propagation. The original HAZ crack gradually progressed due to the service fatigue, while galling damage induced residual tensile stresses into the contact surface between the shank pin and the eye. Once initiated, the fatigue crack could then advance due to the high stress concentration at the root of the crack. Also, lack of corrosive products at the fractured surface suggested that the process of failure must have developed quite rapidly.

The shank material designated as AISI 4140 was studied very carefully and showed that all the elements of chemical composition, with the exception of carbon, were within tolerance. The results are shown in Table 11.1.

Lower carbon content of the shank material resulted in a lower hardness of the forging. The observed range of Brinell number was 210 to 230. However, the real problem was with the microstructure of the weld, where untempered martensite appeared. The area of martensite showed a Brinell hardness of 600, which had probably developed within a very short time after the cooling of the austenite. Assuming that a volume change, associated with martensite transformation, acted together with the inherently brittle untempered martensite, it was possible to postulate that the combined effect generated the original cracks in the HAZ. The evidence of cleavage surfaces in the fractured regions strongly indicated that a substantial portion of the HAZ was as brittle as glass. In other words, even very small hook loads in the presence of a crack could supply sufficient strain energy for crack propagation.

The conventional properties of the forged shank material, as stated in Table 11.2, were satisfactory. However, CVN energy had no relation to the mechanical properties, and it proved to be outside the normal scatter as a function of temperature, as shown in Fig. 11.24. The scatter region shown conforms to the standard CVN impact energy as a function of temperature, while the lower curve was obtained from a precracked CVN geometry, indicating extreme strain rate and notch sensitivity. It was noted at the time that the crane industry

TABLE 11.1 Chemical Composition of Shank (%)

Element	Shank Material	AISI 4140
C	0.32	0.38–0.43
Mn	0.84	0.75–1.00
P	0.016	0.04 max
S	0.014	0.04 max
Si	0.33	0.20–0.35
Cr	1.01	0.80–1.10
Mo	0.18	0.15–0.25

TABLE 11.2   Conventional Properties of Shank
Material

	Test temperature	
	28°F	72°F
0.2% yield point (ksi)	64	63
Ultimate strength (ksi)	117	111
Elongation (ksi)	28	29

continued with the use of AISI 4140 steel in spite of the fact that for many years other manufacturers had had problems with this type of low-alloy steel in generator shafts and other components. At the same time, laboratory tests indicated that the ductile-to-brittle transition temperature for the 4140 alloy could be as high as 100°F. It was quite feasible, then, that only a few crane manufacturers actually knew the low-temperature limitations of their lifting equipment.

This case study resulted in a number of practical recommendations for the purpose of retrofit and quality assurance activities. Magnetic particle inspection (known as Magnaglow) and fluorescent dye penetrant inspection (known then as Xyglow) were suggested for critical crane parts on a yearly basis. In order to reduce the tendency to galling, the hardness of the shank pin would have to be increased. This was consistent with the Modell number, which is a measure of sliding friction and the tendency to galling because it involves the ratio of Brinell hardness and the elastic modulus of the material. The abrasion and galling appear to decrease with an increase in the Modell number.[14] It was decided that no

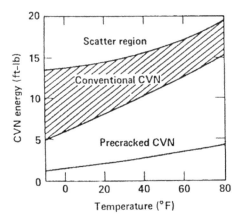

FIGURE 11.24   CVN energy for shank material.

repair welding should be allowed on 4140 steel or on any other alloy steel unless a well-controlled welding procedure could be developed and approved. A standard was also recommended calling for at least 20 ft-lb of CVN energy at $-20°F$ for the retrofit materials involving all the critical parts of the lifting gear. However, this was not always easy to accomplish.

The next case study was concerned with a smaller piece of equipment and a rather obvious type of failure, which relates, however, to a number of examples reported in the literature. The photograph presented in Fig. 11.25 points to the area of failure where the sum of tensile and bending stresses is likely to be at work.[15] The cracks can easily be initiated by the tensile residual stresses in the region of hammer blows. The sequence here is as follows. A hammer blow deforms the material's surface in compression. At the moment the hammer leaves the surface, the material springs back and creates a tensile residual stress. During the process of fatigue, the superposition of a high residual stress and a cyclic tensile stress in service results in the formation of the first crack. The fatigue stresses represent a supply of the elastic strain energy for crack growth until a critical crack length is reached. At this point the final catastrophic failure is propagated in a brittle manner. It was agreed at the time that a nondestructive inspection

FIGURE 11.25   Broken crane hook.

procedure and a limited grinding operation could have removed the cause of such a failure.

One of the well-known cases of residual effects, created by hammer blows in the past, involved the railroads, which required periodic hammer tests of the air tanks on all railroad cars. Metallographic analysis has since shown that the majority of the service failures of these air tanks had been initiated by fatigue cracks created by a combination of residual stresses from the hammer tests and the cyclic stresses accrued in service. Similar conditions were developed in aircraft manufacture, where hammer blows had been used to drive parts into jigs and fixtures. Such requirements and these fabrication practices have now been discontinued.

Careful examination of the fractured hook shown in Fig. 11.25 indicates that this part bears the scars of numerous hammerlike blows, but the only cracks that propagated occurred in the region of the highest combination of tensile and bending stresses. Indentations in areas away from the highest stress field did not develop fatigue cracks, because the applied stresses proved to be lower than the fatigue limit.

The recommendations resulting from this case study were as follows:

1. Perform magnetic particle or dye penetrant on crane hooks at least once a year.
2. Limited grinding operation can remove indentations and the surface layer containing the residual stresses.
3. Tougher steels should be used in critical crane components subjected to low-temperature environments.

Just before the largest ever nuclear underground test conducted in this country (almost a quarter of a century ago), extensive discussions and studies of materials control were taking place with special regard to a crane component holding a load close to one million lb and supporting a crucial experimental package. While the foregoing two cases of hook failure were simply a precursor, the matter of structural integrity and reliability of a large industrial crane became, overnight, the main force driving the crane retrofit. The plane-strain fracture toughness K_{Ic} and the dynamic tear (DT) for defining NDT had suddenly come to the front of scientific and technical deliberations. However, the K_{Ic} parameter still needed a component of nominal stress in order to estimate the critical crack length. Hence the calculation of stress came to the forefront, in case the critical part should be redesigned or modified because of the change of material.

The foregoing brief description covered only the first part of the problem, that is, the programmatic emergency. The second part was more insidious because it dealt with the intricacies of producing the quality material and interactions with industry. The need for the application of fracture mechanics principles to the crane industry was not fully recognized and the design formulas employed by the industry were considered to be of a proprietary nature, making the topic of structural reliability of the existing lifting equipment much more involved and costly. Hence the saga began.

The traveling block of the 750 ton crane was a critical element because it contained side plates supporting the yoke pins. A rough sketch of the block parts, with major dimensions, is presented in Fig. 11.26 (redesigned version).

Since design properties seriously affected the parameters of fracture mechanics, the major portion of the retrofit effort had to be directed toward materials control and fabrication. The use of mild steel plates of 4 in. thickness, as shown in Fig. 11.26, became suspect when the chemical analysis submitted by the manufacturer could not be verified through an independent laboratory process. There was no specification for the yield and ultimate strength or other mechanical properties affecting the design and performance of the crane block. For a maximum lifting capacity of 1,500,000 lb, the highest calculated stress was only 17,300 psi. However, according to the preliminary estimate, the block steel was judged to be close to the quality of AISI-1040 material, with a yield of 30–40 ksi and a marginal fracture toughness at room temperature.

This experience has opened further questions related to the existing hook and block materials forged in either AISI-4140 or AISI-1035 steel, and has resulted in tests run on the samples of quenched and tempered 4140 steel in thicknesses of 1 in. to $5\frac{1}{2}$ in. The tests indicated that this alloy was susceptible to a catastrophic failure in thicker sections, even above 60°F. It was established that such a failure could be precipitated by a small fatigue crack or a defect due to a forging or a weld repair process.

Further digging into the certification process has disclosed that the hanger and sheave pins were made from AISI-8620 steel, which was originally

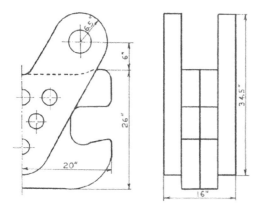

FIGURE 11.26 Major dimension of traveling block.

developed according to an emergency specification during World War II. The emergency standard was designed to replace the AISI-4320 steel. In the final analysis the substitute material had a coarser microstructure and a lower plane-strain fracture toughness than that of a forged 4140 steel. The additional study has also indicated that the original frames of the traveling blocks have, at times, been made from such marginal materials as ASTM-A7 and A36, as well as AISI-4150. The chemical requirements of A7 were absolutely minimal and the material was later replaced by A283. Although the basic minimum mechanical properties were specified for a material thickness up to 4 in., there was still no specific requirement for fracture toughness. In this regard the AISI-4150 steel proved to be even worse than AISI-4140, particularly in a welded state. Repair welds in these materials have been responsible for a number of catastrophic failures in industry.

Special comment is warranted about a popular "garden variety" structural steel known as A36. This is a hot-rolled carbon steel with very poor fracture toughness, especially in 2 in. and thicker sections. It can also have poor resistance to crack propagation at a service temperature as high as 120°F. This steel, of course, can be refined by suitable provisions in mill practice involving normalizing and tempering, although such an improvement can involve a substantial increase in cost. As it is, A36 is a still widely available and inexpensive steel that can be used in critical applications provided rather low nominal design stresses or redundant load paths are maintained.

The final preparations for utilizing the large crane in a high-cost and highly scientific important experiment required the assurance that fracture resistance of the critical lifting components was reflected by the use of a minimum value of 25 ft-lb of CVN energy at a temperature of $-20°F$ in order to retain acceptable toughness up to 70°F in the field. Various metallurgical questions continued to be raised that prompted laboratory tests on the material taken from the hook. The results were clearly disturbing, as shown in Fig. 11.27.

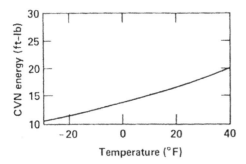

FIGURE 11.27 CVN data for a large crane hook.

Other components of the traveling block were made from the 4140 for-gings, which were also known to be of poor fracture resistance, particularly in thicker sections. These developments precipitated the final technical decision to redesign and to fabricate the entire traveling block using the Navy-grade material known as HY-80 in a forged condition. The new parts included a double hook, carrier plates, pins, bail, and the yoke. The two carrier plates and the double hook are shown in Fig. 11.26, as the main parts of the block system. Introduction of HY-80 to solve critical problems of structural integrity in the field marked the special phase of the crane retrofit program. It was hard to dispute the remarkable degree of fracture toughness found in HY-80, which could be as high as that shown in Fig. 11.28.[16]

In general, steel having a yield strength higher than about 180 ksi falls into the category of a premium strength. The quench and temper (Q&T) or quench and aging (Q&A) processes are necessary to control the appropriate strength levels. However, the resistance to fracture correlates rather poorly with the increase of strength. It appears that the metallurgical techniques designed to increase yield strength have the effect of reducing the number of dislocations. This effect alone compromises the ductility and toughness. For high-strength steels (Fig. 11.28) such a reduction can be quite significant. The term "fracture resist-ance" often is used as the ratio of fracture toughness K_{Ic} to the yield strength of the material S_y. This is a convenient normalization, although toughness and resistance to fracture can be considered as synonymous.

As the need for fracture control and fracture-safe operations increased in the 1970s during the underground nuclear tests, crane retrofit became a permanent

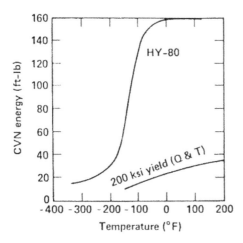

FIGURE 11.28 Comparison of CVN energies.

feature of the fieldwork. There certainly was a justified concern that the lifting equipment and fixtures traditionally supplied for the construction industry were not necessarily fabricated from fracture-tough materials. Users were beginning to make changes in manufacturers' ratings of cranes, because, for instance, the industrial cranes were not specifically designed for dynamic loading. However, there were no provisions for crane derating in relation to the level of fracture toughness of the crane materials. The goal of the retrofit program was to upgrade the fracture resistance of the critical crane components instead of the additional derate. This was not always a clear-cut proposition, but it was certainly a move in the right direction.

The criticality issue was defined as the failure of the major component upon which the crane system would allow the load to drop. The task was therefore to recognize the more critical parts first, such as all crane hooks and the eye connections in the primary load path, because the industrial standards pertaining to cranes did not have any specific requirements for fracture-tough materials. In terms of fracture mechanics the aim would be to select a material that could guarantee that a brittle failure would not occur over a specific range of section sizes, flaw dimensions, and operating conditions within a certain stress range. Specifically, the stress should not exceed the yield strength. Also, the cost of procuring better materials was, at the time, considered to be a prudent choice.

Since stress analysis has always played an important role in retrofit decisions, it is well to point out certain configurational details and the role of analytical tools in making technical decisions. The accepted modus operandi was that, in the great majority of cases, the finite-element methodology had to be augmented by the appropriate closed-form solutions.

One of the unique areas of crane technology was the "medium-range emplacement rig" (MRER), covering load capacities between 300 and 630 tons. Out of the general category of manufactured components, such as crown blocks, traveling blocks, or connectors, the configuration of a double hook presented an interesting stress analysis problem because of the presence of a central bore for inserting a large-diameter pin, as shown in Fig. 11.29. This special configuration was seldom available at the time, although German machine design practice had included this concept, without, however, provision for a pin. The locations marked as A, B, and C in Fig. 11.29 relate to the points at which the critical stresses were calculated using the established principles of advanced strength of materials and the theory of elasticity. Point A refers to the maximum filet stress, B represents contact stress, and C relates to the stress in the matching pin. The effect of the bore diameter on the three different stress fields is given in Fig. 11.30. Stress curve A includes bending, axial, and shear components. Curve B features the highest stress at the bore, when the bore diameter is 7 in. Curve C deals with the pin, modeled as a beam in double transverse shear and bending. The follow-up calculations, using a finite-element model,

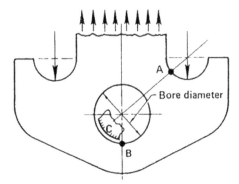

FIGURE 11.29   Double-hook concept.

involved the hook and eye areas separately, showing significantly lower stresses. However, the finite-element model was considered for one bore diameter only. In the end, the choice of material was governed by fracture resistance and the validity of NDT in relation to the service temperature. This material featured the minimum yield strength of 100 ksi and an elongation of 15%.

The important phase of the retrofit of large cranes, approaching one million lb working capacity, involved the so-called elevator links (often referred to in the

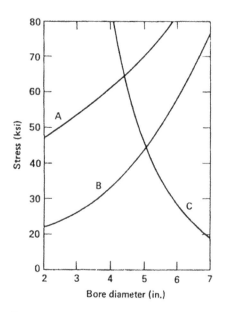

FIGURE 11.30   Critical stresses in double-hook design.

crane industry as bails). The problem of structural integrity centered around the analysis, manufacture, and testing of these critical components. The specific example illustrated in Fig. 11.31 refers to a specially developed weldless link, fabricated from a single piece of high-grade steel, hammered and drop forged to enhance the ultimate tensile strength. The location of the minimum diameter of the link cross-section of $3\frac{1}{2}$ in. is shown in Fig. 11.31. The preliminary analysis of stresses under 400 kips load indicated a maximum compressive, contact stress of 85.2 ksi. This is a difficult problem area involving the two curved surfaces in contact between the hook and the link. The maximum tensile stress, across the horizontal section, was 67.3 ksi. Both stresses were determined by a finite-element model and referred to the inner boundary of the link. However, these results were still considered to be a crude approximation because a three-dimensional effect was not accounted for. Additional calculations were also made using a closed-form approach based on the classical curved beam theory,[17] which indicated higher stresses calling for redesign, utilizing a fracture-tough material. Since the loads on these elevator links were rather high, other approaches such as

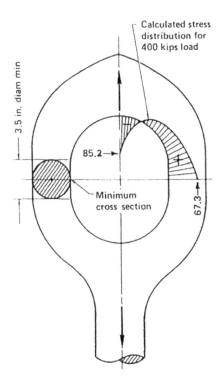

FIGURE 11.31   Part view of elevator link.

low boundary stress or load path redundancy were not recommended. The general picture and scope of the retrofit program was quickly emerging with such components as pins, links, and eyebars denoted as critical. The process was also accelerated with further discovery that the majority of these components were fabricated from the 4140 material, heat treated to a yield level of 100 ksi, and having a very doubtful fracture toughness.

In view of the current emphasis on computer numerical techniques in structural analysis, it may be of interest to recall a few examples of comparison between the solutions using computer modeling and hand-calculated result based on the closed-form approach, recorded during the crane retrofit effort described in this section.

It is interesting to note that finite-element modeling of the swivel eye geometry indicated in Fig. 11.32 was not easy, although this was a seemingly elementary problem of machine design. Yet it was important to verify the critical stress levels for the purpose of fracture control. The areas of modeling difficulties centered around the clearance between the eye and the pin, as well as the details of thread geometry. The preliminary estimate of pull-out stresses is noted in Fig. 11.32 on the premise of zero clearance and a perfectly rigid pin.

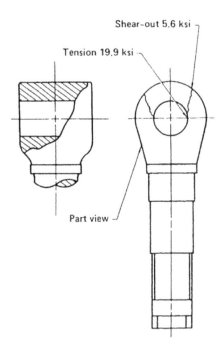

FIGURE 11.32   Swivel eye geometry.

A comparison between the computer and closed-form results is given in Fig. 11.33. Similar correlation of the maximum stress levels for the upper eye of the elevator link, determined for the inner surface, is illustrated in Fig. 11.34. The shape of the link with the preliminary stress distribution is indicated in Fig. 11.31. Neither the finite-element nor the curved-beam solutions accounted for the three-dimensional features of the problem.

The upgraded version of the large crane lifting fixture (double-hook type) is shown in Fig. 11.35. The hook has a 10 in. diameter hole in the pin-version design for the overall load support, and a $4\frac{5}{8}$ in. pin support hole if the elevator links are not used. The analysis neglected the double-curvature effect (pin and hook contact surface). The finite-element model was applied to one-half of the fixture, shown in Fig. 11.35, with the vertical plane of symmetry prevented from moving in the direction perpendicular to this plane. Displacements parallel to the plane of symmetry were allowed.

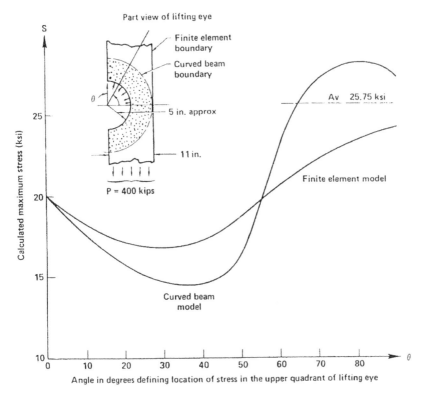

FIGURE 11.33 Comparison of computer and hand calculations for swivel eye modeling.

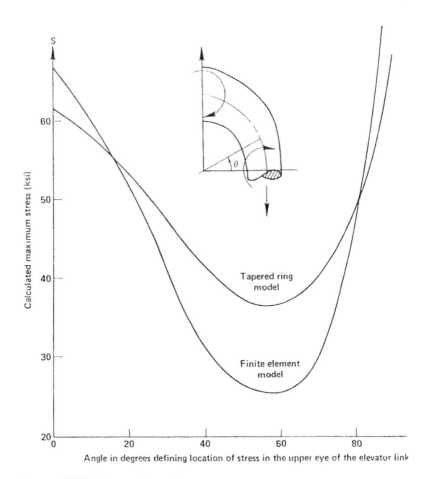

FIGURE 11.34 Comparison of computer and closed-form results for elevator link modeling.

In analyzing the effects of pin contact forces, a cosine load distribution was used as both the upper and lower openings in the fixture. On the premise of a relatively ductile behavior of the fixture material, the use of the von Mises yield criterion, to convert the finite-element stress field to an equivalent uniaxial stress, was justified. The form of the conversion was

$$\sigma_e = (\sigma_1^2 - \sigma_1\sigma_2 + \sigma_2^2)^{1/2} \tag{11.3}$$

where σ_e is the equivalent uniaxial stress, while σ_1 and σ_2 define the principal stresses.

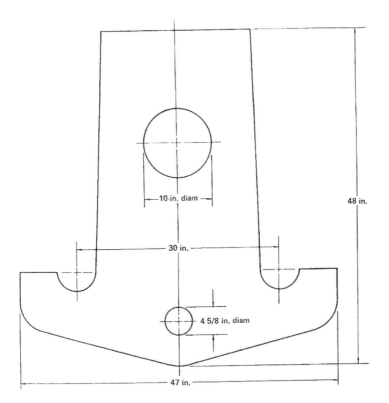

FIGURE 11.35   Large crane lifting fixture.

In the curved-beam model, the closed-form solutions involved estimating bending, normal, and shear stress components separately, and then calculating the maximum principal stresses. The maximum stress criteria were employed in these calculations because the material more likely to be used by the crane industry for large hooks was still the 4140 type, susceptible to brittle behavior. The conventional finite-element codes, however, often presented the results in terms of the von Mises criterion. The specific computer and closed-form curves for the large lifting fixture are given in Fig. 11.36.

Although it was essential to have the best information on the stress distribution in critical crane components, the fracture-safe performance could not be predicted on the basis of conventional material properties alone. This has been proven by many years of crane retrofit effort. Fabrication defects still occurred in various forms such as laminations, seams, voids, inclusions, and slag pockets; these acted as stress raisers, particularly in the notch-sensitive materials, such as steels exhibiting low fracture toughness. In addition, cooling, rolling, and forging

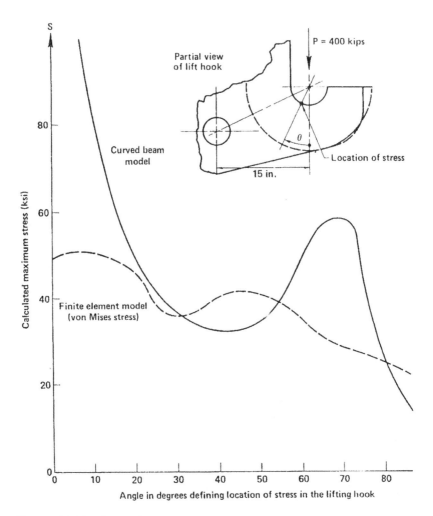

FIGURE 11.36 Comparison of computer and closed-form results for double lift hook modeling.

typically induced residual stresses. Any residual effects in the form of a tension applied to the saddle area of the hook would be unfavorable, and therefore stress relief heat treatment after fabrication was certainly highly recommended.

The examples of evaluation of the critical components discussed so far have been extended to other parts of the cranes such as anchor links, sheave brackets, bridle attachments, links, equalizing rings, and hinge brackets, to mention a few. The task was not mundane because it was, at times, very difficult to locate the material's history and credible service records. The experimentalist and the inspector had often to assume that the fracture resistance properties were totally unknown.

It was, indeed, fortunate that the crane retrofit program could fall back on the use of high-yield materials technology in the yield range 80–130 ksi.[18] The workhorse material for the early phase of crane retrofit was HY-80, suitable for a full range of shapes, forgings, or castings. Typical average toughness properties of HY-80, compared with those of the lower strength construction materials used in the Navy at the time, are shown in Fig. 11.37. The main requirement for this material was a CVN level of 50 ft-lb at a test temperature of −120°F. Separate specifications covered plate, extrusions, rolled shapes, castings, and forgings. However, there were certain limitations involving chemical composition, heat treatment, and welding. One of the more recent developments in this area is the use of the low-carbon and low-alloy steel known as HSLA-80, which has superior weldability.

INTEGRITY OF UNDERGROUND STRUCTURES

This section is concerned with fracture control and material studies related to piping, casing, canisters, and similar components found largely in underground explorations or test facilities for industrial, scientific, and other purposes. The material presented covers the case history of special underground tests where pipe or canister fracture could not be tolerated for reasons of public safety and to prevent the loss of scientific data of national importance. Although the

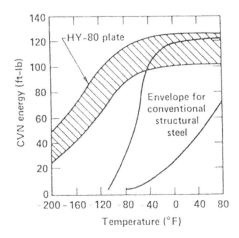

FIGURE 11.37 Typical CVN regions for HY-80 and conventional steel.

emphasis on the latter point may have been diminished during the past few years, the general concern for materials control in other areas has remained strong. Materials science will always be a part of the discipline triad, with fracture mechanics and stress analysis as the two other elements of equal weight.

The primary contributions to the development of specifications and hands-on data for the oil well casing, tubing, and drill pipe have been provided by the American Petroleum Institute (API) during the past 60 years. Recommended properties and design criteria have also been widely publicized in trade and professional magazines, allowing for countless combinations for design of emplacement strings and the associated hardware.[19] Over the years a number of grades of steel have been approved as API standards with minimum yields between 40 and 150 ksi.

The lowest grade material utilized in field operations at the start of underground testing was API J55, having a yield strength of 55 ksi, and there were some concerns about the level of fracture toughness of this material over the range of service temperatures involved. The main concern was that for a relatively low fracture toughness and the yield of 55 ksi, the critical crack size might be too small for normal inspection techniques. Since the NDT and K_{Ic} data for the API J55 grade were not available, a special study was undertaken to break ground for the application of fracture mechanics and materials science as the first step toward the development of fracture control plans for this and other API materials.[20]

The original API grade having 55 ksi yield and 20% elongation was generally qualified as an acceptable structural material. The chemistry, however, was rather poorly defined with respect to sulfur, phosphorus, and carbon contents, which could have had an influence on NDT. The carbon content of 0.40–0.50% was certainly rather high for good toughness characteristics, and apparently the material was produced by the least expensive method and without any special regard for fracture control. Random material samples indicated a spread in NDT of 0–150°F, and all the larger test pieces failed at ambient temperature in an essentially brittle manner.

Although the original case study zeroed in on the J55 casing, it soon became clear how limited were the fracture toughness data available for steels with comparable chemistry and strength levels. For instance, a material with 50 ksi yield and 20% elongation, such as A302B steel, shows variation of K_{Ic} with temperature as plotted in Fig. 11.38, for the two types of processing. The test specimens for this case were taken from a 7 in. thick plate and fatigue cracked during the process of obtaining the values of K_{Ic}. Figure 11.38 clearly shows that, in general, the heat-treated condition creates markedly higher fracture toughness due to the presence of a finer microstructure. For instance, at a service temperature of about 30°F, the dashed line gives the relevant K_{Ic} values as 125 and 55 ksi (in.)$^{1/2}$. Once the K_{Ic} parameter and the working stresses are

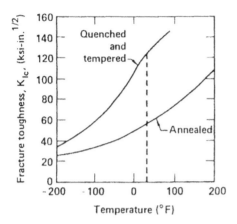

FIGURE 11.38 K_{Ic} variation with temperature for A302B steel.

known, the critical crack size can be determined from the conventional formulas of fracture mechanics. The crack size is then proportional to the ratio squared, K_{Ic}/σ^2, where σ is the nominal applied stress. It is also interesting to note that both curves in Fig. 11.38 tend to flatten out at the lower temperatures. The preliminary estimate of the plane-strain fracture toughness for the API J55 material was only 20 ksi $(in.)^{1/2}$. For a working stress, say, of 40 ksi, a critical crack length calculates to be 0.16 in. In practice, this crack size will not be very easy to detect.

From the materials point of view the carbon content in a conventional steel is responsible for controlling the mechanical strength. However, as the carbon content is increased, NDT shifts toward the higher temperatures, while the energy required to produce a fracture decreases. For normalized, air-cooled, iron–carbon alloys, the effect of carbon content is illustrated in Fig. 11.39. The characteristic curves and their relative displacement should not be regarded as sufficiently

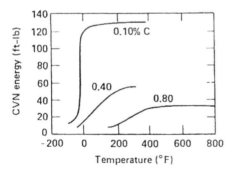

FIGURE 11.39 Effect of carbon content on CVN energy.

accurate for indexing the NDT values. These curves are only intended to convey the idea that the general effects of carbon content on the amount of energy to fracture and the location of NDT are significant. At the same time, the alloying elements such as nickel, added for solid solution effects, and aluminum with molybdenum, introduced for grain refinement, should definitely enhance fracture toughness. The virtual lack of these control elements in API J55 explains why the fracture toughness of this material was always marginal.

The case studies of API J55 indicated rather clearly that new efforts should be made to adopt other API grades such as N-80 and P-110 in the quenched and tempered condition. Any thought of retaining J55 for even partial applications in the field was finally abandoned with the failure of a J55 casing. This particular case should be recorded as another example of dealing with the conventional mechanical properties instead of the basic parameters of fracture mechanics.

After the failure of the J55 casing, normal inspection could not pinpoint any specific irregularities, and it was concluded that a microscopic defect must have acted as the crack initiator, produced most likely by a weld spatter. Furthermore, since a very small defect caused brittle failure, the fracture toughness of this material must have been very low indeed. It thus became apparent that normal structures made of J55, or similar materials, had to be designed to a much lower stress to survive.

Indeed, further examination of the fractured region of the casing disclosed that the failure was due to a rapid propagation of a brittle crack that originated at the outer surface near a weld in the HAZ. There was no flaw other than that due to weld spatter, and there was no indication that a crack deeper than 0.02 in. could have been present. The cracked surface had no evidence of shear lips, and it was clear that the fracture was transgranular and typical of a coarse-grain material that failed with limited energy absorption. The austenitic grain size was of the order of 0.15 mm. It should again be emphasized that while the conventional mechanical properties were still quite good, there was no clue as to the low fracture toughness and high NDT during the ductile-to-brittle transition.

With the end of this particular case study and the increased concern for fracture-safe designs, the API grade P-110 was selected for complying with the upgrade of fracture resistance criteria. However, it was determined that even this improved material had not been qualified using rational and quantitative fracture resistance test procedures.[21]

The test samples of the P-110 pipe were obtained from the three industrial sources that complied fully with the specifications in force at the time. The mechanical properties obtained from the manufacturers indicated a reserve of yield strength over the specified level of 110 ksi between 4 and 14%. The average ratio of yield to ultimate strength was 0.87, and the average elongation was about 24%. The pipe specified for the research was $9\frac{5}{8}$ in. in outer diameter, with a wall thickness of 0.435 in.

The work conducted by the NRL was centered around the determination of the complete temperature transition relationships in terms of DT and CVN parameters. It also involved technical assistance in the selection of materials from the NRL data bank and in the characterization of new materials with respect to tensile properties and fracture resistance. The main thrust of the NRL program was to determine fracture resistance properties of P-110 and 4130 materials. Emphasis was also placed on the safe use of the CVN energy criteria for specifications and the extension of fracture resistance criteria for section thicknesses smaller than $\frac{5}{8}$ in. The educational effort was centered around the strictly practical aspects of fracture mechanics under a general subject of fracture-safe assurance for engineering structures.

The results of this case study were presented separately for the DT and the CVN energy as a function of temperature. All the data were for the longitudinal specimens with the actual fracture in the circumferential direction. The true nature of the temperature transition of fracture toughness under the maximum constraint condition that a crack can build up in the pipe is best defined by the DT energy curve shown in Fig. 11.40. The cross-section of the test piece is given in the diagram. The point denoted by (C) is used here as the criterion for the elastic–plastic fracture state. The code letters A, J, and S designate the manufacturing sources supplying the P-110 casing.

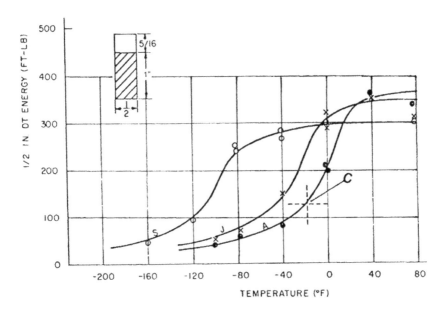

FIGURE 11.40 Dynamic tear vs. temperature for P-110 casing (from Ref. 21).

The complete temperature transition relationship for the CVN energy is illustrated in Fig. 11.41. The code letters A, J, and S designate the industrial sources as in Fig. 11.40. The CVN test sample imposes an undefined constraint condition that is less than the maximum for this size of pipe, so that the transition temperature characteristics are displaced in a more unpredictable manner.

It may be noted that the differences between the DT and CVN transitions are significant. For instance, the midpoint of the transition curve (S) in Fig. 11.41 is about $-120°F$ at 25 ft-lb, while a similar midpoint in Fig. 11.40 for the (S) curve is at $-100°F$ and 150 ft-lb. The comparison of the midtransition region for the (J) curve gives $-90°F$ for the CVN energy and $-30°F$ for the DT test. The interpretation of the CVN data for the (A) curve becomes even more complex because of the vertical displacement of the curve.

The relationship between the CVN and DT energies is given in Fig. 11.42. There is a significant scatter of the results between the two dashed lines representing the upper and lower boundaries of the scatter band. Assuming that the CVN criterion can be selected from the upper boundary, the relevant value may be estimated from a simple expression, such as

$$CVN = 0.16DT + 21 \tag{11.4}$$

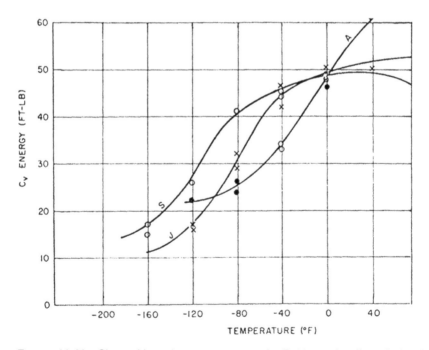

FIGURE 11.41   Charpy V-notch vs. temperature for P-110 casing (from Ref. 21).

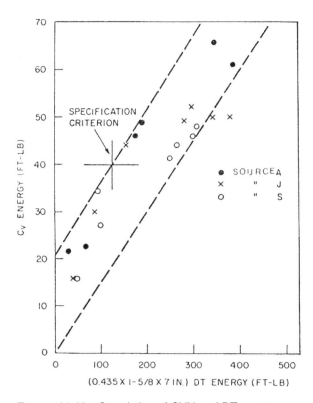

FIGURE 11.42   Correlation of CVN and DT energy.

The NRL case study has shown that a significant level of confidence can be achieved when a DT value of 125 ft-lb corresponds to the API grade P-110 pipe with 0.435 in. wall. The CVN value from Eq. (11.4) calculates to be 41 ft-lb. Many years after this study, the industrial grade C-110 appeared to meet the original API P-110 requirements. The specific goal of the C-110 material was to be even more restrictive than P-110. For instance, the C-110 had to be quenched and tempered to assure a minimum of 90% martensitic structure, and it was stress relieved by heating to at least 1000°F. The carbon content was between 0.20 and 0.35%, with the phosphorus and sulfur limits well below those for the API grades. In comparing the strength levels, the C-110 had a maximum yield about 10% lower and a higher elongation. The minimum transverse CVN value of the C-110 material was specified as 22 ft-lb at room temperature up to a wall thickness of 0.75 in. The special point (+) in Fig. 11.42 denotes the proposed specification criterion.

As an extension of the case study of the API-grade materials, one high-strength steel, successfully used in sour gas well operations, was characterized by the NRL with reference to fracture resistance criteria suitable for use in a specification. The material was known as AISI 4130 steel, fabricated often as underground piping of 8 in. outer diameter and $\frac{5}{16}$ in. wall thickness. The test specimens, cut from the pipe samples in the longitudinal direction, had to be of undersize type for both DT and CVN experiments. These specimens were tested over a range of temperatures from $-160°F$ to $+80°F$, in order to develop the temperature transition criteria. The transition characteristics for 4130 steel in terms of DT energy were established as shown in Fig. 11.43. This chart was prepared for circumferential crack propagation under a longitudinal stress. The specification criterion for general yielding in the presence of a through-thickness flaw is indicated by a dashed line just above the 100 ft-lb level. Since the test specimens for Fig. 11.43 were $\frac{5}{16}$ in. thick, it was necessary to determine the equivalent DT energy to the 500 ft-lb level normally obtained from a $\frac{5}{8}$ in. DT specimen. The conversion from $\frac{5}{8}$ to a $\frac{5}{16}$ in. test specimen results was accomplished with the help of the following expression:[21]

$$(DT)_E = P_{RF}(d)^2(B)^{0.5} \tag{11.5}$$

where $(DT)_E$ = equivalent DT energy (ft-lb), P_{RF} = plastic resistance factor (ft-lb/in.$^{2.5}$), d = net section depth (in.), and B = specimen thickness (in.).

The cross-sectional geometry of the test specimens used in the study of 4130 steel is shown in Fig. 11.44. The relevant dimensions and proportions of the cross-section are $a = 0.6B$, $d = 2B$, $B = 0.3125$ in. (7.9 mm), and $d = 0.625$ in. (15.8 mm). Hence, from Eq. (11.5)

$$\frac{P_{RF}}{(DT)_E} = \frac{1}{d^2 B^{0.5}}$$

$$= \frac{1}{0.625^2 \times 0.3125^{0.5}}$$

$$= 4.58$$

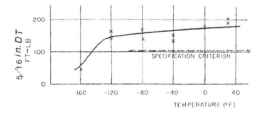

FIGURE 11.43 DT energy for 4130 steel (100 ksi yield) as a function of temperature.

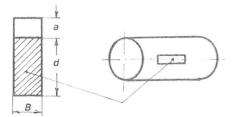

FIGURE 11.44 Geometry of longitudinal test specimen.

and the equivalent DT energy for the subsize specimen is

$$500/4.58 = 109 \, \text{ft-lb}$$

This number is represented in Fig. 11.43 by the dashed horizontal line described as the specification criterion. The Charpy V-notch energy, as a function of temperature for longitudinal (L) and circumferential (C) directions of crack propagation, is shown in Fig. 11.45. The yield strength of the material is 100 ksi, and the specification criterion, denoted by (SC), is given as 60 ft-lb. The upper curve was derived from the circumferential crack propagation under longitudinal stress. The lower curve, on the other hand, represents a longitudinal crack propagation under a conventional hoop stress. Both curves were obtained from the 7.5 mm test specimens.[21] The specification criterion (SC) indicated as 60 ft-lb provides assurance that a through-thickness crack will not propagate catastrophically at elastic stress levels. Both transitions of the DT and CVN

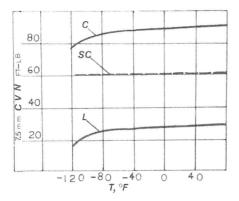

FIGURE 11.45 CVN energy for 4130 steel (100 ksi yield) as a function of temperature. (C), (SC), and (L) have no dimensions and merely serve as diagram markers.

energy (Figs. 11.43 and 11.45) indicate clearly that the material studied was on the upper shelf at temperatures as low as $-80°F$.

Continuous material studies and the related experience were shaping materials selection philosophy in terms of the fracture mechanics parameters and their impact on design of underground systems. Certain projects of national importance had primary concerns for safety and reliability driving the technical and operational activities, at times in conflict with schedules and economy. This was unavoidable, however, because of the involvement of a new science, fracture mechanics, in day-to-day operations. The applications of linear elastic fracture mechanics (LEFM) were particularly slow in coming in the various branches of industry affecting the procurement and certification of new and the well-established engineering materials. The obvious choice, then, was to select and maintain certain "workhorse" types of materials and design procedures that would represent a minimum of conflict with the fracture control philosophy.

In the forefront, of course, was the high-yield family of structural steel in the yield range of 80–130 ksi,[18] which was particularly suitable for welding thick components and developing impressive forgings. Regardless of composition and heat treatment, the modulus of elasticity and weight density remained fixed at 30×10^6 psi and $0.284 \, lb/in.^3$, respectively. The important mechanical properties included good ductility, toughness, and resistance to low-cycle fatigue. The role of good ductility is to improve metal formability, and resistance to fracture at a local stress raiser and deformation under dynamic conditions. The minimum acceptable elongation for HY-80 steel has been established at 20%. The transition zone is similar to that of a typical structural steel because of the ferritic microstructure. Steels with high upper-shelf toughness tend to absorb energy by deformation, while low upper-shelf toughness signifies the material's susceptibility to slight deformation and longer crack extension.

The developments in high-yield technology, however, had some constraints. For example, an upper limit of 0.18% for carbon assures good weldability. The phosphorus limit of 0.04% reduces detrimental effects on ductility and toughness. The limit for sulfur is 0.05% in order to minimize the formation of iron sulfide, which liquefies under normal rolling and forging temperatures. On the other hand, adding manganese forms manganese sulfide, which has a higher melting point and acts as a strengthening agent for the ferrite–pearlite matrix. The upper limit of manganese in this regard is 1.3%. Carbon and manganese increase the tensile strength, while normalization enhances the toughness.

The "workhorse" type of high-yield technology is still HY-80, which has, however, a few limitations that should be kept in mind. The final tempering temperature should not be less than 1100°F and the microstructure at midthickness of the plate should contain at least 80% of martensite. The use of this material should also be matched by developments in electrode materials for reliable welds. This is not an easy task because of the variable conditions of structural constraints, joint preparation, weld accessibility, cooling rate, transformation, and shrinkage. In general, the chemical composition and microstructure of electrodes have a strong effect on the properties and performance of the weldment. In order to provide the maximum resistance to weld cracking, the electrode carbon should be limited to 0.10%, with the phosphorus and sulfur peaking at 0.03% each. The maximum silicon content should be 0.60% in order not to decrease the resistance to hot cracking. Other limits include molybdenum 0.75%, nickel 1.0%, and vanadium only at 0.2%. The limitations for the electrode coatings have a 0.2% maximum water content by weight and 0.1% for the submerged arc welding process. It is a good practice to select the weld metal to overmatch the strength of the base metal, as well as to match the ductility and toughness. However, it is very difficult to match fracture toughness of HY-80 steel.

One of the major problems in welding is cold cracking. To minimize this concern, "low-hydrogen" ferritic electrodes have been developed, without which the welding of HY-80 steel would be very costly and perhaps impossible under field conditions. The explanation of cold cracking is complicated and is really in the province of metallurgy rather than conventional engineering.[22] This topic is not a new one and it is destined to drag along for a few more years. Nevertheless, it is well to keep in mind some of the principal variables and interactions surrounding this area of concern with special regard to hydrogen content and internal stresses, such as:

- Restraint and rigidity of structures;
- Internal pressure due to hydrogen;
- Thermal gradients;
- Volumetric changes in metallurgical transformation.

To mitigate welding stress effects, controlled preheating and other techniques can be applied. However, the use of uniform heat may not be quite self-evident because of other possible effects, as roughly illustrated in Fig. 11.46. The problem is that any corrective measures must be related to the weld size, temperature distribution, and control, as well as to shrinkage strain and the cooling rate. All of these elements represent a complex behavior affecting the process of cracking.

Essentially, as the weld size increases, the cooling rate drops, while the shrinkage strain increases. This particular combination should be beneficial, resulting in fewer cracks until the strain becomes dominant. This highly complex behavior can also be related, say, to crack density (number of cracks per unit area) as a function of heat input, as sketched in Fig. 11.47. A similar interpretation is possible if we express number of microcracks per square inch of steel surface as a function of kilojoules per inch of weld. For example, the area of 100 microcracks per square inch might correspond to a heat input of about 30 kJ.

The last topic covered in this brief account of HY materials involves the qualification of fracture toughness of a welded plate using the "explosion

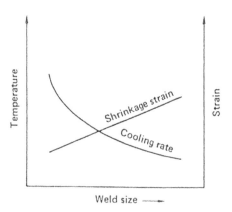

FIGURE 11.46 Shrinkage and cooling of welds in HY-80.

bulge" technique.[23] In this test an explosive is detonated a given distance from the plate (such as shown previously in Fig. 4.13). The explosion produces a uniform force on both supported and unsupported areas of the specimen. The mode of failure indicates whether the metal is brittle or ductile at a specific temperature. The fracture can clearly select the weakest path because the dynamic pressure places equal demand on all portions of the metal. An ideal bulge test can produce a full hemispherical bulge without failure. Investigations utilizing "explosion bulge" have shown that this technique can also be used to determine dynamic

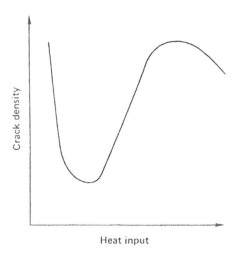

FIGURE 11.47 Effect of heat input on microcracking.

resistance to crack propagation. The lower bound of these experiments is given in Fig. 11.48.

In summary, HY-80 steel has proven to be tougher and more resistant to fracture than any other structural steel in a particular range of strength/weight ratio. Its only disadvantage is the rigorous control required in fabrication.

The application of HY-130 steel to underground hardware was viewed with some doubt because of the potential hydrogen cold-cracking phenomena, requiring stricter fabrication controls during welding and nondestructive testing. For instance, shielded metal arc (SMA) welding and a covered electrode had to be developed for use with HY-130 steel. The HY-130 weldment is more sensitive to stress corrosion, and it shows the fatigue life at comparable stress levels to be about one-third the life of similar HY-80 weldments. Otherwise, HY-130 is adequate for large welded structures, and plate elements can be formed to curvature, retaining a high degree of fracture resistance after cold forming.

In addition to experience with the API and high-yield materials it is necessary to mention such basic materials as 4330 (vanadium-modified) steel at the high range of yield and A537 structural steel covering the lower level of strength. These two additional steels have been selected on the premise of their proven resistance to fracture.

The lower range of the yield was certainly filled by long-standing experience with the normalized version of steel under the full name of ASTM A537 Grade A. The recommended CVN value was 35 ft-lb at 70°F and 25 ft-lb at −75°F, for a plate thickness less than about 0.6 in. The maximum acceptable NDT at the time was −60°F, determined by means of a drop-weight test

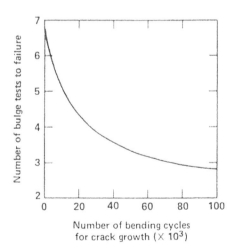

FIGURE 11.48 Bulge technique in testing fatigue of HY-80 steel.

(DWT) according to the existing standard.[24] The dynamic tear (DT) tests conducted on the A537 plate are shown in Fig. 11.49.

Typical mechanical properties for A537 are given in Table 11.3. Essentially, there are two classes of this material, and the maximum plate thickness supplied by industry is 4 in. The ratio of yield to ultimate strength is reasonably low, so that good ductility may be expected. This characteristic alone does not help to predict fracture toughness, and in relative terms it is easier to judge the toughness by making a direct comparison with other materials.

One of the more sophisticated materials covering the higher level of strength is 4330 (vanadium-modified) steel produced from ingot, forging, or bar configuration. Its production requires fine grain, fully killed, and vacuum degassed electric furnace quality to give the properties shown in Table 11.4. The ultimate strength of this material is seldom specified, although it should be on the order of 140–170 ksi. The NDT temperature was observed to be between −60 and −70°F. In the past experience, critical components such as pins, shafts, axles, and shackles were made from this material, and it is still customary to extend the use of 4330 (vanadium-modified) to hooks and flanges.

It should be stated, however, that there were special cases, in relatively recent practice, of low elongation and CVN values of this material.[25] A metallographic examination showed a significant amount of manganese sulfide stringers oriented in a manner to lower the aforementioned properties. The CVN fractured surfaces indicated some evidence of a large-grain pull-out that may

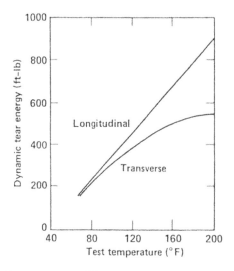

FIGURE 11.49 DT energy for A537 steel.

TABLE 11.3   Strength Characteristics of A537 Steel

	Normalized	Quenched and Tempered
Ultimate strength (ksi)		
Thickness less than 2.5 in	70–90	80–100
Thickness 2.5–4.0 in	65–85	75–95
Yield strength (ksi)		
Thickness less than 2.5 in	50	60
Thickness 2.5–4.0 in	45	55
Elongation in 2 in. (%)	22	22

TABLE 11.4   Properties of 4330 (Vanadium-Modified) Steel

Yield strength (ksi)	120–150
Elongation (%)	16
Reduction of area (%)	40
Minimum CVN (ft-lb)	
Longitudinal	35
Transverse	25

have contributed to lowering the CVN energy. The cases of high sulfide content in the past were resolved by reheat-treating with subsequent retesting.

The construction steel known as A36, although popular in many applications, has never been regarded as a fracture-tough material. The relevant CVN energy level of 15 ft-lb at 10°F is consistent with the bridge steel specification for such materials as A572, A440, A441, A242, and A588,[11] covering the yield range 35–65 ksi. However, in spite of its popularity, any use of A36 in critical applications should be discouraged.

LESSONS FROM STRUCTURAL FAILURES

It is normal to be horrified by unexpected structural failures of catastrophic proportions caused by insidious crack behavior in engineering materials or by plain negligence in design. It is also normal to expect that every incident of fracture and destruction is an opportunity to learn about all quirks of nature and the practice of humans dealing with natural disasters, large or small. However, history is full of examples showing that the pace of learning is alarmingly slow for many scientific or economic reasons, and that at times we never learn. And sometimes the pace is slow because free enterprise is protected by proprietary rules and because we have to survive in a litigious world. A typical example can be quoted from past records concerning the failure of a high-pressure gage, which caused lost-time injury. When the user sent a "speed message" regarding the incident to the manufacturer, the "speed reply" was simply: "we will not give out the information on the heat treatment and fabrication of our product since we consider this confidential proprietary information."

The purpose of this section is to discuss a few cases related to structural failures in order to gather some information on material and crack behavior as a practical matter of helping in design. The first general problem is concerned with the slow learning process for scientific reasons in the area of hydrogen embrittlement.[26]

Hydrogen affects structural metals by reducing the strength and the capability of plastic deformation. Embrittlement problems continue to occur in welded structures because of a quick diffusion process at normal temperatures, and even after thousands of investigations this problem is poorly understood. Large internal blisters and cracks can be developed in steel as a result of sulfide corrosion, and ferrous metals in general become embrittled under stress. Many hydrogen problems occur during the manufacture of iron and steel when the hot metal cools and transforms. Most molten metals readily absorb large quantities of hydrogen, creating a problem with large steel castings, although this process can be mitigated with vacuum degassing equipment. Other sources of hydrogen involve inorganic or organic electrode coatings, or the storage of chemicals such as is the case in the petrochemical industry. The latter is the most damaging and it represents irreversible embrittlement, sometimes referred to as "hydrogen attack." To improve the resistance of low-alloy steels to this attack, additions of titanium and vanadium can be very effective. Also additions of chromium and molybdenum in somewhat larger quantities are helpful.

Especially insidious and spectacular is the so-called delayed failure or static fatigue that occurs when a part is operating for long periods of time under relatively small stresses until a sudden fracture takes place. High-strength steel structures are particularly susceptible to this type of behavior. This behavior can also be induced by transformation, quenching, or residual stresses. Research indicates that a brittle crack can actually grow during "static fatigue," and that the failure occurs when the ligament of the specimen can no longer support the service stress. Plane crashes have been traced in the past to fractures of high-strength steel landing gear. Also, the industry in general reported failures of high-strength and high-hardness parts such as springs, lock washers, and bolts, after a longer period of service.

The classical Griffith theory[27] originally proposed for fracture in glass filaments can be extended to high-strength steel, embrittled by hydrogen, assuming the presence of rather small cracks. The Griffith stress depends on the "square root of surface energy." The effect of hydrogen in promoting the brittle behavior is to decrease the surface energy and to allow crack growth at lower stresses. This is only one of several theories of the hydrogen effect. Unfortunately, the mass of experimental data supporting various theories, from Griffith to dislocation, is apparently contradictory, because of the enormous complexity of the many interacting parameters.

The foregoing brief summary of the nature of hydrogen embrittlement is based on the state-of-the-art information compiled a quarter of a century ago, and one would normally assume that many lessons have been learned in the interim on how to avoid any major disasters. As recently as seven years ago, the National Safety Council used, as a warning about hydrogen menace, the case of a catastrophic failure of a large steel tank near Chicago, with disastrous results. Apparently, the tank had been built and operated properly, and it was inspected regularly until the failure in 1984. According to the investigators, hundreds of similar tanks are in operation all over this country and the world, and many contain "corrosive hydrogen sulfide." The failed tank was 60 ft high and $8\frac{1}{2}$ ft in diameter, with 1 in. thick steel walls. The pressure in the tank was 200 psi, with a stress at failure of about 10 ksi, suggesting a high factor of safety even for a "garden variety" steel such as A36. Since a lot of energy was still contained just before the explosion, and since the vessel's rupture released a sudden vaporization of the liquefied gases, the top portion of the tank, weighing 20 tons, landed more than half a mile away, creating havoc, while the remaining portion of the tank and the facility were consumed by fire.

Chemical and mechanical tests have shown that the tank materials exceeded specifications and that the welds were stronger than the base metal. Extensive cracking was evident in the HAZ near the repair welds, which were susceptible to hydrogen damage. One of the main cracks propagated through nine-tenths of the wall and eventually continued to grow right around the tank. The final, near instantaneous fracture was triggered because of hydrogen embrittlement. It is quite obvious from this finding that unusual care should be exercised in making welded repairs and modifications to these vessels because field welding invites susceptibility to hydrogen damage. In view of the past experience with the theory and practice related to hydrogen cracking, it has to be concluded that the "lessons" aspect of this type of incident continues to be hampered largely by economic factors.

The Chicago case also brings into focus the degree of theoretical difficulty in dealing with the unusual aspects of pressure vessel failure, which may or may not have received full attention from researchers and experts in fracture mechanics. There is no easy way for the design practitioner to develop a simple approach to a problem if the solution requires, say, a new or modified theory related to fracture. A rather splendid example of such a situation is provided by the paper of Finnie et al.[28] dealing with the effect of thermal stresses on the timing of initiation and propagation of fracture, such as experienced in the Chicago disaster. According to this work, a large drop in temperature caused by the expansion of the propane–butane fluid and high surface heat transfer resulted in thermal stresses in the tank, which increased the stress intensity factor sufficiently for crack growth. The time sequence required for initiation and propagation of the through-wall crack was shown to be in agreement with the field observations. The delay between the two phases of this process was 15 min, as noted in the original records of the investigation. The study has not only shown the theoretical justification of the thermal effects, but it has also provided a practical explanation of the sequence of events of the tank failure for engineers and analysts. This is a rather unique and welcome contribution.

The next case is concerned with a high-pressure gage failure; the tubular part of it is shown in Fig. 11.50. The investigation was started with visual examination of the ruptured tube, and the basic assumption was made that high-pressure hydrogen was present. At the same time, a metallurgical analysis was present. At the same time, a metallurgical analysis was initiated to see if the preliminary investigation and tentative conclusions were on the right track.

There were a number of observations concerning this incident that did not point directly to the immediate impressions. Gas analysis three months after the failure indicated limited presence of hydrogen, but indicated that this could have diffused out of the steel rather quickly at room temperature. Other high-pressure gauges of the same design have been in hydrogen service for several years without a problem. The gauge that fractured at 30 ksi was actually proof tested at 75 ksi at the manufacturer's plant. Also, the tube's bore was noted to be concentric.

The macrostructure is indicated as location "1" in Fig. 11.50, with the "chevron" markings pointing to the origin area of the crack. The chevron markings along with the shear lip suggested that the steel known as 431 stainless had some ductility. There was also a high-stress area at the opposite side of the tube, marked as location "2."

The extensive analysis of microstructure yielded a number of metallurgical observations and provided a basis for a comparison of the various tube samples. It was concluded that the fractured tube was improperly heat treated and had rather low resistance to cycling loading. Most likely, the absorbed hydrogen contributed to the failure. This particular high-pressure tube was made of very poor-quality material, and the rupture could have occurred with any fluid in the identical environment.

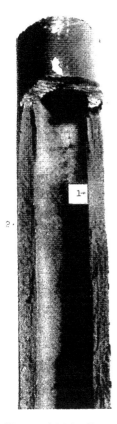

FIGURE 11.50 Part view of failed pressure gage tube.

The basic lesson here was that all pressure systems using high-pressure tube gages (known as bourdon tubes) should be designed with the philosophy that the failure of the pressure gage will take place. The tube must be made to appreciably deflect in operation and to survive a number of stress cycles higher than those in any other part of the system. Finally, off-the-shelf high-pressure gages should only be employed with nonhazardous fluids. This leads to the second case of failure of bourdon-type tubing made of stainless steel.

Service of austenitic 300 stainless steel in a hydrogen atmosphere is normally considered to be quite satisfactory. However, this case is concerned with a 316 stainless steel tubing (bourdon type) that was designed to work in a hydrogen environment under high pressures and to be subject to possible local effects of plastic deformation. The mechanical properties of the material were certified as follows.

Ultimate tensile strength 95–125 ksi
Tensile yield strength 65 ksi (minimum)
Elongation (minimum) 18%

The ratio of yield to ultimate strength was between 0.5 and 0.7, indicating good ductility to take care of local plastic effects. However, the entire problem was not straightforward. The microstructure indicated a significant difference of the character of the surface between the outer (OS) and inner (IS) locations, as shown in Fig. 11.51. The outer edge had shown the normal cold-worked, equiaxed grains, while at the inner edge the cold-worked grains were elongated in the radial direction. There was also clear evidence of longitudinal cracking driven by the hoop stress at the inner surface (IS). The cracks eventually progressed to about two-thirds of the way through the wall before the final catastrophic failure. The sequence of the mechanism of failure was then becoming a little clearer, although the cracking of plastically deformed stainless steel in a hydrogen environment in general was still in question.In conclusion on this case, it was reasonable to assume that the normally used sinking operation created laps or folds of the material on the inside surface of the tube, which acted as stress concentrations leading to localized yielding. If at the same time the protective oxide layer was broken, the hydrogen atmosphere would establish a clear passage to the now exposed grains. Although rare, this was a clear case of hydrogen embrittlement of stainless steel.

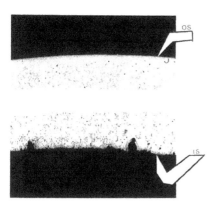

FIGURE 11.51 Surface effects in stainless steel bourdon tube (150×).

The next case in the category of incidents caused by a steel fracture is concerned with developing certain criteria for existing equipment and structures, in order to prevent a potential failure. Such a problem is often very difficult because of insufficient design and inspection data. Many vessels, for instance, are operating without any established limits of service temperature and with little or no information related to the basic parameters of fracture mechanics. The design information is often of the proprietary nature; or it simply does not exist. In either case, the responsible engineer is still required to develop a rationale for his recommendations in the form, say, of limiting the temperature and operating pressure.

The particular case in point involved three large gas storage vessels, made from Kaisaloy 3 steel, having an ultimate strength of the order of 70 ksi. The vessels were used to store helium and represented a large amount of bottled-up energy. Although the information was sketchy, it was agreed that the chemical and mechanical properties of Kaisaloy 3 would be similar to those of A537, Class 1, steel. This type of steel was known to be used for auto bumpers, trucks, tractors, bridges, railroad cars, and earth-moving equipment. There was no stated application in the pressure vessel industry, and the available information included a special comment that the elongation and reduction in area might vary with thickness and rolling temperature. The thickness of the Kaisaloy 3 cylindrical section was 3.18 in., and, for conservative reasons, the thickness of the vessel heads was not used in the course of final deliberations.

A few years prior to the review of helium storage vessels the NDT temperature for the A537, Class 1, steel was established to be 0°F, using 1 in. and 2.5 in. thick material. Although these tests were limited in scope, this was the only information on the nil-ductility temperature available at the time. Using, then, 0°F as the NDT and following the rules of the American Society of Mechanical Engineers (ASME) Pressure Vessel Code for the lowest permissible service temperature, consistent with the 3.18 in. thick vessel, the minimum operating temperature was selected as +40°F.

The nominal applied stress for the case at hand was established in relation to the recommended factor of safety. For a non-ASME-code vessel, containing a large amount of stored energy, the safety factor should have been at least equal to 4. Hence the applied stress became

$$\sigma = 70/4$$
$$= 17.5 \, \text{ksi} \quad (121 \, \text{MPa})$$

The next point of contention was the level of fracture toughness, which was taken as $K_{Id} = 62$ ksi (in.)$^{1/2}$, utilizing the work of Pellini.[29] Hence, using the basic LEFM relation, the critical crack size a_{CR} was estimated from

$$a_{CR} = 0.32\left(\frac{K_{Id}}{\sigma}\right)^2$$
$$= 0.32(62/17.5)^2$$
$$= 4\,\text{in. (102 mm)}$$

Since $4 > 3.18$, the critical flaw has met the leak-before-break criterion, and $+40°F$ was recommended as the minimum operating temperature for the helium storage tanks. There was, however, a possibility that below $+40°F$, a brittle fracture could develop at stresses around $0.5\,S_y$ or less. In such a case, a factor of safety of 4 would not be enough.

The main lesson to be learned from this case is that recertification of existing equipment or structures presents an extremely difficult problem dealing with the three legs of the "discipline triad" — materials, fracture mechanics, and stress analysis — with very limited support of pertinent experimental data.

As a follow-up on the statement of general importance of the experimental data in any investigation or any design function, it may be of interest to refer to a study dealing with a well-known and established reputation of the fracture tough material designated as HY 130 steel. This study was undertaken to compare the test values of the DT, CVN, plane-strain fracture toughness K_{Ic}, ultimate strength in tension S_u, yield strength S_y, and elongation with some of the requirements cited in conventional material specifications. The tests were conducted in accordance with the ASTM procedures: at room temperature and $0°F$. All the properties were recorded for the test samples from the longitudinal and the transverse rolling directions. The reason for distinguishing between the two directions was explained as the potential existence of elongated stringers of impurities that could significantly reduce the fracture toughness and elongation of the material. A summary of the average test results from this study is given in Table 11.5. It is of interest to note that the conventional mechanical properties were relatively unchanged between the various test conditions. However, the situation with the fracture mechanics parameters was quite different, particularly with respect to the rolling direction. According to the established material specifications, the CVN value at $0°F$ should have been 55 ft-lb of energy, and certainly not less than 40 ft-lb. The expected DT energy at $0°F$ was 500 ft-lb. The tests indicated significantly higher CVN and DT results, for the test samples consistent with the essentially axial direction. At the same time, the tested material did not quite conform to the material specification in the transverse rolling direction, due to the anticipated effects of elongated stringers of impurities. The study

TABLE 11.5 Summary of Average Test Values for HY 130 Steel

Condition	DT (ft-lb)	CVN (ft-lb)	K_{Ic} (ksi in.$^{1/2}$)	S_u (ksi)	S_y (ksi)	Elongation (%)
Transverse						
0°F	274	51.5	181	149	140	16
Room temp.	617	59.2	182	146	137	16
Axial						
0°F	1285	118	323	149	138	22
Room temp.	1330	125	315	146	136	21

Source: Lawrence Livermore National Laboratory.

also points to the fact that even with careful test procedures, it may be difficult to zero in on the correct magnitudes of fracture parameters. This tends to make the task of recertification of existing equipment and structures somewhat problematical.

The last case selected for this chapter deals with the failure investigation of a pressure vessel system used to conduct experimental studies of rock behavior under very high confining pressures. A part view of the vessel is given in Fig. 11.52. The recorded pressure at failure was 76.5 ksi. The material was designated AISI Type S5, with a yield strength of 176 ksi, an ultimate tensile strength of 200 ksi, and an elongation of 13%, at a Rockwell hardness of 44. When the hardness goes to about 51, the corresponding mechanical properties become 225 ksi, 257 ksi, and 9%, respectively. The strength ratio for both cases of hardness is 0.88.

The approximate location of crack initiation (CI) in Fig. 11.52 points to the onset of the threaded area of the main cylinder, also shown in the approximate

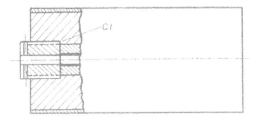

FIGURE 11.52 Part view of pressure vessel system.

sketch of Fig. 11.53, including the manner of loading and the details of thread design. The pressure envelope is shown by arrows, and the resulting axial load due to internal pressure is transmitted by the cylindrical closure by means of a threaded joint. The dashed lines indicate the relevant position of the closure and the plug during normal operation of the pressure vessel system. The thread detail in Fig. 11.53 belongs to the main cylinder.

It soon became obvious that the fracture must have been initiated near the root of the first thread, since in a typical threaded joint of a conventional design the lion's share of the load is carried by the first two or three threads.[30] However, the exact location and the size of the crack responsible for initiation of the fracture was difficult to determine. According to the best guess based on the metallurgical examination, the initiation area appeared to be about 0.010 in. in diameter, so that for a yield stress of 175 ksi (correspondingly roughly to a Rockwell number of 44), the apparent plane-strain fracture toughness can be

$$K_{Ic} = 175(\pi \times 0.005)^{1/2}$$
$$= 22\,\text{ksi}\,(\text{in.})^{1/2} \quad (24\,\text{MPa}\,\text{m}^{1/2})$$

The extremely low fracture toughness indicated that the S5 steel was very notch sensitive, consistent with the observations that the fracture was very flat and it was difficult to locate the initial crack. The flat fracture also signals the limited energy absorption during crack propagation.

A number of useful conclusions and lessons learned from this case study can now be summarized:

- S5 steel is not a fracture tough material and can only be used for shrink-fit pressure vessel liners in the manner specified for tungsten carbide or a similar low-toughness material.

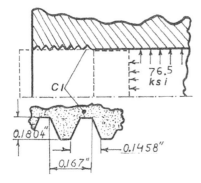

FIGURE 11.53 Details of loading and thread design in pressure vessel.

- Existing S5 vessels of similar design and under similar service pressures can only be operated remotely and from approved man-rated enclosures.
- Design of such vessels should include estimates of service performance based on the established parameters of fracture mechanics and stress analysis.

It should be stated for the record that the metallurgical examination of the S5 steel indicated a carbon content of 0.55, manganese 0.80, silicon 2.00, and molybdenum 0.40%. The microstructure contained significant amounts of ferrite resulting from the particular heat treatment, suggesting a degraded performance in comparison with the properly quenched and tempered structures. The proper applications of this material, however, can include punches, chisels, hammers, pneumatic tools, stamps, concrete breakers, and shear blades.

The information on specific case studies and the related experience presented in this chapter so far has been concerned with typical failures of engineering structures caused by the initiation and propagation of brittle fracture of a material such as steel. The parameters of LEFM such as flaw size, fracture toughness at service temperature, and the applied nominal stress are normally considered in modern design and failure analysis, where the main objective is to develop fracture control plans and procedures to provide assurance of structural integrity of engineering products. In other words, the principles of fracture mechanics, materials science, and stress analysis are utilized directly to monitor and minimize some of the self-destruct tendencies of conventional structural materials.

A special case related to brittle behavior of materials is taken here from the field of biomechanical technology dealing with the selected features of microsurgical blading intended for the incision of the plaque layer inside the artery.[31] This is indeed a special situation where the mechanism of brittle fracture is induced intentionally for the purpose of lowering the pressure loading during the dilation of the artery. In essence, the artery as a whole can be modeled as a pressure vessel consisting of an inner, brittle cylinder representing the plaque and the outer cylinder, which is the healthy part of the artery showing elastic behavior and significant elongation under stress. We therefore have a nested cylinder, which requires a total higher pressure for the dilation process and which is a function of the geometrical and physical properties. Hence, if the structural integrity of the plaque cylinder can be downgraded, the resultant internal pressure for the dilation can be lowered, minimizing, in medical terms, the trauma. This is essentially accomplished by the radial motion of the microsurgical blade, which creates a notch across the field of hoop tension. In a more general sense, the highest stress concentration can be expected when the notch depth is large while the corner radius and the notch angle are small. The approximate shape of the notch and location inside the artery are given in Fig. 11.54. The features of the plaque cut should be similar to those of a fine saw cut, where the notch dimensions can be of the order of 0.100 in. in length, 0.020 in. in depth, and 0.010 in. wide, without specifically defining the corner radius μ.

The notch effect can be described as follows. Because of the presence of the notch, the relevant hoop stress is markedly increased. For a brittle material the high tensile stress can lead directly to failure by sudden fracture due to separation of atoms, resulting in cleavage. This type of fracture behavior is certainly consistent with the brittle nature of plaque, particularly in the case of a very small radius μ, which is expected to be 0.000125 in. The conventional stress concentration factor can be plotted as a function of the h/μ ratio, as shown in Fig. 11.55. When $\mu = h$, and μ represents a semicircular notch, the factor k tends to a standard theoretical value of 3. The use of a rather small notch radius μ leads to the establishment of a triaxial stress system in the plaque in contact with the cutting edge. The crack created by this mechanism develops a local plastic zone and advances ahead of the cutting edge, destroying the fracture integrity of the plaque cylinder. The triaxiality of the stress field alone contributes to brittle behavior even in a ductile material.

The foregoing study assumed the nested cylinder configuration to be of concentric geometry. In the real world, however, we may have a partially blocked artery, which can be modeled with the help of eccentricity e, according to the sketch in Fig. 11.56. This dimension defines the shift between the two centers of curvatures assuming two circular patterns. The ratio of the maximum hoop stress σ_h at point (A) to the dilating pressure P can be stated as

$$\left(\frac{\sigma_h}{P}\right) = \frac{2R_o^2[R_o^2 + R_i^2 - e(2R_i + e)]}{(R_o^2 + R_i^2)[R_o^2 - R_i^2 - e(2R_i + e)]} - 1 \tag{11.6}$$

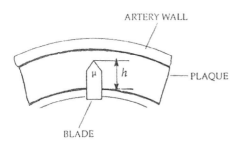

FIGURE 11.54 Notch shape and location in artery plaque.

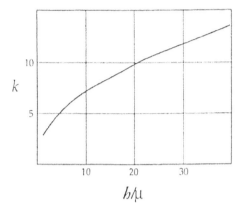

FIGURE 11.55 Stress concentration factor at small radii in plaque.

When e tends to zero, Eq. (11.6) reduces to the classical Lame formula. Also the geometry restricts the degree of eccentricity to

$$e < 0.5R_i \tag{11.7}$$

The argument for increasing the notch effect to encourage brittle behavior should apply equally well to the concentric and eccentric models.

As shown by the various case studies, application examples, and the related experience gathered in this chapter, there are few limitations on the use of practical fracture mechanics. The potential field of applications is staggering in its

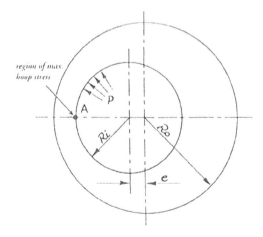

FIGURE 11.56 Model of eccentric blockage by plaque.

variety, from the broken hull of the *Titanic* to plaque in a human artery, where the mechanism of fracture and the concepts of stress will always be a matter of extra-ordinary subtlety and challenge.

SYMBOLS

a	Crack size, in. (mm)
B	Material thickness, in. (mm)
CVN	Charpy V-notch, ft-lb (mm-N)
d	Depth of net section, in. (mm)
DT	Dynamic tear, ft-lb (mm-N)
$(DT)_E$	Dynamic tear (equivalent), ft-lb (mm-N)
e	Eccentricity, in. (mm)
FTE	Fracture transition elastic, ksi (MPa)
FTP	Fracture transition plastic, ksi (MPa)
h	Depth of notch, in. (mm)
k	Stress concentration factor
NDT	Nil-ductility transition, °F
P	Internal pressure, ksi (MPa)
P_{RF}	Plastic resistance factor, ft-lb (in.)$^{-2.5}$[(mm-N) (mm)$^{-2.5}$]
r	Average radius in. (mm)
R_o	Outer radius, in. (mm)
R_i	Inner radius, in. (mm)
S_u	Ultimate strength, ksi (MPa)
S_y	Yield strength, ksi (MPa)
t	Wall thickness, in. (mm)
ε	Elongation
μ	Notch corner radius, in. (mm)
σ	Applied nominal stress, ksi (MPa)
σ_1, σ_2	Principal stresses, ksi (MPa)
σ_e	Equivalent uniaxial stress, ksi (MPa)
σ_h	Hoop stress, ksi (MPa)

REFERENCES

1. Pellini, W.S.; Puzak, P.P. *Fracture Analysis Diagram Procedures for the Fracture-Safe Engineering Design of Steel Structures*, NRL Report 5920; U.S. Naval Research Laboratory: Washington, DC, 1963.
2. Babecki, A.J.; Puzak, P.P.; Pellini, W.S. *Report of Anomalous "Brittle" Failures of Heavy Steel Forgings at Elevated Temperatures*; ASME Publication: New York, NY, 1959; Paper 59-Met-6.
3. Greenberg, H. How to distinguish brittle from tough steel. Met. Progr. **1957**, *71* (6).

4. Lemcoe, M.M.; Pickett, A.G.; Whitney, C.L. *Cyclic Pressure Tests of Large Size Pressure Vessels*, Progress Reports, Project No. 733-3; Southwest Research Institute: San Antonio, Texas, 1962.

5. Kooistra, L.F.; Lemcoe, M.M. Low cycle fatigue research on full-size pressure vessels. Weld. J. **1962**, *41* (7).

6. Blake, A. *Fracture Control and Materials Technology for Downhole Emplacement*, UCRL-53398; Lawrence Livermore National Laboratory: Livermore, CA, 1983.

7. Pellini, W.S. *Principles of Structural Integrity Technology*; Department of the Navy, Office of Naval Research: Arlington, VA, 1976.

8. Robertson, T.S. Propagation of brittle fracture in steel. J. Iron Steel Inst. (London) **1953**.

9. Brown, W.F.; Srawley, J.E. *Plane Strain Crack Toughness Testing of High Strength Metallic Materials*, ASTM Special Technical Publication, No. 410; American Society for Testing and Materials: Philadelphia, 1967.

10. Lange, E.A. Fracture toughness measurements and analysis for steel castings. AFS Trans. **1978**.

11. Rolfe, S.T. Fracture and fatigue control in steel structures. Eng. J. Am. Inst. Steel Construction **1977**.

12. Lange, E.A. *Dynamic Fracture-Resistance Testing and Methods for Structural Analysis*, NRL Report 7979; Naval Research Laboratory: Washington, DC, 1976.

13. Landon, P.R. *Metallurgical Report on the Broken Drill Shank from the Nevada Test Site*, Internal Report, PMD Note No. 155; Lawrence Livermore National Laboratory: Livermore, CA, 1965.

14. Blake, A. *An Experimental Investigation of Abrasion of Metals in Sea Water*, Laboratory Report No. 578; British Electricity Research Laboratories: Leatherhead, England, 1955.

15. Landon, P.R. *Deetz Broken Crane Hook*; Private communication, 1966.

16. Lange, E.A. *Fracture Toughness of Structural Metals*, NRL Report 7046; Department of the Navy, Naval Research Laboratory: Washington, DC, 1970.

17. Blake, A. *Design of Curved Members for Machines*; Robert E. Krieger Publishing Co.: Melbourne, FL, 1979.

18. Heller, S.R.; Fioriti, I.; Vasta, J. An evaluation of HY-80 steel as a structural material for submarines. Part I and II. Nav. Eng. J. **1965**.

19. *Oil Country Tubular Products Engineering Data*; ARMCO Steel Corporation: Middletown, OH, 1966.

20. Goldberg, A. *Does J55 Pipe Casing Have Sufficient Toughness*? Internal Report, PMD Note No. 308; Lawrence Livermore National Laboratory: Livermore, CA, 1968.

21. Lange, E.A.; Cooley, L.A. *Fracture Control Plans for Critical Structural Materials Used in Deep-Hole Experiments*, NRL Memorandum Report 2497; Naval Research Laboratory: Washington, DC, 1972.

22. Voldrich, C.B. Cold cracking in the heat-affected-zone. Weld. J. **1947**, *12* (3).

23. Hartbower, C.E.; Pellini, W.S. Explosion bulge test studies of the deformation of weldments. Weld. J. **1951**, *30* (6).

24. *Standard Method for Conducting Drop-Weight Test to Determine Nil-Ductility Transition Temperature of Ferritic Steels*, ASTM E-208; American Society for Testing and Materials: Philadelphia, 1975.

25. Landon, P.R. *4330V Material Specification*; Private communication, 1982.

26. Rogers, H.C. Hydrogen embrittlement of metals. Science **1968**, *159* (3819).

27. Griffith, A.A. The phenomena of rupture and flow in solids. Phil. Trans. Royal Society **1920**, *221*, 163–198.

28. Finnie, I.; Cheng, W.; McCorkindale, K.J. Delayed crack propagation in a steel pressure vessel due to thermal stresses. Int. J. Pres. Ves. Piping **1990**, *42*.

29. Pellini, W.S. Analytical design procedures for metals of elastic–plastic and plastic fracture properties, Part II. *Proceedings of Joint U.S.–Japan Symposium on Application of Pressure Component Codes*, Tokyo, Japan, 1973.

30. Blake, A. *Design of Mechanical Joints*; Marcel Dekker: New York, 1985.

31. Blake, A. *A Mechanical Evaluation of the Dilatation of Atherosclerotic Disease with the Barath Surgical Dilatation Balloon System*, IVT Technical Report Series, Vol. 2, No. 2; Interventional Technologies Europe: Republic of Ireland, 1993.

12

Introductory Fracture Control

HISTORICAL PERSPECTIVE

It is probably not necessary to go back to the Viking ships and the American horse buggy to realize that flexibility and amount of resilience are the essential elements of structural quality. Without these features the structure is simply not able to absorb the energy of external forces. Of course, the structure can be grossly overloaded and then all bets are easily off, especially when a flaw or a crack is present. The material breaks, usually in tension, as the crack spreads across the direction of loading and the stored-up strain energy is potentially available to propagate this crack. Such a statement, of course, suggests that we are dealing with the self-destructive mechanism of the material because of the presence of strain energy in a resilient structure. The only question remains is what portion of the stored energy is directly converted into fracture energy. And it is not surprising that modern fracture mechanics is more concerned with the conversion of energy into fracture than with the traditional forces and stresses acting on the structure. This situation, however, points clearly to the complexity of the fracture control process, where any oversimplification of this problem may be dangerous. As the conventional design principles, developed over many years of practical experience, the history of fracture mechanics teaches that any fracture control options depend heavily on the engineering judgment in combining the three main branches of technical knowledge involving materials science, fracture mechanics, and stress analysis. The term "discipline

419

triad" used in Chapter 11 was simply intended to emphasize that the foregoing three elements are of equal importance and comparable complexity. While we have to live with cracks and stress concentrations, the triad of knowledge may be the best source of methodology for controlling the menace of fracture. It is, however, the process of integration of the specialized fields that requires many book volumes to cover the entire topic of fracture control. This chapter can hardly do more than scratch the surface.

History shows that imperative needs often drive the development of new branches of science and engineering, and the case of fracture control is no exception, although it took more than a century to crystallize the knowledge of brittle behavior and the basic characteristics of possible measures of preventing or minimizing the brittle fracture. The catastrophic failure of a structural material usually takes place prior to plastic deformation, and the crack propagates at very high velocity, leaving a flat, fractured surface. The occurrence of brittle fracture may be less frequent than that of fatigue, yielding, or buckling, but it can be more destructive, particularly in engineering systems such as tanks, pressure vessels, ships, bridges, or airplanes.[1]

Some of the earliest reports dealt with structural failures in the late 1800s such as a 250 ft high steel standpipe during a hydrostatic acceptance test, riveted gas holders, or tanks for holding water or oil. What was especially disappointing was that the materials used in the construction had met tensile strength and elongation requirements. The most notorious case of the early 20th century was the failure of a giant tank holding molasses, involving a significant number of fatalities and injuries in addition to property damage. The extensive litigation and technical audit that followed concluded that the tank failed by overstress, and observed that the testimony of engineers and scientists was almost equally divided as to the causes and the state of knowledge of brittle fracture. It is difficult to say, even today, whether such a statement is still valid.

In the period before World War II, a number of welded bridge components failed soon after entering service because of brittle fracture. All investigations confirmed that the sudden failures were initiated in defective welds, and Charpy impact tests have proved that the majority of bridge steels at the time were brittle at the prescribed working temperature.

Brittle failures continued in spite of past experience. A large number of ships were built during the war and it was not until an unusual number of ship failures were reported that brittle fracture was finally recognized as a problem of major proportions. By 1946, 20% of all merchant ships developed cracks, some rather extensive, leading to a number of Liberty ships being broken completely in two. The fracture normally started at square hatch corners or square cutouts. Subsequent design changes involved rounding and strengthening of the hatch corners, removing square cutouts, and adding crack arresters in the form of rivets.

It is fascinating to follow the accounts of a sequence of events and the changes in design philosophy of those times because history repeats itself with or without the lessons to be learned. For instance, until the 1940s, metal structures were, in most cases, fabricated with the aid of rivets and bolts. This type of construction assured that at least a local fracture could be isolated from the rest of the structural member, and a total collapse could thereby be prevented. The arrival of the monolithic nature of structures based on a series of weldments was not fully appreciated with reference to crack propagation. In a sense, the structural continuity provided a degree of assurance that even a small fracture initiation could progress without impedance, followed by a sudden split of the entire hull of a ship. Hence, again, it was time to swing back to design improvements based on the theory of crack arresters, restrictions on the chemical composition of ship steels, and general improvements in fabrication techniques, which, alas, were not the complete solution to the problem.

This problem was made more difficult because only a limited amount of data was available relating the metallurgical effects to the fracture resistance of steels. Hence metallurgists needed more time to develop better steels, while designers had no firm basis for correlating the crack size with the stress level among the conditions of fracture initiation. Not enough attention was paid to fracture properties of the welds and there was enough inertia to prevent the abandonment of riveted and bolted structures combined with the ductility criteria. Although experience with the welded ships negated the ductility criterion, the general appeal and value of the riveted and bolted connections should never be underestimated.[2]

While the science of fracture mechanics in general and the applications of linear elastic fracture mechanics (LEFM) were honing the specific areas of fracture methodology from the 1950s through the 1970s, brittle fractures were still occurring at various intervals and with familiar severe consequences. For instance, between 1951 and 1953 two relatively new welded cargo ships and a tanker broke in two. A year later another welded tanker, fabricated from improved steel and designed using up-to-date techniques, suffered a similar fate. Between 1960 and 1965 another 10 failures were recorded and a number of unpublished fracture events, with the special case of a large tank barge (584 ft long), continued through 1972. The last case was unusually disturbing because a one-year-old vessel of this size and design suddenly broke almost completely in half while sitting in port with calm seas. Investigation has shown that the Charpy V-notch (CVN) notch toughness of the material was acceptable, but it was marginal when measured by the dynamic tear (DT) method. However, the primary reason for failure was the overload by improper ballasting. Hence the human factor contributed to the actual brittle fracture.[3]

Although the problems encountered in the shipping industry were in the forefront of attention, the aerospace industry and other manufacturing areas had their share of material failures related to brittle behavior. For instance, two

Comet aircraft failed catastrophically in the 1950s because of very small fatigue cracks initiated from the rivet holes near the window openings. This aircraft failed at high altitude, and subsequent tests included a full-scale simulation of the pressure differential that induced the appropriate stress field in the fuselage. Other aircraft failures involved landing gear and rocket motor cases because of undetected defects or growth of subcritical cracks triggered by fatigue or stress corrosion. Also the manufacturers of special heavy equipment in the 1950s and later years were exposed to several failures of steam turbines, generator rotors, and other components such as those described briefly in Chapter 11. These incidents induced the manufacturers to conduct extensive studies of brittle fracture in order to develop approaches to fracture control.

Barsom and Rolfe[1] selected the case of the Point Pleasant Bridge disaster,[4] which well reflects the concerns of the 1960s and 1970s for structural integrity of countless bridges in this country and elsewhere. The case of Point Pleasant, although at a high price, appears to represent a turning point in the bridge-building industry in the direction of recognizing the necessity of paying attention to brittle fractures in bridges. However, it is hard to understand at times why in so many instances in the modern world we seem to wait for a major disaster before making a preemptive decision.

The foregoing bridge collapse came as a complete surprise and the subsequent investigation of the eyebar suspension chain defined the cause of the bridge collapse as a cleavage fracture in the lower limb of the eye of the eyebar. As expected, later and extensive use of fracture mechanics and metallurgy resulted in a number of conclusions and lessons, which can be summed up as follows:[5]

- The initial crack on the surface of the hole in the eye was caused by stress-corrosion, triggered by hydrogen sulfide under fatigue conditions.
- The growth of the crack to a critical size developed under normal working stress.
- The eyebar steel had very low fracture toughness at service temperature.
- High hardness of eyebar steel was inviting stress-corrosion cracking.
- Close spacing of joint components prevented normal application of antimoisture coating.
- High design load resulted in yield stress level at the inside surface of the eye.

This and other cases of brittle fracture in steel bridges called for corrective measures in fabrication, materials, and design, which were compiled in a special publication.[6] The use of fracture mechanics has also shown that it is not enough to have a specified material with the acceptable level of fracture toughness without paying special attention to the complicated interrelationships between the materials, design, fabrication, and loading environment. These interrelationships are particularly important from the practical point of view in the constantly ongoing process in modern engineering, heavily oriented toward the optimization of performance, safety, and cost. All such elements enter the considerations of the planning and execution of fracture control. This process in itself is sufficiently involved even without any mention of the statistical aspects of the variation of fracture parameters.

More than 50 years ago the overall philosophy of fracture control was based on rather simplistic assumptions of lower allowable design stresses, thinner materials, and riveted (or bolted) plate members, which actually performed as crack arresters prior to the onset of fracture mechanics theory and application. Formal fracture-safety guidelines did not exist, and the majority of failures were not catastrophic. In the boiler and pressure vessel industry, relative safety was obtained by continually decreasing the allowable stresses, expressed as a certain percentage of the maximum tensile stress.

The next practical restriction on design was to eliminate as far as possible the stress concentration in the form of square hatch corners and cutouts such as those found in the original Liberty ships. Such measures, supported by the addition of rivets as crack arresters, were bound to reduce the incidence of failures until the World War II shipbuilding program produced large-scale monolithic structures through welding.

The third general type of fracture control was to improve notch toughness of the materials by the assurance of at least 10 to 15 ft-lb of CVN impact energy at the service temperature, which actually did not prevent crack formation but provided a degree of slowing and arresting of crack propagation. Some success in this area was marked by the establishment of the 15 ft-lb transition temperature criterion, which worked as long as the service temperature did not fall below the transition point. During the 1950s this criterion was modified and strengthened by the development of the nil-ductility transition (NDT) temperature, as indicated in the various chapters of this book. The new criterion of fracture control then specified that the selected material should have an NDT temperature lower than the service temperature.

In historical terms the establishment of NDT was an important point on the learning curve dealing with the general effort to develop the methodology for preventing brittle fractures by zeroing in on improved notch toughness. The role of maturing fracture mechanics has been to give lower stress, minimized stress concentrations, and improved toughness guidelines a quantitative meaning, a rare feature in the maze of theoretical and numerical procedures in modern engineering, which is full of qualitative characteristics. The publications intended as a direct help to design engineers, in the form of practical aspects of fracture mechanics, should include tools based on the transition temperature criteria when dealing with materials selection. Although for many years now the literature has been dominated by the use of classical fracture mechanics as the primary research tool, it is well to keep in mind at least some of the pragmatic statements of Irwin.[7] To paraphrase, the goal of fracture mechanics should be to increase efficiency of the fracture control plans with the minimum toughness requirements, inspection standards, state-of-the-art design methodology, and fabrication quality. Proof testing is especially valuable because it often reflects the fracture failure experience in the proof test instead of in service. Finally, the use of a transition temperature approach rather than fracture mechanics is not a disadvantage provided fracture mechanics methods can be employed in design modifications. What it all boils down to is that fracture mechanics should definitely have practical objectives.

Throughout this brief historical account, the experience with brittle fractures prior to 1940 has appeared to be limited because the majority of larger structures such as tanks, ships, and bridges were essentially held together by rivets and bolts. This still makes a lot of sense when we look at the 100-year-old Eiffel Tower, held in place by 2,500,000 rivets, as a monument to structural integrity and reliability.

A rather bewildering exception came to light about ten years ago when maritime experts studied the photographs of the wreckage of the *Titanic*, which sank in the North Atlantic Ocean on its maiden voyage in 1912.[8] The investigators suggested that brittle fracture was the basic cause of the disaster because the steel had inferior fracture resistance at low temperature, judging by the appearance of riveted hull fragments still resting on the ocean floor. It appears that under the impact of striking an iceberg, the "glass-brittle" steel simply shattered. Under such conditions, any defect exceeding critical size can trigger catastrophic crack propagation at extremely high velocities, sometimes as high as 7000 ft/s.

The basic investigation of the *Titanic* disaster is likely to continue for some time after all the plate sections of the hull are recovered. It is quite obvious that reliance on the riveted construction to arrest crack propagation in the *Titanic* case was not sufficient. The combination of the unprecedented size of the ocean liner and the inferior fracture resistance of the structural steel was certainly too much for the level of structural integrity provided for in the hull design. The tentative conclusions also leave little doubt that a better grade of steel showing a lower NDT temperature would not have fractured at a high rate of speed, thus allowing badly needed time for rescue operations.

ANALYTICAL OPTIONS OF FRACTURE CONTROL AND FAILURE REVIEW

It should not be at all surprising that fracture control can be exercised in numerous ways, depending on the various rules and regulations of a particular organization and the type of structural systems involved. There is little uniformity in the approaches and the goals of control philosophy unless there are specifications and requirements with respect to unique products and agencies such as, for instance, commercial airplanes, military aircraft, shipping industry products, nuclear pressure vessels, chemical plants, railroad equipment, or bridges, to mention a few.

Another set of variations in the area of fracture control comes from strictly technical requirements such as classical fracture mechanics, transition temperature technology, inspection methodology, or statistical fracture mechanics. There is simply no end to variations, and there are no simple solutions to fracture control plans, unless a wealth of experience is backing up the particular product.

A separate and very difficult decision to be made is to weigh the cost of a failure against the expense of developing a fracture control plan and executing the entire process of control. It should also be stated that a fracture control plan is a set of recommendations intended for a given structure, and it should not be extended to other structures without a fully justified reason.

The most rudimentary and pragmatic method of fracture control is to inspect and repair the structural component in a timely fashion. The analytical effort is normally difficult to assess, because the analysis plan may not be in phase with the inspection planning and the intervals of inspection. Broek[9] suggests the following list of options for the implementation process:

- Repair detected cracks during periodic inspections.
- Provide some measure of protection when partial failure occurs.
- Develop a technique for a rational decision to retire or replace the part without actual knowledge of crack presence.
- Provide repair after failure in proof test.
- Remove cracks from the surface at periodic inspections.

The foregoing rational measures are greatly complicated when cracks are simply not detectable by inspection. The term "not detectable" does not necessarily imply that the inspection methodology or the equipment is at fault. Some structures may be so large that the size alone defies inspection, or so complicated in form that the inspection process is not feasible. The second major problem is that many repairs are simply inappropriate, as indicated by some of the case studies in Chapter 11 and by examples in industry, including more recent incidents. The solution to repair may introduce more problems than it solves.

As we follow the entire pragmatic plan, the complexity of action increases because the process of repair demands a layer of analytical studies and we slip into the area of statistical fracture mechanics, which transforms into another option of fracture control. Here layer upon layer of complexity surrounds our effort when the statistics of crack detection begins to affect our decision-making capability. Here it is insufficient to deal with the number of crack detections alone because the true probability of detection is affected by the number of inspectors.

Or, in other words, the true probability number can only be approached with a very large number of inspectors at work, and we do not mean 3 or 5, but 100 inspectors. Clearly, this is not a simple problem to manage, and even the probability of deflection characteristics obtained under laboratory control may not be very relevant.

As we move from one aspect of control to another, the theoretical tools of statistical fracture mechanics arrive at the stage of Monte Carlo techniques, more variables, and more functional relationships that require simplifying assumptions that are bound to twist the physical meaning of the problem no matter how elegant the mathematical statistics may be. This is certainly way too far from engineering applications of practical fracture mechanics.

Certain industries and regulatory agencies have developed rules and requirements addressing damage tolerance criteria on the premise that cracks and minor flaws cannot be eliminated entirely, and that the modern era has developed a new generation of fracture-sensitive materials and structures. In addition to this constraint, many structures operate in hostile environments and extreme temperatures in such areas as offshore platforms, chemical processing plants, nuclear plants, and aerospace systems, to mention a few. The designs are being extended to high-performance materials and high operating stresses requiring improved inspection procedures and, above all, refined analytical tools. The strange feature of the entire field of improvements such as better materials, design techniques, and control procedures is that it invites immediately increased stress levels, weight reduction, and cost cutting, so that a vicious circle of progress–risk–cost continues.

Be that as it may, the analytical options of fracture control exist and the damage tolerance is expected to provide the following parameters:[9]

- Relationship between residual strength and crack size;
- Maximum allowed crack size;
- Crack growth time interval;
- Allowable size of a preexisting flaw;
- Time intervals for inspection, proof test, repair, or replacement.

There are a number of options that can improve damage tolerance. Also, in order to maintain relative safety, fracture control decisions should be based on the length of crack growth time. However, the basic question remains: Which option should be considered when the growth time is too short for practical economic reasons? In order briefly to review this topic, assume the following definitions and symbols:[9] H = growth time of a given crack; a_p = maximum allowed length of a flaw; σ = applied nominal tensile stress; β = Broek[9] notation for a function in stress intensity factor expression for a generic configuration, a number of which are discussed in Chapter 3 of this book.

The use of a material with higher fracture toughness should result in a larger a_p. The effect of this action on the parameter H is expected to be small because most of the service life of the part in question is consumed during the early phase of crack growth. This was shown, for instance, in the Design Problems of Chapter 5. In general, an average reduction in the crack growth rate increases H by a similar amount. Using Broek's[9] example, rate reduction by a factor of 2 calls for H to increase by a factor of 2.

Selecting a more sophisticated inspection procedure should reduce detectable crack sizes and have significant influence on H. This should also call for fewer inspections and have other benefits related to the replacement life and proof test interval.

Assuming that crack growth rates are proportional to a third or fourth power of the stress intensity factor, a modest reduction in stress will result in a significant increase of the parameter H. A modest stress reduction seldom leads to design changes in the structure but it can affect the stress intensity factor. The net reduction in β appears to be as good as lowering the applied stress σ.

Provision of redundance and crack arresters represents a very direct and practical approach to the improvement of damage tolerance. A well-designed structure for this purpose should have multiple load paths, particularly if the stressed members can transfer the load in shear rather than tension. Similarly, a significant improvement can be attained through the incorporation of doublers and crack arresters. All the foregoing features improve the overall design and increase the parameter H. The only problem is that seldom are damage tolerance assessments made during the early design phase.

The concept of damage tolerance reaches a special level of importance in the fields of commercial and military aviation, for obvious reasons of safety and economy. Tolerance here is clearly defined as the ability of the aircraft structure to sustain a certain amount of damage in the form of cracks and to operate without catastrophic consequences until the particular component can either be repaired or replaced. Damage tolerance, of course, can be maintained more easily if we have the option to incorporate fail-safe design features such as multipath loading and crack arresters, requiring significant analytical and design effort. However, it is generally understood that this type of control may not be completely effective if a larger damage is not fully attended to. It is therefore vitally important that repairs are followed with the appropriate inspection procedures, and the analysis of crack growth from a presumed initial crack size throughout the service life of the component. In essence then, the remaining structural integrity of the component must be based upon the probability of fracture. The remaining structural integrity is usually referred to as the "residual strength," consistent with the so-called limit load, which can only occur once in the aircraft life. The crack length a_p corresponding to the residual strength is not the same as the critical crack length defined in the classical fracture mechanics as

$$a_{\mathrm{CR}} = \frac{1}{\pi}\left(\frac{K_{\mathrm{Ic}}}{\sigma}\right)^2 \tag{12.1}$$

In this expression a_{CR} represents one-half the length of the crack. When this symbol denotes the total critical length, we get Eq. (2.20). Unfortunately the customary notation of $2a$ for a crack length in general suggests caution in juggling the factor of 2.

It should be stated for completeness of the entire requirement under the most stringent rules of commercial aviation that the probability of the limit load coinciding with the occurrence of the maximum allowed crack length is extremely small. At the same time, high residual strength and large a_{p} assures long inspection intervals. This should help to achieve a damage tolerant structure at a much lower cost. The reader interested in the detailed comparison of commercial and military requirements for aircraft may wish to consult the material compiled by Broek.[9] Damage tolerance requirements of other regulatory agencies (Lloyds of London, Veritas, American Bureaus of Shipping, or American Society of Mechanical Engineers) are largely of the preventive type and demand little or no analysis.

The use of fracture mechanics and damage tolerance analysis reduced the total number of failures in this century in conjunction with some improvements of design methodology and quality control. Although, at this time, the number of structural failures is relatively low, even with the best analytical techniques, all failures can hardly be eliminated. However, failure analysis continues to be an important element supporting new structural developments.

An indispensable phase of the failure analysis process is a quantitative fractographic evaluation of the fracture mechanism, and it helps to arrive at solutions pointing the way toward the prevention of future failures.

Experience tends to indicate that in many cases the failures are caused by design and fabrication shortcomings, and only a few cases by materials alone. Material selection, for instance, is a design function; hence, the first order of improvement is through design and production, with due input from fracture mechanics technology.

The definition of load-bearing criteria involves the conventional factors of safety (such as, in most cases, 1.5 to 3.0), and therefore the structures would seldom be subject to true overload failure. Cracks often tend to develop in service when design allows notches and discontinuities, and when quality control is at fault. The effects of residual stresses or temperature may not be accounted for in early design, allowing some form of crack initiation and propagation. And in a sense, the presence of a crack may be regarded as a partial failure. Although the crack by itself may not constitute an immediate fracture, the mechanisms of fatigue, creep, or stress-corrosion can easily lead to cleavage or intergranular separation. The insidious part of this mechanism is that the fracture can take place under normal operating stresses. Generally, we do not need to have the maximum stresses under overload conditions for the fracture to occur. Gross abuse, extreme underestimation, or very poor design will seldom combine to create the worst service environment.

It is well to point out that failure analysis, which is a very significant portion of engineering effort, can be obscured by the differences in definitions of brittle fracture used by, say, fractographers, metallurgists in general, or design engineers. Does brittle fracture have to involve some plasticity? And when should the fracture be regarded as ductile? What is the real difference between the fracture of an unnotched bar in tension and plastic deformation confined to the fracture path of the crack? When the overall plasticity is limited, the fracture is normally considered to be brittle and yet the fracture mechanism may appear to be ductile. The only safe way from the purely engineering point of view is to assume that most fractures in service caused by the cracks are brittle, regardless of the subtle differences in describing the mechanism of fracture. At least there is one point difficult to dispute: that the main culprit is a crack or defect of some kind without which the fracture is not expected to occur. And from the pragmatic point of view, it is not the crack per se but the fundamental causes of cracking that should command our attention. These include material and fabrication defects, poor choice of material and heat treatment, as well as questionable design. All these elements, of course, constitute the framework of quality control.

The more obvious causes of defects derived from the manufacturing can include blunt tools, overheating, welding operations, carburizing, nitriding, and surface hardening, which can induce volumetric changes in the surface layer leading to residual stresses that are extremely difficult to assess at the time of design analysis. Another area that compromises the design quality includes hidden stress gradients found in bolted joints where the centrally located bolts transfer minimal loads, causing overload and cracking of the highly stressed regions near the outer bolts. This mode of failure is likely to develop under elastic service conditions much before the plastic deformation can assure even load sharing by the bolts.

The case of secondary displacements is possible in shrink-fit assemblies when, say, the load is transferred from the shaft to the shrunk-on part, which at the beginning of the shaft torque transfer can cause local movement, fretting, and possible fatigue cracks. Similar situations can arise under bolt heads at bolt shafts. The analysis of the origin of crack mechanism is essential in this process, and it represents a rather unique challenge because we often deal with hidden secondary effects, very small defects and cracks in more brittle materials, as well as unanticipated stress and strain fields.

   Knowledge of fracture control would be incomplete without the contribution of fractography, which helps us to understand mechanisms of cracking. Fatigue damage develops under nominal elastic stresses but, as stated previously, a fatigue crack cannot be initiated without plastic deformation, no matter how small and local. The growth mechanism, characterized by a series of blunting and sharpening effects, creates a formation of distinct lines on the fractured surface known as fatigue striations. Since one striation corresponds to one fatigue cycle, the rate of crack growth can be deduced from the measurement of the striation spacing.

   Other lessons of fractography show that a stress-corrosion crack tends to follow the material's grain boundaries because the chemical content of the boundary is different from that inside the grain. A microstructure showing dimples can be utilized at times for a qualitative evaluation of the local stress field, depending, however, on the angle of view. The dimpled surface diffuses light and it appears dull gray, in contrast with the glittering surface of cleavage. The appearance of a cleavage depends on the rate of loading, temperature, and stress. With sufficient plastic deformation to relieve stresses, cleavage is not expected to show up. In some alloys cleavage fracture cannot be induced under the majority conditions of stress and strain. Of special interest in damage tolerance assessment, the cracks transverse to the hydrostatic tension (under plane-strain conditions) indicating a high stress field. The fracture topography (projectional displacements) can be used in conjunction with fracture mechanics analysis to define the size of the crack tip opening displacement (CTOD). Knowing this parameter, the toughness of the material can be calculated from the following set of equations:

$$\text{CTOD} = 2h \tag{12.2}$$

$$(S_y)\text{CTOD} = \frac{K_c^2}{E} \tag{12.3}$$

and

$$K_c = [E(S_y)\text{CTOD}]^{1/2} \tag{12.4}$$

where h = height from surface topography, in. (mm), S_y = yield strength, ksi (MPa), E = elastic modulus, ksi (MPa), K_c = plane-stress fracture toughness, ksi (in.)$^{1/2}$ (MPa m$^{1/2}$), and CTOD = crack tip opening displacement, in. (mm). For instance, taking a measurement from a micrograph of a fractured aluminum surface, $h = 0.00053$ in., yield strength of 65 ksi, and elastic modulus of 10,000 ksi, the fracture toughness follows from the foregoing expressions.[9] Combining Eqs. (12.2) and (12.4) gives

$$K_c = [2hE(S_y)]^{1/2} \tag{12.5}$$

and substituting

$$K_c = [2 \times 0.00053 \times 10,000 \times 65]^{1/2}$$
$$= 26.2 \, \text{ksi (in.)}^{1/2} \qquad (28.8 \, \text{MPa m}^{1/2})$$

Other formulas for fracture toughness expressed in terms of the CTOD parameter are given below.

Plane strain:

$$K_{Ic} = 1.26[(\sigma_f)(E)\text{CTOD}]^{1/2} \qquad (12.6)$$

Plane stress:

$$K_{Ic} = 1.1[(\sigma_f)(E)\text{CTOD}]^{1/2} \qquad (12.7)$$

where

$$\sigma_f = 0.5(S_y + S_u)$$
$$S_u = \text{tensile strength, ksi (MPa)} \qquad (12.8)$$

The combination of fractography and the CTOD parameter provides a useful correlation with the help of a topographic measurement, h.[10]

While the striation counts in fatigue cracking are a source of useful information, there are certain limitations of this fractographic technique. Fractographs show a very small portion of the crack and the photographing process is rather slow, so that in many actual investigations the striation counts are limited to a few dozen fractographs. This type of limited study does not provide sufficient information.[9] However, in a proper use of this technique, where each striation spacing is obtained from an average of numerous measurements, the numerical integration of the rates can be illustrated as a continuous crack propagation curve as a function of the number of fatigue cycles, as it is normally done with the help of linear elastic fracture mechanics (LEFM). An example of a curve based on the fractographic technique is shown in Fig. 12.1.

The elementary design formulas and calculation examples given in Chapter 3 can be used to estimate the toughness from the crack size at fracture in order to indicate the rough order of magnitude of the toughness for the purpose of comparing the estimate with the material specification. If the design stresses are not known, the toughness can be taken from standard handbooks and the nominal stress levels calculated. One way to proceed is to develop the residual strength information, and on this basis the stress at fracture can be selected. The residual strength, of course, must be dependent on the crack size because we are dealing here with the strength of the ligament, and it must therefore be finite. A typical characteristic is shown in Fig. 12.2. It should be added that the ligament portion of the part in question is assumed to be crack-free for the

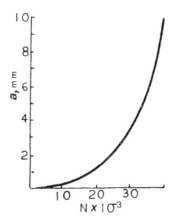

FIGURE 12.1 Crack growth curve (N, number of cycles; from Ref. 9).

purpose of establishing the maximum load the part can carry. The calculation is based on the premise that the maximum stress is consistent with the ultimate strength of the material with or without the design factor of safety (or ignorance). At the same time, crack growth occurs with time, as shown, for instance, in Fig. 12.1. It is also of interest to note that the fatigue striations indicate the direction of fracture propagation, as shown in Fig. 12.3 by the arrow. The chevron marks on the fractured surface represent a high-speed fracture of a material of low toughness with high stress driving the crack.

The information presented so far may be useful in determining the inspection intervals for similar components as a part of the fracture control plan. The analysis for this purpose is based on the load history, applied nominal stress,

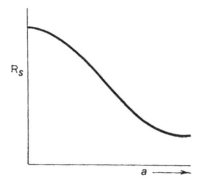

FIGURE 12.2 Approximate residual strength diagram (R_s, residual strength).

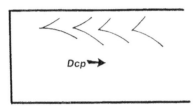

FIGURE 12.3 Chevron marks on cracked surface (Dcp, direction of crack propagation).

and the crack growth rate. A change in loading or environment during the crack growth can have a significant effect.

It is important to recall that the stress intensities are additive as long as the stress systems involved are of the same mode. For more details on the use of fracture mechanics in conjunction with the tools of fractography, the reader is referred to specialized literature.[11,12]

PRACTICAL ELEMENTS OF FRACTURE CONTROL

In order to optimize the performance and safety of modern-day structures within a rigid envelope of economics, design engineers have to start from elementary predictions of service loads and environmental conditions, followed by stress calculations. The next logical step is to compare these results with the potential critical stresses in the more likely failure modes of structural behavior. The structures are finally sized through an iterative process to guard against failure modes that deserve primary attention. Ideally, good structural design demands that all possible modes are evaluated in order to avoid any catastrophic surprises. How close we can get to the ideal is a matter of a number of circumstances such as time, skills, and cost that can be assigned to the design project. The choice of the failure modes and the sequence of modes deserving special attention are not always clear. In a more conventional approach, the general yielding and instability modes may well appear to be of primary importance, until the issues of subcritical crack growth and unstable crack extension under fatigue, stress-corrosion, and a combined effect of these two environments overshadow everything else. The assumption that the proper choice of materials and design stress levels will take care of all the failure modes of any consequence may not always be true, because we are violating the principle of a "discipline triad," which calls for a balance among the skills of materials science, fracture mechanics, and stress analysis in developing fracture control plans and the specific guidelines. And even under ideal circumstances and with the proper balance of skills, fracture control may not always be totally effective.

According to the general philosophy of Barsom and Rolfe,[1] failure by fracture or subcritical crack growth by fatigue in a hostile environment should be selected as the dominant failure modes on the premise that more conventional modes of yielding and instability are also pertinent to the major assault upon the problem of fracture. This approach is certainly consistent with the basic idea of practical fracture mechanics and its relation to engineering design.

The object of fracture control guidelines is to minimize the threat of brittle fracture. The types of elements of fracture control normally included in the plan deal with improving notch toughness and eliminating stress raisers as well as defining welding and inspection techniques. The integration of such elements into the requirements for a specific structure forms the main portion of a format fracture control plan. When such a plan is being developed for a large structure (airplane, bridge, ship), many complex issues and detailed procedures are at stake that present significant difficulties of formulation, interpretation, and implementation. One of the first tasks is to identify the factors responsible for fracture under prescribed conditions of service and their individual contributions to failure. The design methods should be scrutinized as to their role in partial or total assessment of the structural failure. Out of the total exercise, specific recommendations should evolve for the purpose of ensuring the safety and reliability of the structure against fracture. Such recommendations then would include the elements of material selection, design stress levels, fabrication, and inspection.

The extent of useful life of a component is normally defined by the time needed to initiate and propagate a crack from its subcritical state to a critical dimension. Hence the material selected for a particular component will be judged by the characteristics of initiation, subcritical propagation, and unstable crack propagation. The unstable portion of crack propagation, which is the final stage of useful life, depends on toughness, crack size, and the level of the applied stress. The total effect, then, depends on three factors, and a marked increase in failure possibility can only be blamed on a single factor that is significantly different, say, from the other two. This is a considerable constraint. For instance, if the structural material selected for the job is tough enough to prevent brittle fracture while design and fabrication characteristics happen to be poor, the structural reliability and safety of the component during its useful life cannot be guaranteed. It should be noted, in general, that the useful life of a structural component depends on a number of variables and any oversimplification, such as the claim that "forgiving materials" showing superior notch toughness can correct any shortcomings committed by the fabricator, inspector, designer, or user, may be spreading a false sense of security.[1]

Another area of misconception concerns the yield strength level residual stresses and weld discontinuities. This philosophy maintains that the primary cause of fracture in a welded structure can be easily traced to the foregoing interior characteristics of the weldment. This theory appears to ignore environmental effects, cyclic history, and stress redistribution, and other effects in the vicinity of the fracture origin. Although residual stresses and weld discontinuities can influence the mechanism of failure, correct appreciation of the preventive action can only be realized after a complete study of all the parameters involved. This is a necessary condition for the development of a reliable fracture control plan.

Various factors are known to contribute to brittle fractures in large welded structures. These include service temperature, material toughness, design, welding, and residual stresses. However, Barsom and Rolfe[1] point to the three principal factors that control the susceptibility of structures to brittle failure:

- Notch toughness at a given service temperature, loading rate, and plate thickness;
- Size of a crack or discontinuity at a point of fracture initiation;
- Tensile stress level, including residual stress.

These factors are interrelated through the concepts of LEFM in such a manner that the combination of stress and crack size defines the critical stress intensity factor for a particular specimen thickness and the rate of loading. This combination of the foregoing parameters constitutes the state of fracture. The goal of a fracture control plan is to ensure that the stress intensity factors K_I throughout the lifetime of a structural component will not exceed the critical stress intensity factors (K_{Ic}, K_c, etc.). This condition is similar to a simple criterion in stress analysis that states that the working stress in a component, σ, should be smaller than the yield strength of the component material S_y. The degree of safety and reliability can be defined as the conventional factor of safety in LEFM or the elementary stress analysis. Such factors can be specified by a code of practice within a generic class of structures and applications. In a critical environment, such as for instance a nuclear power plant, the relevant fracture control plan would be restricted to the specific structure, providing assurance of extremely low probability of service failure. Under these conditions even a minor failure will not be tolerable. In other applications, occasional failures during fabrication or service of a component may be permitted where the consequences of a failure might be minimal and where it would be more efficient and economical to replace a failed component. An extreme illustration where a service failure can be tolerated might be the repair or replacement of plates of the loading bed of a dump truck.

As implied previously, good engineering design practice should use all possible modes of failure such as buckling, yielding, or corrosion that will apply to a variety of structural components in bridges, buildings, pressure vessels, or aircraft. The purpose of this section is to summarize practical aspects of technical information and design guidelines pertaining to prevention of failure by fracture or subcritical crack growth leading to material fracture.

A number of national institutes and regulatory bodies have developed information on the probable loads and service conditions applicable to design life of specific structures. These may concern highway bridges,[13] pressure vessels,[14,15] aircraft structures, and other components and systems. From the point of view of LEFM it is necessary to establish the rate of loading in order to decide on the type of a controlling toughness parameter (K_{Ic}) or (K_{Id}). Other dynamic loadings caused by wind, seawaves, or earthquakes are based on specific field measurements, experience, and the various code requirements. Bridges, buildings, rotating machinery, or structural components in aircraft are subject to fatigue loading of constant or variable amplitude, and regardless of the type of loading, fatigue action can induce subcritical crack growth by various means as stated in Chapter 5. The point is that starting with even the smallest initial flaw, the potential for larger cracks is always there as long as the structure is subjected to load cycling.

Various environmental effects such as cavitation, corrosion, or stress-corrosion can encourage crack growth. Also, the transition temperature approach to practical fracture mechanics suggests that temperature can have a significant influence on the fracture of structural components. This is especially prevalent in some of the low- and medium-strength steels exhibiting brittle-to-ductile transition.

Estimating initial flaw sizes for analytical purposes may not be straightforward because the quality of fabrication is known to control the character and the overall dimensions of the initial cracks and defects. And last but not least, it should be observed that the inherent fracture toughness of the structural material can be strongly affected by the variations in chemical composition and heat treatment. This puts a special burden on the designer because prior to the selection of the material, all possible conditions of design options, fabrication, loadings, service requirements, reliability, safety, and cost should be examined. In a sense we expect that the particular structural design of larger and more complex systems will be optimized, an assignment of a tall order. Specifically, in the case of an anticipated brittle fracture, the critical ratio of the stress intensity factor to the yield strength of the material should be selected for the appropriate rate of loading, temperature, and the material thickness. However, if the overall weight of the proposed structure is too high, the allowable design stress must be increased so that, for the same factor of safety, a higher yield strength of the material should be used. Here, therefore, we arrive at the point of a typical conflict in design where the two following requirements exist:

- High critical ratio of stress intensity to yield strength;
- High yield strength of the material.

The first requirement is of the "prevention of failure" type, while the second is performance oriented. The evaluation of the first requirement is based on

LEFM, while the second requirement comes from the traditional approach using stress analysis on the premise of "perfect" fabrication and the no-flaw condition. Since the two criteria are obviously in conflict, and as long as this analysis is made prior to the final decision concerning the material, there is some room for a compromise. Barsom and Rolfe[1] provided an excellent example of a practical evaluation of the various fine points of such a compromise in selecting the material for a specific pressure vessel. This evaluation indicated certain advantages of the fracture mechanics approach to the problem. In this context, several comments may be in order that directly relate to the task of developing a fracture control plan in general.

It has been generally recognized that the K_{Ic}/S_y or K_{Id}/S_{yd} ratio is a convenient index for assessing the relative toughness of materials used in engineering design. Since we normally expect that modern structures should be able to tolerate yield stress levels in the immediate vicinity of a structural discontinuity, the critical crack size can also be related to the yield stress at its tip. It follows from the fundamental LEFM expressions that, indeed, the critical crack size is directly proportional to $(K_{Ic}/S_y)^2$ or $(K_{Id}/S_{yd})^2$. For a structural steel the parameters K_{Id} and S_{yd} may be estimated using Eqs. (4.8) through (4.10). The use of one of the high ratios indicates good fracture toughness and, consistent with proper economic considerations, represents a desirable design condition.

The main question remains, however: How high should such ratios be for the best performance of a large structural system where a complete monitoring inspection of crack initiation and propagation cannot be assured? It does not take very long to figure this out because of a great number of factors and variables involved in the entire process. The type of structure, frequency and access for inspection, load path redundancy, quality of fabrication, design life, probability of failure due to overload, consequences of failure, and total cost show clearly that no simple answer can be found. The only positive and rational statement that one can make is that some flaws, local yield stresses, and plane-strain conditions are likely to exist in all structures. Under such conservative assumptions the K_{Ic}/S_y ratio can serve as an indicator of the "relative safety" of a structure against brittle behavior.

Another vital question may be posed in relation to a case where the material cannot be changed because of existing codes and practices or simply due to economics. This case is similar to that discussed in Chapter 6 in connection with the "lower-bound design stress" used to certify existing equipment or "one-time usage" type structures handling hazardous materials. If the material cannot be changed, only a limited fracture control plan can be executed by restricting the operating temperature or reducing the design stress levels as the main variables. It will also help to review all service conditions and design loads that affect performance.

Many factors that can affect the resistance to fracture of large welded structures can be tied to the three main parameters mentioned before — stress, toughness, and crack size. The corresponding symbols pertaining to this discussion are:

Stress (σ) nominal
                   (σ_{des}) design
                   (S_y) yield
Toughness (K_I) stress intensity
                   (K_{Ic}) plane strain
                   (K_c) plane stress
                   (K_{Id}) dynamic
                   (K_{Iscc}) corrosion
Crack size (a) general
                   (a_0) initial
                   (a_i) arrested
                   (a_{CR}) critical

The calculation of the nominal stress, which is always featured in the fundamental formula of fracture mechanics, can be rather simple using elementary equations $(P/A$ or $M/Z)$ or can be unusually complicated involving the classical solutions of the theory of elasticity, depending on the geometry of the various structural shapes such as plates, shells, or box girders. Such problems can be worked using "closed-form" solutions[16] for simpler geometries or computer codes based on the "finite-element method" (FEM) for more complex shapes.

In the specific case of welded construction, the residual stresses at the discontinuities are generally assumed to be of the order of yield strength. The ductility of the material is expected to redistribute the peak stresses. There is, of course, the possibility of a brittle fracture, so that knowing the fracture toughness of the material and localized yield stresses, the critical crack size can be determined and compared with the maximum potential flaw size that can be detected using modern inspection techniques.

The effect of local residual stresses on crack growth under plane-strain and plane-stress conditions is illustrated roughly in Fig. 12.4. The crack a_0 was assumed to initiate in the presence of yield level residual stress S_y and to arrest quickly on the exit from the residual region, over a distance (A) to give the initial crack length a_i for the start of the fatigue crack growth. The irregular curve (D) corresponds to a possible crack extension from a_0 to a_i. The crack growth due to fatigue in the region of plane strain is given by (B). Similarly, the fatigue crack growth in plane stress is denoted by (C). Hence we can define the critical crack lengths for plane strain and plane stress by (B) and (C), respectively. The effect of temperature on the plane-strain fracture toughness is shown approximately in Fig. 12.5.

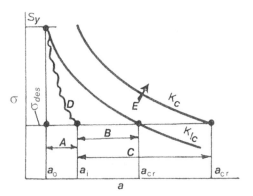

FIGURE 12.4 Growth of cracks under various constraints. (A)–(E) have no dimensions and merely serve as diagram markers.

The conservative approach related to Fig. 12.4 is to calculate the crack size a_0 based on S_y rather than the design stress σ_{des}. Under fatigue loading, the crack grows out of the residual stress zone to become the crack size at the design level of stress. All this points to the fact that the "critical crack size" defined by the intersection of the horizontal line at σ_{des} and the K_{Ic} curve is not a material property, because it depends on the design stress.

Assuming the design stress σ_{des} in Fig. 12.4, the corresponding critical crack size can be calculated. If this number proves to be larger than the plate thickness, the subcritical crack growth should relax the stress ahead of the crack. This means that the local plane-strain condition transforms into a plane-stress behavior. The corresponding fracture toughness will increase, in line with the general trend toward higher K_{Ic} values as shown by arrow (E) in Fig. 12.4. It was also shown previously in Chapter 4 that K_c should be greater than K_{Ic} or K_{Id}.

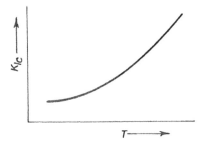

FIGURE 12.5 Effect of temperature on K_{Ic}.

The comparison between the crack growth in a normal and in a corrosive environment is shown schematically in Fig. 12.6. The growth of a crack by fatigue or corrosion fatigue is denoted by (A). The effect of stress-corrosion alone, in terms of crack length, is given by (B). The sketch is approximate and not to scale.

It is possible to use lower values of toughness when tensile stresses are decreased and when compressive residual can be induced by means of shot peening or case hardening, often applied to gear teeth and landing gears. There are also other mitigating circumstances such as crack orientation not in a critical plane or the effect of a decreasing stress field. The effect of temperature on plane-strain fracture is reflected in different values of crack size even under conditions of a constant design stress. The relevant trend is illustrated in Fig. 12.7. The specific level of toughness K_{Ic} can be selected for a given temperature, as shown in Fig. 12.5.

The diagram in Fig. 12.6 suggests that a good design practice is to use the K_{Iscc} limiting curve, in order to assure that failure coincides with the onset of stress-corrosion crack growth. This is certainly a conservative assumption because, once started, the stress-corrosion crack will propagate until complete failure occurs. While we have methods for predicting the number of cycles of fatigue, the rate of stress-corrosion crack propagation is extremely difficult to estimate correctly, because of the large number of variables involved. Many of these relate to the chemistry of the corrosive medium such as concentration of the corrodent at the crack tip, or the temperature.

The three basic variables involving stress, discontinuity, and toughness signify in general three approaches to control of the threat of brittle fracture. The use of fracture mechanics makes such approaches more quantitative by ensuring, say, that $K_I < K_c$, similarly to the process of stress analysis assuring that $\sigma < S_y$.

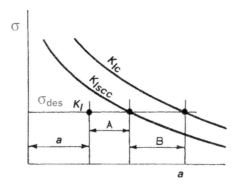

FIGURE 12.6 Crack growth in normal and corrosive environments.

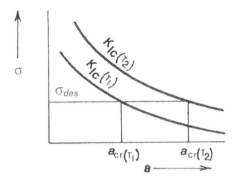

FIGURE 12.7 Effect of toughness and temperature.

In essence we deal with various margins of safety. A rather clear schematic illustration of the reduction of the design stress is given in Fig. 12.8.[1] Using the symbols of reference, the diagram can be described as follows. The initial and new margins of safety are given by (A) and (B), respectively. The original and the reduced design stress levels are denoted by (C) and (D), corresponding to the critical crack lengths a_{CR} involved. In other words, lower design stress allows longer critical crack. Also, the symbols (E) and (F) refer to the original and the reduced stress levels in design. The illustration in Fig. 12.9[1] suggests improved quality of fabrication and inspection. By using the same design stress level and the material's toughness for the particular service conditions, the initial margin of safety is shown as (A), the new margin as (B), and the reduced flaw size a_o at (C) indicating improved quality of the product. Similar graphical illustration can be applied to the case of improved notch toughness while the level of the design stress and the quality of fabrication remain unchanged. In general

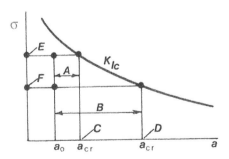

FIGURE 12.8 Reduction in design stress.

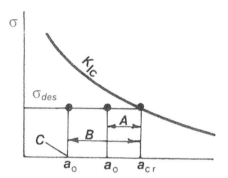

FIGURE 12.9   Reduction of flaw size.

the relative influence of the three basic parameters can be described by the fundamental expression, such as

$$K_{Ic} = (\text{numerical constant}) \times (\text{stress}) \times (a)^{1/2}$$

It is clear from the relationship above that it is easier to control σ and K_{Ic} than a. However, the parameter that strongly governs the rate of growth of the subcritical crack is K_I or ΔK_I, raised to the power of 2 or greater as shown already in Chapter 5. This effect may certainly be more significant to the useful fatigue life than the change in K_{Ic}.

It should be stated in addition to all the material reviewed in Chapters 3 and 5 that the basic analysis of elastic behavior of the stress field near the crack tip involves the single important quantity K, generally accepted as the "stress intensity factor." When the loading causes an opening mode of displacement, this factor is denoted by K_I. Since the form of the fundamental expression is the same for K_I, K_c, K_{Ic}, and so on, it is well to characterize the remaining parameters:

- Nominal applied (gross section) tensile stress σ is assumed to be normal to the plane of the crack and in the crack's vicinity.
- Characteristic dimension of the crack (or flaw), a, such as crack depth in the case of a cracked surface.
- Dimensionless constant whose numerical value is a function of the crack geometry, the ratio of the crack size of the structural part, and the type of loading such as tension or bending.

The fundamental premise of LEFM is that unstable crack propagation occurs when the value of K_I reaches the critical level of the stress intensity factor, denoted as K_{Ic}. There is full agreement as to this concept between LEFM specialists and design practitioners: K_{Ic} is the plane-strain fracture toughness, which is a temperature-dependent material property. In the case of structural carbon and alloy steels this parameter also depends on the rate of loading applied to the flaw. A general schematic illustration of the energy levels for various rates of loading is given in Fig. 12.10. The vertical axis represents the levels of performance in terms of the absorbed energy in tests of notched specimens for the plane-strain (A), elastic–plastic (B), and plastic (C) ranges of behavior. The S-shaped curves designate static loading (E), intermediate loading rate (I), and impact loading (D) toughness characteristics.[1] The methods for control of fracture behavior utilizing lower design stress, improved fabrication techniques, and tougher materials discussed so far, are certainly the important approaches to practical engineering design. Barsom and Rolfe[1] expand this practical design methodology into the area of other effective design characteristics of interest to practitioners concerned with the general topic of fracture control planning. A brief summary of major points follows.

Structural Materials

These are selected for unusual toughness so that the material does not fail by brittle fracture under the most severe operating conditions. The use of HY-80 steel for submarine hull structures, critical components of cranes handling hazardous materials, and special hardware employed in underground nuclear tests provides good examples of this method. This approach, however, may not be the best in fatigue applications.

Multiple Load Path

This should not be confused with the term "redundant" related to statically indeterminate structures. When, however, our system consists of several independent structural parts, then we have the condition of a "multiple load path." Failure of

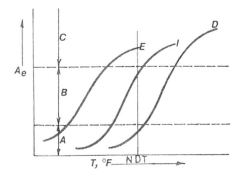

FIGURE 12.10 Notch toughness vs. temperature at various rates of loading (A_e, absorbed energy).

one structural shape transfers the load to the remaining members.[17] Naturally, the factor of safety on toughness should be greater for a "single load path" structural design. Also, the cracks propagating in a multiple load path system may later arrest, although individual components will have to be replaced or repaired. Still, the failure of the entire structure would not be expected.

Crack Arresters

The fail-safe philosophy of crack arrest has been known and practiced extensively in the aircraft and shipbuilding industries. The relevant requirements cover four areas of activities:

- Assurance of high notch toughness;
- Provision of effective local geometry;
- Proper location of crack arresters;
- Design of crack arrest systems as the energy-absorbing and deformation-restricting mechanism.

Control of Crack Growth

The methods normally used in fatigue fracture control also apply to control of crack growth with the help of the stress intensity fluctuation ΔK and the critical stress intensity factor K_{Ic}.

Loading Rate Reduction

The majority of structures under normal conditions operate at loading rates that are slow to intermediate, and where the lower range of notch toughness may still be acceptable. Experience shows that relatively few older structures were subjected to brittle fractures. Hence control of the loading rate is rather effective.

The main problem in developing a fracture control plan is not the application of the foregoing technical rules but, simply, the economic decision. Since the potential for design overloads and nature-made unexpected loads always exists, the goal of sound engineering is to optimize structural performance in relation to economic reality. How much basic notch toughness should we have in the system if the material toughness can be specified directly in terms of K_{Ic} or K_{Id}? It quickly turns out that a material specification using K_{Ic} or K_{Id} parameters is often totally unrealistic, based on economic as well as technical considerations. It is therefore not surprising that the essence of practical fracture mechanics can only be concerned with the simplest concepts of LEFM, fracture-safe design based on the transition temperature, elementary material tests, case studies of failed structural systems, and the related experience. And it is also not surprising that material requirements for nuclear pressure vessels (American Society of

Mechanical Engineers, ASME, Code, Section III) and steel bridge members (American Association of State Highway and Transportation Officials, AASHTO, Guide Specification) are written in terms of the CVN impact results and NDT temperature criteria.[18–20]

Investigations of the effect of temperature on the fracture toughness of carbon and alloy steels have shown a significant increase of toughness over a relatively narrow temperature range. This study included K_{Ic}, K_{Id}, K_{Ia}, and similar parameters, with K_{Ia} denoting crack arrest fracture toughness. The lower-bound curve of all the experimental results was designated as the K_{IR} curve, shown in Fig. 12.11. This parameter is known as the "reference fracture toughness" selected as the conservative criterion, indexed to NDT temperature by the ASME Code.[20] An analytical expression for the curve in Fig. 12.11 is

$$K_{IR} = 26.777 + 1.223 \exp \{0.0145[T - (RT_{NDT} - 160)]\} \qquad (12.9)$$

where RT_{NDT} is the reference temperature for NDT, in °F. For instance, taking $T = RT_{NDT} = 0$, Eq. (12.9) gives $K_{IR} = 39.2$ ksi (in.)$^{1/2}$. If necessary, the K_{Id} parameter can be estimated using formulas in Chapter 4. In general, the relationship between the dynamic yield strength S_{yd} and K_{Id} is

$$K_{Id} = (\text{numerical constant}) \times S_{yd} \qquad (12.10)$$

In Eq. (12.10), the constant factor varies between 0.5 and 0.78. The approximate value of 0.6 is often used in preliminary estimates.

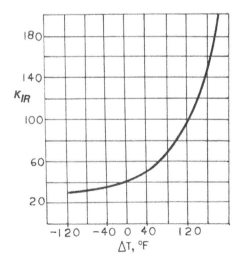

FIGURE 12.11 Reference stress intensity factor K_{Id}, [ksi (in.)$^{\frac{1}{2}}$].

Fracture control criteria for nuclear pressure vessels were defined by the ASME Code a little over 30 years ago. It was necessary to establish the degree of strength reduction in the presence of defects, the influence of the rate of applied loading, and residual stresses caused by welding. The widely adopted procedure since the 1940s under the name of "transition temperature" and the tests based on the CVN, and drop-weight techniques, continued to be in demand. Loading on the structure was permitted only at a temperature higher than the nil-ductility transition (NDT) by the increment determined from service experience, model tests, and engineering judgment. This was therefore an eminently practical approach, representing materials behavior, quantitative influence of fracture mechanics, and stress analysis. When the material was subjected to significant neutron radiation, the transition temperature was increased while the fracture toughness of ferritic steels suffered a drop. This situation required the development of a systematic surveillance program in accordance with the methods of the American Society for Testing and Materials (ASTM).

The conservative practice of K_{IR} was followed next by the procedure of a postulated defect size, in vessel shell or head regions remote from stress raisers, defined by the ASME Code (Section III, NB-3113).[20] The shape of the defect was taken to be semi-elliptical, located on the surface. Although buried flaws are generally more difficult to detect, they can only produce about half the level of stress intensity for a given size of the defect.

With a conservative assumption of the defect size and geometry it is possible to calculate the two stress intensity factors based on the general primary membrane stress due to pressure and the thermal stress caused by the maximum expected thermal gradient through the vessel thickness during operation. Effects of residual stresses are excluded from the calculation, largely because of the conservative assumptions and the reduction of peak residuals through heat treatment, service, and radiation effects. Hence the sum of the calculated applied stress intensity factors can be compared with the reference stress intensity factor K_{IR}. From this, the governing operating condition can be stated as follows.

$$K_{IP} + K_{IT} \leq K_{IR} \tag{12.11}$$

where K_{IP} = stress intensity factor for pressure, and K_{IT} = stress intensity factor for temperature.

The developments leading to Eq. (12.11) for fracture control planning in the field of nuclear pressure vessels did not include any conventional factors of safety because of the use of conservative assumptions in selecting K_{IR} and the flaw size. If required, for instance, any additional conservatism can be achieved by multiplying the normal stress intensity term K_{IR} by a factor of 1.5 to 2.0 and the thermal component K_{IT} by 1.0 to 1.3 for the usual operating conditions. For other activities such as a hydro test, less conservative safety factors may be appropriate. For more details on a condensed version of the ASME Code, Section III for nuclear pressure vessels, consult Barsom and Rolfe.[1]

In spite of a very fine operational record of steel bridges, material scientists and designers continue to emphasize the fracture mechanics methodology as a tool for maintaining the safety and reliability of bridge structures. In addition, AASHTO has recognized the need for a more formal fracture control plan with special regard to notch toughness, welding, and inspection criteria, applicable to critical tensile components. It is also gratifying that following a research effort on the use of A36 and A572 steels in bridge-related industry, AASHTO adopted the CVN impact toughness requirements for a total of five ASTM steel designations. These included A588, A514, and A517 steels suitable for bridge construction. It appears that the decision to go with the CVN rather than K_{Ic} requirement was affected by at least two factors. One factor was the result of tests indicating that fracture did not occur under plane-strain conditions. The other was a simple recognition of the fact that K_{Ic} tests were costly and difficult to run, while there was a good correlation between the K_{Ic} and the CVN test results. The AASHTO decision, then, was driven by practical considerations that are hard to dispute. The tests indicated plainly that the fracture toughness of the steel increased rapidly from plane-strain to fully ductile behavior, consistent with the sharp fracture toughness transition characteristics of the CVN results, derived from intermediate strain rate and impact loading. A description of detailed procedures for developing the CVN fracture toughness requirements, imposed on all primary tension members of bridge steels, is available for further reading.[1] These requirements are tied to the minimum operating temperatures, which are linearly related to the CVN test temperatures. To minimize the problem of dealing with a great variety of testing and service temperatures, the three zones of service and the corresponding test temperatures were established for all bridge steels having a yield strength of 50 ksi or less.

In dealing with materials having yield strength values significantly higher than 50 ksi, it was determined that the temperature shift between the low strain rate (static) and impact decreased as the yield strength increased, as shown approximately in Fig. 12.12. The temperature shift was measured roughly from the NDT temperature established by the drop-weight test.[1] It was also shown that for yield strengths of bridge steel higher than 65 ksi, a temperature reduction of 15°F corresponded to a yield strength increment of about 10 ksi. For other details of specifications, the interested reader is directed to the formal regulatory document of AASHTO.[19]

It should be added that during the verification tests — which used cover plates and transverse stiffeners fabricated from A36 and A572 Grade 50 steels, and subjected the simulated bridge members to fatigue — the loading conditions were of the severe type, involving a combination of temperature, strain rate, and prior fatigue. The tests suggested that members containing fatigue cracks up to

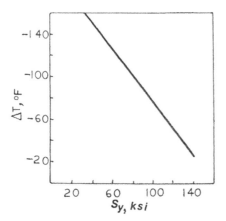

FIGURE 12.12   Effect of yield strength on temperature shift.

0.4 in. deep could survive 100,000 cycles at the maximum allowable design stresses, assuming sufficient fracture toughness of the materials. In actual service
cracks of such a magnitude, and even larger, could be tolerated on the premise
of the redundancy built into the design of various bridges and because of fracture
arrest on the cracks entering the decreasing stress fields. In addition, there is only
a very low probability that the most severe combination of the lowest service
temperature and the maximum strain rate will occur at the same time.

Engineering use of temperature transition principles has so far become a
cornerstone of formal fracture control planning involving structural requirements, material selection to be consistent with the lowest service temperature,
and certification of the product in relation to the original specification. The latter
point, in particular, cannot be assured without a standardized fracture test. The
relevant test specimen is cut to specific dimensions from the part under consideration, and it must assure the specimen quality in terms of the constraint and the
transition temperature. The plane-strain transition is expected to take place
immediately above the NDT temperature.

The most important feature of the true fracture control plan is its indepen-
dence of any technical decision governed by an opinion. The certification process
supporting the true control plan must also be based on analytical methods that are
fundamentally rational. In addition, however, this process should be statistically
sound. And true control appears to be governed by purely random events dictated
by random metallurgical effects. The foregoing overtones of statistical fracture
control certainly indicate that designers cannot predict structural reliability of a
given structure by direct calculation. The task of a fracture control plan is,
then, to show that the specified structural performance is not a stochastic event.

However, since there seems to be no definitive boundary between probabilistic and stochastic solutions, we have to deal with bracketing conditions. Hence, design solutions can only be totally deterministic or potentially stochastic.[21]

LIMITATIONS AND ERRORS

This section is concerned with general comments related to potential limitations and errors based on service experience, affecting design, fabrication, and inspection procedures. Modern technology dealing with large and complicated projects can hardly tolerate the type of consequences resulting from the disasters of the first half of the 20 century, as well as some more recent service fracture failures, as indicated in the various chapters of this book. There is a definite need for comprehensive fracture control plans and yet there is an obvious lack of applicable past experience in this area because of increasing economic constraints. There is a rather serious trend in modern living toward economic limitations and errors of judgment in the various areas of technology where the industry would pay for liability and litigation rather than for a comprehensive planning of fracture control and product reliability derived from practical engineering knowledge.

Some of the specific technical limitations and potential errors, within the framework of practical fracture mechanics, can be discussed along the two basic lines of approach to fracture analysis and control. The first is fracture-safe design based on the transition temperature criteria, where we attempt to circumvent the problem by designing the structure to operate above the transition temperature. This puts the structure into the ductile region of behavior.

The other approach is based on the concepts of linear elastic fracture mechanics through the application of stress intensity factors. This leads directly to the plane-strain fracture toughness, governed by the ASTM methodology, and it essentially applies to higher strength materials exhibiting brittle behavior. This is a relatively simple matter until we come across a high-toughness material that requires thick test specimens and high load capacity testing rigs. In general the K_{Ic} evaluation eventually became a costly procedure requiring quality equipment and trained personnel.

Structural integrity cannot be derived from abstract considerations because crack inspection requires metallurgical knowledge in such categories as the state of fracture tied to the crack size and localized metallurgical damage defined as the sensitivity of the material to embrittlement. There is also some evidence that a combination of certain metallurgical factors makes practical control of minute critical cracks infeasible. An example of such a situation may be the case of a standard-grade C-Mn, pearlitic steel, subjected to arc strikes that triggered unexpected failures of large structures.[21] The failure due to a sudden development of microscopic cracks represents a serious limitation on the inspection methodology, because there is essentially no visible flaw until the very moment of failure. Such conditions are consistent with the plane-strain behavior of steels of intermediate strength and titanium alloys during gas-metal arc welding, and in the case of high-strength steels subjected to stick-manual arc welding. Since it is impossible under such conditions to control the presence of metallurgical notches (local embrittled sites), fracture control can only be done through improved welding procedures. This limitation does not apply to plane-stress behavior.

It should also be noted that in addition to arc strikes, high residual stresses in the damage areas due to welding can reduce the critical cracks to microscopic sizes under plane-strain conditions, with very little tensile ductility remaining.

Since the local volume of the embrittled material, consistent with the microscopic cracks, is also very small, any inspection technique can do very little, and the basic problem of control becomes that of monitoring metal hardness, quality, and impurities on a microscopic scale. For the critical crack sizes of 0.2 in. (5 mm) or higher, the inspection techniques become more realistic. The embrittled volumes of metal increase and the number of likely origins of fracture becomes smaller, at least on a statistical basis.

The foregoing brief evaluation of some of the practical limitations shows clearly that the use of plane-strain conditions poses a serious problem of certification of structural integrity, because of a fast fracture potential.

The uniformity of design, fabrication, and certification of engineering products has developed over the past 25 years into a number of codes, rules, and standards on the grounds of analytical methods for various failure modes, which now go beyond the traditional practices of overload, buckling, or plastic instability. It is now possible to rationalize the mechanism of fracture and the crack growth failure modes. This does not mean, however, that long-established engineering practices have been changed or ignored in the modern environment. The new considerations start with fracture and logically follow through with crack growth, which requires a separation of the certification rules for low- and high-strength materials. For instance, the ASME Code deals with fracture, crack growth, and neutron damage for commercial nuclear power plants. However, this code is independent of U.S. federal standards. Also, since 1972, the code has included fracture mechanics principles for characterization and analysis. The code has become a matter of legal responsibility for the designers, manufacturers, and users of the relevant components and systems, and the process of licensing is described as "certification by analysis" rather than "certification by rule." This is an important limitation because the licensing is based on deterministic analyses and not on opinion.

Legal requirements for certification in the form of codes, rules, and standards for other branches of industry can be published now in the *Federal Register* and thereby become law. In this manner, both industry and regulatory agencies can adhere to rational procedures. The use of modern certification principles has been particularly helpful in the development of high-performance machines and structures, such as high-quality aircraft. Also, in the case of a structural failure resulting in legal procedures, compliance with the *Federal Register* can help to prove the use of best practice and engineering knowledge.

Engineering interest in practical fracture mechanics is best defined in terms of the K parameter in LEFM because of its dependence on the flaw geometry and the elastic stress system in the vicinity of the crack. Except under unusually brittle conditions, the fracture is initiated at the crack tip. The critical level of K is, of course, the K_{Ic} parameter. Also, the larger the plastic zone size, the tougher is the material. And in general the size of the plastic zone is directly proportional to the term $(K_{Ic}/S_y)^2$. This also means that for a very small plastic zone, a very small amount of energy has to be used to develop the unstable crack, and the material is brittle.

The first limitation in this scenario is, however, that the K_{Ic} value on its own does not translate into the physical meaning of fracture toughness if the yield stress is not specified. For instance, taking K_{Ic} as 50 ksi (in.)$^{1/2}$ and the two values of yield stress S_y as 30 and 200 ksi, the ratio of the small to large plastic zone size is obtained as follows:

$$(K_{Ic}/S_y)^2 = (50/200)^2 = 0.0625$$
$$(50/30)^2 = 2.7777$$

and

$$\frac{0.0625}{2.7777} = 0.0225$$

or $1/0.0225 = 44.4$; that is, one material is very tough and the other material must be rather brittle. It also follows from this elementary illustration that materials of different yield strength will have the same toughness if their $(K_{Ic}/S_y)^2$ ratios are identical. In the foregoing example, the high-yield material would need to have $K_{Ic} = 333.3$ ksi (in.)$^{1/2}$. However, independently of this discussion, the K_{Ic} should be considered a fundamental parameter of fracture mechanics and a material's property. The physical meaning of plane strain is a "maximum triaxial constraint" to plastic flow. It also follows from this statement that the plane-strain plastic zone size cannot be decreased by increasing the depth (or size) of a sharp crack, so that the corresponding value of K_{Ic} must be regarded as the minimum.

The stress intensity parameter K depends on the size and geometry of the crack, and the stress level opening the crack. The K_{Ic} level can be reached by a large crack and low stress, or by a small crack size and high stresses. As the triaxial constraint increases, the stress at the crack tip will eventually exceed yielding. In a brittle material, small cracks can be severe enough for stress intensity to reach the K_{Ic} level. In a more ductile material, larger cracks would be required to reach similar stress levels.

There are three conditions for the development of instability under plane strain. These are: a large volume of metal surrounding the crack, a large crack to prevent lateral contraction, and sufficient crack depth to reach the critical K level. The physical aspects of the foregoing statements are reflected in the expression limiting the thickness:

$$B \geq 2.5\left(\frac{K_{Ic}}{S_y}\right)^2$$

The maximum crack depth and the depth of the uncracked ligament are both taken as $0.5b$. If the ligament depth is insufficient, the volume of metal surrounding the crack will be too small to maintain the plane-strain constraint, so that K_c rather than K_{Ic} conditions will apply. This limitation will exist even if the above thickness requirement for B is met.[21]

One of the problems with the development of fracture control planning is that crack (or defect) geometry and size have to be postulated. However, potential crack locations and geometrical discontinuities can be anticipated, so that the overall influence of the "assumption" is somewhat lessened. Further improvement in maintaining some degree of preciseness in fracture control may be found in the area of inspection with regard to detectable crack configurations. Some combination of visual and x-ray techniques may be best, although it is still necessary to assess several scenarios of damage development in relation to the shortest life and projected inspection intervals.

The establishment of the initial defects in welded structures may, at least partially, be successful if we can identify porosity, lack of fusion, or lack of penetration. If such a defect can pass quality control inspection at hand, then the defect can be regarded as an initial crack for the purpose of planning fracture control procedures. In the case of judging the initial shape and size of the defects in castings, the assumptions are more delicate and they are limited by the particular quality control experience.

One of the natural limitations in the area of crack control is illustrated by the case quoted by Broek[9] of a military airplane. Of a total of 2000 holes in a wing subjected to a fatigue test, 6% were found to be cracked and crack sizes noted. Subsequent analysis provided crack growth curves and flaw sizes between 0.02 and 0.05 in. for the equivalent initial defects. However, the efforts to correlate such defects, say, with the hole quality expressed in terms of roundness, reaming, burrs, or scratches were inconclusive, so that the extrapolated flaw sizes (0.02 to 0.05 in.) appeared to have no special bearing on the initial quality control. The practical lesson from this points to the ever-present detectable cracks that are tied to the current technology, and that in the end may lead to arbitrary assumptions. Two special words, "assumption" and "arbitrary," are emerging from this discussion as shadows over the honest efforts to create a degree of precision in crack assessment in spite of the various obstacles.

Every methodology, including the elementary version of practical mechanics and the two closely related branches of materials science and stress analysis, are bound to have certain shortcomings and intrinsic limitations in dealing with Mother Nature. However, the worn phrase "garbage in, garbage out," reserved in many cases for the computer world, may also apply to the field of fracture mechanics. The input and the assumptions are ever present in design and research activities, and in the deliberations of causes and consequences.

There are various sources of error, which can be put in specific categories. For the purpose of this chapter dealing with limitations and errors related to practical aspects of fracture mechanics, the material is classified under four headings:

- Uncertainties and assumptions
- Interpretations
- Inaccuracies
- Shortcomings

The foregoing subdivision of the entire area of potential error sources should, hopefully, assist the practitioner to zero in on the particular problem without, however, more detailed examination of the reference material. Hence the prime intent of this section is to suggest a brief overview. For a more comprehensive treatment of this topic, the interested reader is advised to study the work of Broek.[9]

Uncertainties and Assumptions

Errors due to input data depend on the environment where misinterpretation and incorrect use of information can make a significant difference, particularly in the area of fatigue crack growth. Questions can arise in dealing with the level of constraint, equation fitting, data scatter and inaccuracies, and special parameters involving retardation or changing environment. Parameters such as K_c, K_{Ic}, J_R, and da/dN are probably in the forefront of dispute. The term J_R denotes the fracture energy of a nonlinear material, which is not considered in this book. The errors in a_p, defining the permissible crack size, are moderate and not likely to exceed 20%. A small change in a_p is not too significant in fracture control, unless the a_p term itself is rather small. Misinterpretation of scatter or equation fit can introduce an error factor on the order of 1.5 to 3. Similar factors can be found in dealing with flaw geometry (circular vs. elliptical), and the initial crack shape can be quite different from the elliptical geometry.

Interpretations

In the first place, all load histories involve a degree of approximation, no matter how much care is taken in assuring correct geometry and other details of the analysis. Load history is an exercise in statistics. The technique known as "clipping" reduces the magnitude of the highest cycling loads, with no cycles omitted. "Truncation," on the other hand, reduces the number of very small amplitude cycles, which only have a limited effect on the entire spectrum. The decisions on clipping and truncation should be made by the appropriate experts to assure correct sequencing. Any misuse of the foregoing techniques can introduce another layer of errors.

Inaccuracies

The causes of error in determining stress intensity can be encountered in actual load values, stresses, and the overall geometry factor (sometimes called the correction factor), which enters the formula for the stress intensity factor K. It is a multiplication factor that goes together with $\sigma(a)^{1/2}$. In the work of Broek,[9] this factor is known as β, where the parameter K is given as

$$K = \beta\sigma(\pi a)^{1/2}$$

The crack growth is often defined as the 3rd or 4th power of K. Hence the errors in estimating the fatigue life are proportional to the errors in β and σ to the 4th power. So a 10% error in stress corresponds to a factor

$$(1.1)^4 = 1.46$$

The errors in stress come from the estimate of loads and the limitations of stress analysis. No matter how confident the stress analyst, the error in stress concentration, load distribution, and eccentricities is not going to be much lower than 10%. This error is not any lower in computer, finite-element modeling with the various approximations of boundary conditions and three-dimensional cases interpreted in two dimensions. The error in β may be of secondary importance when inaccuracies of loads, stresses, and flaw shapes are included in the overall analysis of design limitations.

Shortcomings

As stated previously, a small error in LEFM is possible in the case when small cracks cause collapse of relatively small structures. For long cracks and large structures, however, error due to the LEFM procedure is very small, as shown by plots of crack length vs. life in cycles. Some schools of thought suggest using the J-integral approximations approach for representing (da/dN) data. According to Broek,[9] the use of J in fatigue crack growth analysis is of no practical interest. The representation of data by K is simpler and well justified. Also, there is no advantage in switching to elastic–plastic fracture mechanics (EPFM), particularly for smaller cracks, because the error will not happen to be markedly lower. The foregoing considerations do not address the error due to scatter.

In general, the errors due to computer modeling are caused by integration schemes, rounding of numbers, and special equations for retardation phenomena and the state of stress. The retardation phenomenon occurs when one single high stress is interspersed in a constant amplitude fatigue. Further comments are given in Chapter 5. Although sophisticated retardation models are available, all computer codes contain individual assumptions, so that retardation calibration parameters are not transferable between the various codes. Numerical integration is normally used in calculating crack growth. In most applications, single precision is more than sufficient. Several calculational examples in computer work using single and double precision are available.[9] In the case where all computer errors in a complex problem can be compounded, the total factor on estimated fatigue life can be 2.7 and 137,000! This can hardly be defined as an error.[9] Although such extreme compounding of the errors is highly unlikely, it is obvious that the result can be influenced much more by an "assumption" rather than a "shortcoming." Although good computers are close to perfect, the result is still a function of input and assumptions, and the final decision should be made on a case-by-case basis.

Considering the various limitations discussed so far, it is relatively easy to lose confidence in practical fracture mechanics as a design tool. Indeed, no method or tool is without limitations, and certainly fracture mechanics is of very little help if the user has no basic information on loads, stresses, and material properties. The same, of course, can be said of material science or stress analysis, and other branches of engineering methodology.

It is also easy to question the limitations when using well-defined terms such as LEFM or EPFM. LEFM is designed specifically to deal with plane strain, and yet the procedures are essentially the same for plane stress, plane strain, and a transitional state of stress. Of course, if the toughness becomes unusually high, certain interpretations are in order, so that other branches of engineering methodology may be more appropriate. And what happens to fracture strength when small cracks face the plane-strain environment or elastic–plastic conditions? And in the mathematical sense, what happens to fracture stress if $a \to 0$? Is LEFM still conservative when fracture strength becomes infinite? And what becomes of ASTM thickness standards when fracture toughness is very high? And so on. Broek has provided a brief philosophical look at the problem of misconceptions[9] that ties in very well with the topic of limitations and errors; that discussion also teaches that no branch of engineering methodology is perfect, and that from a practical point of view, this methodology is not likely to improve in the climate of the modern economy in this country and elsewhere. More papers are published every day, with fewer practical results and less pragmatic wisdom. Without waiting for technical improvements, currently available tools of fracture mechanics, in unison with material science and stress analysis, can go a long way toward the solution of engineering problems, in spite of limitations and errors. And, collapse and fracture conditions in fracture control will still be competing 100 years from now.

SYMBOLS

A	Cross-sectional area, in.^{2} (mm^2)
a	Crack length or depth, in. (mm)
a_{CR}	Critical crack length, in. (mm)
a_i	Length of arrested crack, in. (mm)
a_0	Initial crack length, in. (mm)
a_p	Permissible length of crack (or flaw), in. (mm)
CTOD	Crack tip opening displacement, in. (mm)
da	Crack increment, in. (mm)
dN	Number of stress cycles
E	Elastic modulus, ksi (MPa)
H	Growth time of crack, hr
h	Topographic height, in. (mm)
J	J integral, lb/in. (N/mm)
J_R	J integral for nonlinear material, lb/in. (N/mm)
K	Stress intensity factor, ksi (in.)$^{1/2}$ [MPa (m)$^{1/2}$]
K_c	Plane-stress fracture toughness, ksi (in.)$^{1/2}$ [MPa (m)$^{1/2}$]
K_I	Stress intensity factor (Mode I), ksi (in.)$^{1/2}$ [MPa (m)$^{1/2}$]
K_{Ia}	Stress intensity factor, arrest, ksi (in.)$^{1/2}$ [MPa (m)$^{1/2}$]
K_{Ic}	Plane-strain fracture toughness, ksi (in.)$^{1/2}$ [MPa (m)$^{1/2}$]
K_{Id}	Dynamic fracture toughness, ksi (in.)$^{1/2}$ [MPa (m)$^{1/2}$]
K_{IP}	Stress intensity at pressure, ksi (in.)$^{1/2}$ [MPa (m)$^{1/2}$]
K_{IR}	Reference stress intensity, ksi (in.)$^{1/2}$ [MPa (m)$^{1/2}$]
K_{Iscc}	Stress intensity at stress-corrosion, ksi (in.)$^{1/2}$ [MPa (m)$^{1/2}$]
K_{IT}	Stress intensity at temperature, ksi (in.)$^{1/2}$ [MPa (m)$^{1/2}$]
M	Bending moment, lb-in. (N-mm)
P	Concentrated load, lb (N)
RT_{NDT}	Room temperature NDT, °F
S_u	Ultimate strength, ksi (MPa)
S_y	Yield strength, ksi (MPa)
S_{yd}	Dynamic yield strength, ksi (MPa)
T	Temperature, °F
Z	Section modulus, in.^{3} (mm)^3

β	Parameter in stress intensity equation
β	General symbol for thickness, in. (mm)
ΔK_{I}	Stress intensity factor fluctuation, ksi $(\mathrm{in.})^{1/2}$ $[\mathrm{MPa}\,(\mathrm{m})^{1/2}]$
σ	General symbol for stress, ksi (MPa)
σ_{des}	Design stress level, ksi (MPa)
σ_{f}	Flow stress, ksi (MPa)

REFERENCES

1. Barsom, J.M.; Rolfe, S.T. *Fracture and Fatigue Control in Structures*, 2nd Ed.; Prentice-Hall: Englewood Cliffs, NJ, 1987.
2. Blake, A. *Design of Mechanical Joints*; Marcel Dekker: New York, 1985.
3. Marine Casualty Report. *Structural Failure of the Tank Barge I.O.S. 3301 Involving the Motor Vessel Martha R. Ingram on 10 January 1972 Without Loss of Life*, Report No. SDCG/NTSB, 1974.
4. *Collapse of U.S. 35 Highway Bridge, Point Pleasant, West Virginia*, NTSB Report No. NTSB-HAR-71-1; National Transportation Safety Board: Washington DC, 1968.
5. Bennett, J.A.; Mindlin, H. Metallurgical Aspects of the Failure of the Pt. Pleasant Bridge. J. Test. Eval. **1973**.
6. Fisher, J.W. *Fatigue and Fracture in Steel Bridges — Case Studies*; John Wiley: New York, 1984.
7. Irwin, G.R. Fracture Dynamics. In *Fracturing of Metals*; American Society of Metals: Cleveland, 1948.
8. Witherell, C.E. *Mechanical Failure Avoidance*; McGraw-Hill: New York, 1994.
9. Broek, D. *The Practical Use of Fracture Mechanics*; Kluwer Academic Publishers: Dordrecht, 1988.
10. Broek, D. Correlation Between Stretched Zone Size and Fracture Toughness; *ICF Conference*, Munich, 1973.
11. Ryder, D.A. Elements of Fractography; AGARDograph 155-71, 1971.
12. Broek, D.; Some contributions of electron fractography to the theory of fracture. Int. Met. Rev. **1974**, Review 185 9.
13. American Association of State Highway and Transportation Officials. *Standard Specifications for Highway Bridges*; AASH TO: Washington, DC, 1977.
14. Royer, C.P.; Rolfe, S.T.; Easley, J.T. Effect of strain hardening on bursting behavior of pressure vessels. *Second International Conference on Pressure Vessel Technology: Part II — Materials, Fabrication and Inspection*; American Society of Mechanical Engineers: New York, 1973.
15. Lemcoe, M.M.; Pickett, A.G.; Whitney, C.L. *Cyclic Pressure Tests of Large Size Pressure Vessels*, Progress Reports, Project No. 733-3; Southwest Research Institute: San Antonio, Texas, 1962.
16. Blake, A. *Practical Stress Analysis in Engineering Design*, 2nd Ed.; Marcel Dekker: New York, 1990.
17. Madison, R.B. Irwin, G.R. Fracture analysis of Kings Bridge, Melbourne. J. Structural Div. ASCE **1971**, 97, No. ST9.

18. PVRC Ad Hoc Task Group on Toughness Requirements. PVRC recommendations on toughness requirements for ferritic materials. WRC Bull. **1972**; No. 175.
19. AASHTO. *Guide Specifications for Fracture Critical Non-Redundant Steel Members*; AASHTO: Washington, DC, 1978.
20. American Society of Mechanical Engineers (ASME). Boiler and Pressure Vessel Code, Sect. III; ASME: New York, 1986.
21. Pellini, W.S. *Principles of Structural Integrity Technology*; Department of the Navy, Office of Naval Research: Arlington, VA, 1976.

13

Design Considerations

BASIC CHARACTERISTICS

The main assignment during the application of practical elements of fracture mechanics to design is to protect the machines and structures from unsuspected and uncontrollable structural failures. For this purpose it is prudent to assume that all engineering materials contain some flaws and irregularities, which at the appropriate stress levels can initiate the process of fracture. In particular, high-strength and low-toughness materials can be subject to brittle behavior, and therefore a part of the designer's effort should be devoted to estimating the correct level of working loads that can be sustained without causing crack propagation. In order to assure such a condition it is necessary to select the appropriate design formulas and the certified mechanical properties. Such input is discussed in various portions of this book. It is also shown that a compound parameter such as plane-strain fracture toughness involves the applied stress and the square root of the crack size with the specified numerical multiplier. This parameter remains of primary interest. It is also proper to comment here that although literally thousands of K formulas have appeared in the handbooks and technical papers over the past 40 years or so, it is surprising how often we tend to rely on a simple expression such as $[\sigma(a)^{1/2}]$, multiplied by a suitable constant term. Also the "theory and practice" shows that as long as this product is kept below, say, the K_{Ic} magnitude for the selected material, the presence of crack a for the case at hand should be harmless. In a sense this situation leads to a comparison with the case in stress analysis where the structural element is considered to be safe and the working stress is less, by some acceptable margin, than the yield strength of the material. Therefore, in terms of the elementary symbols, one can state that

$$\sigma(a)^{1/2} \times (\text{constant}) < K_{Ic} \qquad \text{Fracture mechanics}$$

and

$$\sigma < S_y \qquad \text{Stress analysis}$$

It is clear that the S_y and K_{Ic} terms must represent material properties.

In conventional design the key material property has varied over the years, starting perhaps with the tensile strength S_u, progressed to yield stress S_y, and later to the concept of 0.2% proof stress or even microyield, depending on the codes of practice or a degree of sophistication. Similarly, the factors of safety have also varied, depending on the application and weight criticality, say, from 1.5 to maybe as high as 20. The prime function of the factor of safety, of course, is to account for the unknown effects such as fabrication, assembly, and service. Unfortunately, none of the factors of safety recognized the potential of brittle or fast fracture, often at working stresses well below the specified design level. In a strict sense, however, the onset of a brittle failure still requires the presence of a stress concentrator of some sort to trigger the crack propagation. Defects in the form of cracklike discontinuities in the areas of high local stresses will normally serve as crack starters. In terms of fracture mechanics, cracklike defects can involve solidification cracking of welds and castings, lamellar tearing around inclusions, hydrogen cracking in heat-affected zones, and subcritical crack propagation in fatigue or stress-corrosion. These characteristics therefore should constitute a part of the input for the selection of formulas and properties.

The original works of Griffith,[1] Irwin,[2] and Orowan,[3] which formed the basis of modern fracture mechanics, utilized the concepts of surface energy and the critical strain energy release rate, which correlated with the later definitions of the fracture mechanics parameters under plane-strain and plane-stress conditions. These involve theoretical and experimental aspects of fracture mechanics studied in this country and elsewhere, as well as guidelines for control of material toughness specifications. Current techniques deal more readily with the stress intensity factors, transition temperature criteria, J-integral approach, and the crack tip opening displacement (CTOD) methodology. The emphasis in this book, however, is on linear elastic fracture mechanics (LEFM) and the fracture-safe design (transition temperature) applications. The goal here is to transmit the information via selected worked examples, case studies supported by experience, and the balanced approach to design using the elements of fracture analysis, materials behavior, and stress analysis.

The use of the fundamental LEFM parameter K_{Ic} permits setting up the key relationships between the critical crack size and the maximum allowable stresses under elastic conditions as long as the local plastic effect at the crack tip is not significant. When the volume of plastically deformed material is appreciable, the LEFM approach can be invalid and the problem may require CTOD or J-integral application. Under the conditions of gross plastic instability, the onset of failure is governed by stiffness, flow stress, and geometrical parameters, in the immediate vicinity of the crack tip. Unfortunately, the CTOD parameter obtained in the laboratory may show little relevance to the failure of a large structure.

The applied nominal stress around the stress concentrations involving residual stresses may not be easy to specify since stress analysis of locations near the residual stresses is likely to be highly complicated.

Finally the evaluation of closely spaced defects is not a simple matter because the cluster problem has not been well understood, particularly in the areas of poor access and awkward geometrical configurations. Also, the calculated size of the critical defect may prove to be smaller than the resolution limit of the conventional inspection equipment. However, despite such limitations LEFM is still a powerful analytical tool. Hence the specific mission of this text is to emphasize the point that structures, being far from perfect solids, can still operate safely within approved critical crack growth limits developed with the help of equations and tips from practical fracture mechanics.

SELECTED FORMULAS AND DEFINITIONS

The stress field around the edge of an elliptical hole originally described by Inglis' formula, Eq. (2.3), has shown that for a very small minor half-axis in relation to the major half-axis, the stress concentration factor can become rather high. In fact, when the corner radius, approaching the condition of a sharp crack, tends to zero, the theoretical stress concentration factor becomes infinite and loses its meaning as an analytical method. At this point the Griffith theory comes to the rescue with the following expression defining the stress at fracture:

$$\sigma = \left[\frac{2\gamma E}{\pi a(1 - v^2)} \right]^{1/2} \tag{13.1}$$

This formula defines an energy balance under conditions of linear-elastic behavior, where the term (2γ) repesents the work of fracture equal to twice the surface energy. Hence the relationship is established for the plane-strain case between the stress at fracture and crack length if the material's work of fracture is known. In this expression a denotes the half-length of the crack and E is Young's modulus.

This theory was later modified by Irwin and Orowan to make a correction for the presence of a small plastic zone at the crack tip:[1-3]

$$\sigma = \left[\frac{EG_{Ic}}{\pi a(1 - v^2)} \right]^{1/2} \tag{13.2}$$

In this type of formula G_{Ic} represents the critical strain energy release rate in the opening mode of the crack, and v is Poisson's ratio. In metals, the G_{Ic} parameter signifies the amount of plastic work to be done before crack extension. Under plane-stress conditions, the formula reduces to

$$\sigma = \left(\frac{EG_{Ic}}{1 - v^2} \right)^{1/2} \tag{13.3}$$

Equations (13.2) and (13.3) can be used to calculate the maximum applied stress when a component has a crack of known length, on the premise that strain energy release rate is available for the case at hand.[4]

The basic relationships between the various parameters of fracture mechanics have been established during the process of extending linear-elastic theory into the elastic–plastic region of fracture mechanics. These relationships are helpful in the application of the working formulas to design.[5] In this case, in addition to the G_{Ic} parameter we can have a J_{Ic} term, which denotes a measure of fracture toughness determined at the instant of the initiation of crack growth in metallic materials.[6] Hence for plane strain

$$K_{Ic} = \left(\frac{EG_{Ic}}{1 - v^2} \right)^{1/2} \tag{13.4}$$

$$K_{Ic} = \left(\frac{EJ_{Ic}}{1 - v^2} \right)^{1/2} \tag{13.5}$$

and for plane stress, the approximate formula is

$$K_c = (EJ_c)^{1/2} \tag{13.6}$$

If the energy release rate is defined by G and if it signifies the elastic energy per unit crack surface available during a minute crack extension, then using the plane-strain conditions, the appropriate formulas become[2,3]

$$G = \pi \sigma a^2 / E \tag{13.7}$$

and

$$G = \frac{K_I^2}{E}(1 - v^2) \tag{13.8}$$

or

$$GE = K_I^2(1 - v^2)$$ (13.9)

$$GE = \pi \sigma^2 a$$ (13.10)

$$\sigma = \left(\frac{GEa}{\pi}\right)^{1/2}$$ (13.11)

and, from Eqs. (13.8) and (13.10), or Eqs. (13.7) and (13.8), we obtain

$$\sigma = \frac{K_I(1 - v^2)^{1/2}}{(\pi a)^{1/2}}$$ (13.12)

From Eq. (13.4)

$$EG_{Ic} = (1 - v^2)K_{Ic}^2$$ (13.13)

Hence, combining Eqs. (13.2) and (13.13), the original Griffith expression becomes the conventional LEFM formula for plane-strain fracture toughness

$$K_{Ic} = \sigma(\pi a)^{1/2}$$

Similarly, taking

$$K_c = (EG_c)^{1/2}$$

and combining this relation with the Griffith formula, Eq. (13.3), yields

$$K_c = \sigma(\pi a)^{1/2}$$

As stated previously in dealing with various aspects of LEFM, the level of the nominal stress near the crack tip in a structural member is given in terms of the single parameter K, known as the stress intensity factor. In looking back on the history of the development of fracture mechanics, it has to be fully recognized that a quantitative assessment of structural integrity of the machine and structural components was possible because of the science of stress analysis. Using the equations of elasticity, Westergaard[7] and Irwin[8] were able to define the stress and displacement fields in the immediate vicinity of crack tips, assuming three modes of deformation. The most practical solution was Mode I, treated in more pragmatic books and papers. The stresses and deformations were given in terms of K_I and polar coordinates, with regard to x-, y-, and z-directions. An example of dealing with crack tip stresses utilizing a stress analysis technique[4] is presented here as a practical introduction to the specific problems of stress intensity factors, a number of which are included in Chapter 3 of this book, along with elementary calculations. Practical situations in the real engineering world cannot be modeled using simple configurations, so that the majority of design formulas involve approximations to the stress distribution and geometric shapes, as well as corrections accounting for the presence of free surfaces. These corrections are concerned with those factors, used in formulas for stress intensity, where free edge notches are perpendicular to the applied tensile stresses. Examples of "free-surface correction" include the constant factor of 1.12, Eq. (3.4) or M_K, in Eq. (3.23) and Fig. 3.19.

Energy release as a function of crack tip stresses can be, in its simplest form, discussed with the help of the sketch in Fig. 13.1. The stress field in the vicinity of the crack was taken to be

$$\sigma = \frac{\sigma_a}{\left(1 - \frac{a^2}{x^2}\right)^{1/2}} \tag{13.14}$$

Near the tip of the crack, $x = a$, while Eq. (13.14) gives an infinitely high value of the stress shown by the trend of the curve in Fig. 13.1. Also, as the distance x is made large, the stress approaches an asymptotic value of σ_a. It may now be of interest to follow the derivation of the stress intensity factor, starting from Eq. (13.14).[9]

From Eq. (13.14)

$$\sigma = \frac{x\sigma_a}{(x^2 - a^2)^{1/2}} \tag{13.15}$$

In order to be consistent with the notation used by Westergaard[7] and Irwin,[8] the distance ahead of the crack tip is given as

$$r = x - a \tag{13.16}$$

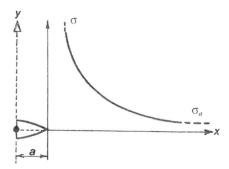

FIGURE 13.1 Stress field near crack.

Combining Eqs. (13.15) and (13.16) yields

$$\sigma = \frac{(r + a)\sigma_a}{[(r^2 + 2ra + a^2) - a^2]^{1/2}}$$

$$= \frac{(r + a)\sigma_a}{(r^2 + 2ra)^{1/2}} \tag{13.17}$$

when $r \ll a$

$$(r + a) \rightarrow a$$

and

$$(r^2 + 2ra) \rightarrow 2ra$$

Hence, simplifying Eq. (13.17) gives

$$\sigma = \left(\frac{a}{2r}\right)^{1/2}\sigma_a \tag{13.18}$$

Multiplying numerator and denominator by $(\pi)^{1/2}$, Eq. (13.18) transforms into

$$\sigma = \frac{\sigma_a(\pi a)^{1/2}}{(2\pi r)^{1/2}} \tag{13.19}$$

or

$$\sigma = \frac{K}{(2\pi r)^{1/2}} \tag{13.20}$$

According to the solution by Westergaard, the primary interest in the stress field along the y-axis, Fig. 13.1, can be represented by the following relation in the fundamental tensile Mode I, opening the crack:

$$\sigma = \frac{K_1}{(2\pi r)^{1/2}}\cos\frac{\theta}{2}\left[1 + \sin\frac{\theta}{2}\sin\frac{3\theta}{2}\right] \tag{13.21}$$

For the stress perpendicular to the crack plane (in the y-direction, Fig. 13.1) to be a maximum, $\theta = 0$, which substituted in Eq. (13.21) gives

$$\sigma = \frac{K}{(2\pi r)^{1/2}} \tag{13.22}$$

Equations (13.20) and (13.22) are essentially identical in form, where $K = K_I$ in the first mode, defining the stress intensity factor and having the units of ksi (in.)$^{1/2}$ or MPa (m)$^{1/2}$. Other units in scientific notation give MN (m)$^{-3/2}$.

The effect of approximation using $r \ll a$ can be illustrated for the likely ratio $(a/r) = 50$, as follows:

$$r = 0.02a$$

$$x = r + a$$

$$= 1.02a$$

From Eq. (13.15)

$$\sigma = \frac{1.02a\sigma_a}{(1.02^2 - 1)^{1/2}a}$$

$$= 5.075\sigma_a$$

From Eq. (13.18)

$$\sigma = \left(\frac{a}{2r}\right)^{1/2}\sigma_a$$

$$= \left(\frac{a}{0.04a}\right)^{1/2}\sigma_a$$

$$= 5.000\sigma_a$$

and the error is

$$\left(\frac{5.075 - 5.000}{5.075}\right) \times 100 = 1.48\%$$

The case of a plate with finite width $(2w)$ and a centrally located crack $(2a)$ long, as shown in Fig. 3.1, appears often in design calculations. The two well-known problem formulations in estimating the stress intensity factors and fracture toughness of the material are based on "secant" and "tangent" functions. Chapter 3 provides a design curve for a quick estimate when (a/w) ratios fall below about 0.7, based on the "secant" solution. The designer also has the "tangent" type formula available for the calculation, such as

$$K_I = 1.4\sigma[w \tan(1.57a/w)]^{1/2} \tag{13.23}$$

The corresponding "secant" formula is

$$K_I = 1.77\sigma[a \sec(1.57a/w)]^{1/2} \tag{13.24}$$

To check the agreement between the two solutions, the ratio of tangent to secant solutions R_K follows directly from Eqs. (13.23) and (13.24).

$$R_K = 0.8[(w/a) \sin(1.57a/w)]^{1/2} \tag{13.25}$$

Assuming the maximum ratio $a/w = 0.7$, consistent with the approximate design curve plotted in Fig. 3.2, we obtain

$$R_K = 0.8[(1/0.7)\sin(1.57 \times 0.7)]^{1/2}$$

The variation of R_K with the (a/w) ratio is shown in Fig. 13.2. It appears that below the 0.1 ratio, there is virtually no difference between the two formulations of the stress intensity levels. In all, there is good agreement between the two formulas, although the majority of workers in the field regard the "secant" formulation as the more accurate of the two.

The design analysis of the stress intensity factor for a double-edge notched, infinitely wide plate under uniform tensile load is extremely simple, as evident from the following.

$$K_I = 2\sigma(a)^{1/2} \tag{13.26}$$

However, this situation deteriorates rather quickly as the panel width dimension becomes finite. In this instance the first practical remedy is to look for an approximation such as employing "secant" correction, Eq. (3.2), in conjunction with Eq. (13.26). This should give

$$K_I = 2\sigma[a\sec(1.57a/w)]^{1/2} \tag{13.27}$$

However, more rigorous solutions are available in the literature[10–12] by accounting for panel length to width ratio effects. Another possibility (still without the influence of panel length) involves the corrections based on "tangent" and "sine" formulations.[4] The relevant formula for the stress intensity factor for two

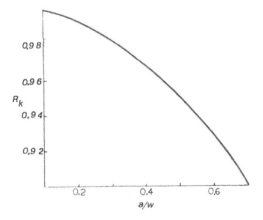

FIGURE 13.2 Ratio of calculated stress intensities (trigonometric).

symmetrical edge cracks in a panel of finite width is

$$K_I = 2\sigma[a \sec(1.57a/w)]^{1/2} \tag{13.28}$$

The ratio of Eqs. (13.28) and (13.27) provides a check on the two solutions involving mixed trigonometric functions. This gives

$$R_L = 0.7 \left(\frac{w}{a}\right)^{1/2} \left[\sin(1.57a/w) + \frac{\sin(\pi a/w)}{10 \sec(1.57a/w)}\right] \tag{13.29}$$

The choice between Eqs. (13.27) and (13.28) is open because neither equation involves the length of the panel. The more precise solutions featuring the effect of length-to-width ratio of the panel on the stress intensity also include the influence of the ratio of crack length to panel width on the free-surface correction factor.[5]

The variation of R_L with the (a/w) ratio is illustrated in Fig. 13.3. Although in this case, the discrepancy between the two formulations is a little greater, this is not surprising due to the increase in overall complexity of the problem.

In addition to formulations based on the trigonometric functions, the factors for the calculation of stress intensities have been derived in the form of polynomials. For instance, using the case of a double-edged crack in a panel of finite width and a polynomial solution,[13] the stress intensity formula becomes

$$K_I = \sigma(a)^{1/2}\left[1.99 + 0.38\left(\frac{a}{w}\right) - 2.12\left(\frac{a}{w}\right)^2 + 3.43\left(\frac{a}{w}\right)^3\right] \tag{13.30}$$

In comparing this result with Eq. (13.27), the following ratio is obtained.

$$R_P = \frac{0.995 + 0.19(a/w) - 1.06(a/w)^2 + 1.715(a/w)^3}{[\sec(1.57a/w)]^{1/2}} \tag{13.31}$$

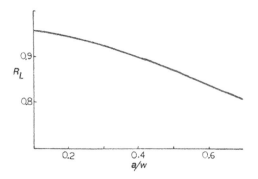

FIGURE 13.3 Ratio of calculated stress intensities (mixed functions).

The comparison of the "secant" solution with the polynomial function is illustrated by plotting the R_P ratio as a function of (a/w) (Fig. 13.4).

The case of a single-edge cracked panel of a finite width included in Chapter 3 was based on combining the tabulated results accepted by the American Society for Testing and Materials (ASTM) Standard[12] and the free-surface correction, but without the effect of the length-to-width ratio of the panel. This effect was evaluated in some detail by Broek[13] on the premise that the (a/w) ratio for the single-edge cracked panel did not exceed 1.0, because cracks of this type, larger than the half-width of the panel, did not hold special technical interest. An example of a specific formula for $(L/w) = 4$, length to half-width ratio of panel with a single-edge crack can be stated as follows:

$$K_I = \sigma(\pi a)^{1/2}\left[1.12 - 0.115\left(\frac{a}{w}\right) + 2.64\left(\frac{a}{w}\right)^2\right.$$
$$\left. - 2.72\left(\frac{a}{w}\right)^3 + 1.90\left(\frac{a}{w}\right)^4\right] \tag{13.32}$$

In checking Eq. (13.32) against the design curve in Fig. 3.6, the differences, for all practical purposes, were found to be insignificant.

The topic of through-cracks emanating from circular holes was briefly discussed in Chapter 3. This section includes additional comments and formulas since the interaction of a crack and a hole continues to be of interest and relates to the conventional idea of stress concentration. Cracks have a habit of starting in high-stress fields caused by structural discontinuities such as fastener or access holes. In fact, based on some aerospace records,[14] about 30% of crack origins were traced to bolt and rivet holes.

As far as the calculation of stress intensity is concerned, it is fortunate that at least elementary through-the-thickness cracks starting from holes can be analyzed with a degree of success, unless these are corner cracks and surface flaws subjected to interference load transfer or residual stress systems. Uncertainties

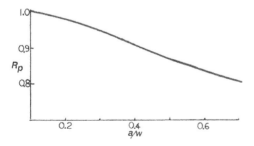

FIGURE 13.4 Ratio of calculated stress intensities (polynomials).

still persist in the areas of flaw development and anisotropy as well as mechanisms of fracture and fatigue. The reverse problem, of course, is also involved when the cracks are approaching the holes and where the expected effect is an arrest of a fast-moving unstable crack.[15]

The stress intensity factor for two symmetrically located cracks (Fig. 3.8) in a very wide plate with a circular hole can be calculated from

$$K_I = 1.77\sigma(a + r)^{1/2} \tag{13.33}$$

In this equation, the term $(a + r)$ represents the effective half-length of the crack emanating from a hole. The formula often used in this case is simply written as

$$K_I = \sigma(\pi a_e)^{1/2} \tag{13.34}$$

and it can be used in many applications for the symmetric case. The formula for the asymmetric case is

$$K_I = 1.25\sigma(2r + a)^{1/2} \tag{13.35}$$

The results obtained from Eqs. (13.33) through (13.35) correlate best with the exact solutions where (a/r) ratios are higher than 0.2. The definition of the effective length a_e is indicated in Fig. 13.5. Cases (1) and (2) represent the asymmetric (single-crack) and symmetric (double-crack) configurations. Another way of calculating the stress intensity for the two cases given in Fig. 13.5 is to follow the least square interpretation of Bowie's work,[15,16] which yields the following two expressions. For the symmetric configuration

$$K_I = \sigma(a)^{1/2}\left[\frac{1.217}{0.277 + (a/r)} + 1.673\right] \tag{13.36}$$

and for the asymmetric single-crack case

$$K_I = \sigma(a)^{1/2}\left[\frac{1.548}{0.325 + (a/r)} + 1.200\right] \tag{13.37}$$

The foregoing equations apply to plates of infinite width.

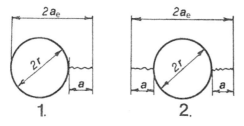

FIGURE 13.5   Notation for crack hole configurations.

The selection of formulas throughout this book is governed by the immediate design need to make the first estimate of the stress intensity factors related to the simplest geometry, stress conditions, and environment. Any treatment of the analytical aspects and applications of fracture mechanics to design can only be superficial because there are so many parameters involved in any rigorous treatment of the basic problems that still await solutions. The comments that follow immediately are based on a number of reviews and observations,[5,13,17] some of which may be more relevant to research than to design. The state of the art points to the fact that certain problems in fracture mechanics may only be dealt with in a speculative manner, no matter how much LEFM practitioners would like to see them in a more deterministic fashion. The final report is still not out on a number of practical issues.

The calculational effort in the area of stress intensity factors in a two-dimensional world has improved, although the formulas for crack behavior in fastener-filled holes, interference fasteners, and cracks approaching holes require constant revisiting and refinement.

Economics holds down any progress, including finite-element effort on entering any three-dimensional assessment of crack behavior, and there is little hope of seeing closed-form solutions and empirical formulas for general applications in more than two dimensions.

It is well at this point to mention the role and effect of fasteners, which are so often underrated in the importance and complexity areas of performance. Welding joints, in spite of the progress, have not been able to overshadow many applications where a bolt or a rivet has maintained its influence on numerous branches of industry. Unfortunately, the theoretical problem of a crack in a bolted joint is not something for which the analyst can invent a general formula. All that is really known is that if the fastener is a loose fit, and when there is no load transfer from the hole surface, there will be an insignificant effect on the crack emanating from this hole. However, with a tight fit, load transfer, and stress redistribution, the stress intensity at the cracked hole will be different from that at an open hole (loose fit) condition. In general, while there appear to be some beneficial effects of interference and cold working of the holes, the crack propagation rates are higher. Unfortunately, such trends cannot be generalized, and systematic test data are hard to find. And the problem becomes more difficult as new parameters and effects are recognized as the building blocks of technical decisions.

And what can be learned from the holes themselves as potential crack arresters? At least qualitative analysis and some experiments indicate that the arrest of a fatigue crack at a hole may be counterbalanced by an increase in stress intensity factor and the increase of a defect size in creating the parameter a_e as the hole becomes instantaneously a part of the effective length of the crack. The situation of crack arrest is better in redundant structures, with the advent, however, of new parameters to be considered in basic understanding of the problem.

Although not much has been said in this chapter about the constant problems with the assumed defect sizes and shapes, it is clear from this and other pragmatic publications that the crack size plays an outstanding role in research and design formulations in all aspects of fracture analysis and control. The advent of fracture mechanics means the intended use of crack size as a design parameter. For instance, the airworthiness requirements in modern engineering are aimed at fail safety and damage tolerance through the development of reliable inspection procedures for crack detection on the order of 0.01 in. size or less. It should always be noted that in using design formulas, there is a certain convention in fracture mechanics in referring to crack sizes. For all cracks having "two characteristic tips," the crack length is denoted by $2a$. In cracks with "one tip," the length is called a. While this statement is elementary in nature, the crack length convention becomes important when comparing design formulas and data generated from various sources.

In closing this brief section on the formulas and definitions that are most likely to be of interest to design engineers rather than experts in the field of fracture mechanics, it is well to emphasize the general form and the meaning of the stress intensity factor, which has been the main reason for extensive investigations and research for about 40 years.

The key feature of the stress intensity factor (K_I in the first mode) is that it relates the local stress at the tip of a sharp crack in a structural component to the applied nominal (global) stress away from the crack. The stress intensity K_I represents the Mode I type of loading under which most brittle fractures are likely to occur. The K_I parameter is directly proportional to the applied stress σ and the square root of the crack length a. It is also very significant that the general form of the stress intensity factor is the same for all cases. It says essentially that

$$K_I = \sigma \cdot (a)^{1/2} \times \text{geometry factor}$$

In the latest work of Broek,[13] there is one symbol representing this factor:

$$\beta = \text{geometry factor}$$

For instance $\beta = 1$ for a panel of infinite width while

$$\beta = \left[\sec\left(\frac{\pi a}{W} \right) \right]^{1/2}$$

corresponds to the plate of finite width equal to W. In this book, the width is generally referred to as $W = 2w$. Under these assumptions the function β can have many forms depending on the geometrical details, and it is the only part of the formula for K_I that has to be derived.

$$K_I = \beta\sigma(\pi a)^{1/2} \tag{13.38}$$

The equivalent formula in the latest textbook of Barsom and Rolfe[5] is

$$K_I = \sigma(a)^{1/2} \cdot f(g) \tag{13.39}$$

Whatever symbols and conventions of the geometry factors such as β or $f(g)$ may be preferred, handbooks show many formulas of various degrees of complexity[10,11,18,19] for different generic configurations, and the designer should be advised that selecting any of such formulas for practical problems can prove to be a major task. In the interim, the best advice may be to try the simpler formulations[5] and procedures combining the elements of LEFM, material property data, and stress analysis to arrive at a preliminary technical commitment.

SPECIAL PROBLEMS

The topic of practical fracture mechanics in design is intended to convey the idea that the term "design" means the synthesis of LEFM methodology, materials characteristics, and stress analysis for the purpose of either preventing brittle fracture or reviewing the results of a structural failure. Experience with using fracture mechanics tools with "before" and "after the fact" structural conditions is equally important in learning on the job. And the learning process appears to be the best when it involves even the most rudimentary calculations, interpretation of formulas, and analysis of the results. This brief section deals with some applications of the design formulas from this text as well as other sources. This information is set up in the form of specific design comments in order to have an easier reference.

Design Comment 13.1

This case is concerned with the determination of the stress intensity factor from a standard compact tension (CT) specimen in accordance with the ASTM Test Method.[20] Several specimen geometries are allowed as long as the plane-strain test conditions are complied with where the crack length and the specimen thickness are greater than $2.5(K_{Ic}/S_y)^2$.

The actual input from industry[21] in this example called for the following specific parameters:

$\sigma = 10$ psi
$a = 0.4$ in.
$w = 1.0$ in.
$a/w = 0.4$

$$K = \sigma(a)^{1/2} F_1 \tag{13.40}$$

$$F_1 = \frac{2(2 + a/w) F_2}{(1 - a/w)^{3/2} (a/w)^{1/2}} \tag{13.41}$$

$$F_2 = 0.443 + 2.32(a/w) - 6.66(a/w)^2 + 7.36(a/w)^3 - 2.8(a/w)^4 \tag{13.42}$$

Substituting gives

$$F_2 = 0.443 + 2.32 \times 0.4 - 6.66 \times 0.16 + 7.36 \times 0.064 - 2.8 \times 0.0256$$
$$= 0.704$$

and

$$F_1 = \frac{2(2 + 0.4) \times 0.704}{(1 - 0.4)^{1.5} \times (0.4)^{1/2}}$$
$$= 11.49$$

so that

$$K = 10(0.4) \times 11.49$$
$$= 72.7 \, \text{psi (in.)}^{1/2}$$

The foregoing calculation is next compared with the results based on Eqs. (3.32) and (3.34).

$$K_{\mathrm{I}} = \frac{P}{B(w)^{1/2}} \cdot f\left(\frac{a}{w}\right)_c$$

where

$$f\left(\frac{a}{w}\right)_c = 29.6\left(\frac{a}{w}\right)^{0.5} - 185.5\left(\frac{a}{w}\right)^{1.5} + 655.7\left(\frac{a}{w}\right)^{2.5} - 1017\left(\frac{a}{w}\right)^{3.5}$$
$$+ 639\left(\frac{a}{w}\right)^{4.5}$$

Substituting again yields

$$f\left(\frac{a}{w}\right)_c = 29.6 \times 0.63 - 185.5 \times 0.25 + 655.7 \times 0.10 - 1017 \times 0.04$$

$$+ 639 \times 0.016$$

$$= 7.33$$

and for $B = w = 1$

$$K_{\mathrm{I}} = 10 \times 7.33$$
$$= 73.3\,\mathrm{psi}\ (\mathrm{in.})^{1/2}$$

Hence

$$K/K_{\mathrm{I}} = 72.8/73.3$$
$$= 0.993$$

Considering the differences between $f(a/w)_{\mathrm{c}}$ and the F_1 and F_2 parameters,[22] the agreement between the numerical results is very close. This is an example of dealing with the different forms of the factors expressing the effect of specimen geometry.

Design Comment 13.2

One of the advantages of adopting fracture mechanics methodology is that it is possible to evaluate the structural integrity of a component in service provided the inspection technique can detect and define the extent of a particular crack. This design problem deals with a steel pressure vessel that has a longitudinal crack that can propagate under certain conditions, driven by the internal pressure. The inspection team has defined the defect as a sharp crack, 4 in. long and 0.8 in. deep. The crack is located on the inside wall of the vessel, which has an inner diameter of 48 in. and a wall thickness of 2 in. The vessel material has a yield strength of 80 ksi, and the crack resistance property defined in terms of the J_{Ic} parameter amounts to 260 in.-lb/in.^{2} The task for the design engineer is to evaluate the operational safety of the vessel subjected to 4000 psi internal pressure.

The conventional membrane hoop stress is

$$\sigma = P_i R/t$$

where $t = 2$ in., $R = (48 + 52)4 = 25$, $\sigma = 25 \times 4000/2 = 50{,}000$ psi.

The plane-strain fracture toughness can be estimated from

$$K_{\mathrm{Ic}} = (J_{\mathrm{Ic}} E)^{1/2}$$
$$= (260 \times 30 \times 10^6)^{1/2}$$
$$= 88.3\,\mathrm{ksi}\ (\mathrm{in.})^{1/2} \tag{13.43}$$

The stress intensity factor follows from Eq. (3.23)

$$K_{\mathrm{I}} = 2\sigma M_{\mathrm{K}} \left(\frac{a}{Q}\right)^{1/2}$$

where the front free-surface correction is

$$M_K = 1.3 \tag{13.44}$$

When the crack depth is less than one-half of the wall thickness, it is customary to assume $M_K = 1.0$, so that the simplified formula for calculating stress intensity for a part-through thumbnail crack in a uniform tensile field becomes

$$K_I = 2\sigma(a/Q)^{1/2} \tag{13.45}$$

The parameter Q, known as the flaw shape factor, can be obtained from a standard chart such as that shown in Fig. 3.21. For this procedure, two ratios are required.

$$a/2c = 0.8/4$$
$$= 0.2$$

and

$$\sigma/S_y = 50/80$$
$$= 0.625$$

Hence, the chart gives

$$Q = 1.21$$

and the stress intensity is

$$K_I = 2 \times 50(0.8/1.21)^{1/2}$$
$$= 81.3\,\text{ksi (in.)}^{1/2}$$

Making now

$$K_I = K_{Ic}$$

the critical crack size can be found after rearranging Eq. (13.45).

$$a_{CR} = \frac{Q}{4}\left(\frac{K_{Ic}}{\sigma}\right)^2$$
$$= \frac{1.21}{4}\left(\frac{88.3}{50}\right)^2$$
$$= 0.94\,\text{in.} \tag{13.46}$$

The margins of safety based on the foregoing calculations are

$K_{Ic}/K_I = 88.3/81.3$

$\phantom{K_{Ic}/K_I} = 1.09$

$a_{CR}/a = 0.94/0.80$

$\phantom{a_{CR}/a} = 1.18$

If the design engineer cannot accept such margins, the only reasonable alternative is to certify the pressure vessel for a lower operating pressure. A repair may be next to impossible due to the crack location.

In problems dealing with flaw analysis three basic parameters are involved, as shown in many formulas throughout this book. However, in the majority of cases the flaw size is unknown and it is necessary to assume the flaw aspect ratio $(a/2c)$. Then, for specific values of σ and K_{Ic}, the ratio (a/Q) can be determined, leading to the crack length $2c$. This procedure, of course, can be reversed for the calculation of stress.

Design Comment 13.3

To follow up on the problem of flaw analysis, the specific question may concern the calculation of the tolerable size of surface flaw, such as may be specified in production. For instance, in the case of a water-quenching operation on a steel member 2 in. thick, the thermal stress was calculated to be 20 ksi. Laboratory tests on the mechanical properties indicated 30 ksi (in.)$^{1/2}$ value of K_{Ic} and a yield strength of the material equal to 80 ksi. The maximum specified surface defect was only 0.024 in., which would be very difficult to verify during inspection. Hence the plan is to judge the severity of this flaw on the basis of fracture mechanics. To predict the tolerable defect size, the aspect ratio $(a/2c)$ is assumed to be 0.1. Since the crack depth is certainly much less than the half-thickness of the part, the correction parameter M_K can be taken as unity, and the expression for the calculation follows from Eq. (13.45).

$$\left(\frac{a}{Q}\right)_{CR} = 0.25\left(\frac{K_{Ic}}{\sigma}\right)$$

$$= 0.25\left(\frac{30}{20}\right)^2$$

$$= 0.56 \text{ in.} \tag{13.47}$$

Since

$$a/2c = 0.1$$

and

$$\sigma/S_y = 20/80$$
$$= 0.25$$

from Fig. 3.21, $Q = 1.07$, so that

$$\frac{a_{CR}}{1.07} = 0.56$$

and

$$a_{CR} = 0.6 \text{ in.}$$

Hence the flaw is not large enough to be over halfway through the section, since the steel member analyzed is 2 in. thick. The question is now, however: What would happen if the calculated thermal stress were much higher, say, close to the yield? For instance

$$\left(\frac{a}{Q}\right)_{CR} = 0.25 \left(\frac{30}{80}\right)^2$$
$$= 0.035 \text{ in.}$$

Again, using $a/2c = 0.1$ and $\sigma/S_y = 1$, Fig. 3.21 gives $Q = 0.87$, so that

$$a_{CR} = 0.87 \times 0.035$$
$$= 0.030 \text{ in.}$$

Unfortunately, this flaw size would be critical, because it exceeds the specified production limit of 0.024 in. And the inspection could not actually say if the fracture did or did not take place (difficult to identify).

Design Comment 13.4

It is, at times, fortunate to have the evidence of an embedded flaw and the reliable mechanical properties so that a comparison can be made between the LEFM and the test results. This was indeed the case in a structural failure of a rocket motor casing during the proof test at a high stress of 180 ksi while the material's yield strength was 230 ksi. Quite often with good materials control, the maximum design stress is allowed to go to about 75% of yield. The observed internal flaw after the rupture of the casing was 0.16 by 0.065 in. The plane-strain fracture toughness was determined to be 55 ksi $(\text{in.})^{1/2}$.

The formula for the calculation of the applied stress is rather straightforward as long as the correction M_K is close to unity; that is, the crack depth is equal to half-thickness of the wall t, as shown in Eq. (13.44).

$$\sigma = \frac{0.56K_{\text{Ic}}}{(a/Q)_{\text{CR}}^{1/2}} \tag{13.48}$$

The input for Fig. 3.21 is

$$a/2c = 0.065/0.16 \times 2$$
$$= 0.2$$

and, assuming $\sigma/S_y = 1$

$$Q = 1.07$$

It should be noted that a represents here one-half of the width of the flaw (small axis of the ellipse is $2a$). Hence, using Eq. (13.48) gives

$$\sigma = 0.56 \times 55/(0.0325/1.07)^{1/2}$$
$$= 177\,\text{ksi}$$

and

$$177/180 = 0.98$$

In practical terms, it is difficult to expect a better agreement between the theory and the field measurements.

It is only fair to admit that this entire book is dominated by those applications of linear elastic fracture mechanics where the stress intensity is of prime importance. Also, the types of materials, such as high-strength steels, alloy steels, and aluminum alloys, can be tested using relatively small specimens, which produce valid K_{Ic} results. However, as the field of material applications expands into the lower range of yield strength for structural purposes, down to about 25 ksi (in.)$^{1/2}$ K_{Ic} values, the LEFM standards put limits on the specimen thicknesses, specimen widths, and crack lengths. It becomes necessary to introduce plasticity corrections as the values of the K parameter for a given stress are higher than the elastic theory would indicate.

Design Comment 13.5

The limited amount of plastic deformation requires only a small correction, which can be introduced using the early approach treating the infinite body in plane stress where the size of the plastic zone was regarded as an additional amount to be added to the initial crack length. Obviously, this approximation could only be justified where the amount of plasticity was not too significant.

Suppose wide sheets of aluminum alloy, having a yield strength of 58 ksi, when tested to destruction indicated a central crack of 1 in. at 29 ksi fracture stress and a crack of 0.65 in. at the failure stress of 35 ksi. Based on the LEFM technique, the plane-stress fracture toughness comes from the following simple formula:

$$K_c = \sigma(\pi a)^{1/2} \qquad (13.49)$$

For $2a = 1.0$, $a = 0.5$ in., and

$$K_c = 29(0.5\pi)^{1/2}$$
$$= 36.3 \text{ ksi (in.)}^{1/2} \quad (39.6 \text{ MPa m}^{1/2})$$

and for $2a = 0.65$, $a = 0.325$ in., and

$$K_c = 35(0.325\pi)^{1/2}$$
$$= 35.4 \text{ ksi (in.)}^{1/2} \quad (38.5 \text{ MPa m}^{1/2})$$

Utilizing LEFM theory with a plasticity correction, the formula is

$$K_c = \sigma\left[\pi a\left(1 + 0.5\frac{\sigma^2}{\sigma_y^2}\right)\right]^{1/2} \qquad (13.50)$$

so that the two cases give

$$K_c = 36.3\left[1 + 0.5 \times \left(\frac{29}{58}\right)^2\right]^{1/2}$$
$$= 38.5 \text{ ksi (in.)}^{1/2} \quad (42 \text{MPa m}^{1/2})$$

and

$$K_c = 35.4\left[1 + 0.5 \times \left(\frac{35}{58}\right)^2\right]^{1/2}$$
$$= 38.5 \text{ ksi (in.)}^{1/2} \quad (42 \text{MPa m}^{1/2})$$

As the ratio (σ/S_y) increases, the plastic zone increases and the discrepancy between the purely elastic and the corrected results for plasticity becomes larger. Further increase in the ratio (σ/S_y) makes the use of LEFM less appropriate.[4]

Design Comment 13.6

With further growth of plastic effects, additional methodology should be used for the establishment of valid fracture toughness parameters. One of the approaches that is still recognized is that described in Chapter 6 under the acronym CTOD, representing the crack tip opening displacement. In simple terms, CTOD is the opening of the crack, denoted usually by δ, which characterizes the onset of crack extension. It is expressed in units of length (usually inches or millimeters) and it may signify the degree of crack instability. Under plane-stress conditions for values of (σ/S_y) lower than unity, Eq. (6.23) is quite appropriate for design. For lower values of (σ/S_y), the term "ln sec" can be simplified, so that Eq. (6.23) becomes[4]

$$\delta = \frac{\pi a \sigma^2}{ES_y} \qquad (13.51)$$

Other forms of this equation include

$$\delta = \frac{K_I^2}{ES_y} \qquad (13.52)$$

and

$$\delta = \frac{G}{S_y} \qquad (13.53)$$

Also, by general analogy,

$$G_c = \delta_c S_y \qquad (13.54)$$

where G_c denotes the critical potential energy release rate per unit of thickness. And specifically (for the plane-strain condition), we can have

$$G_{Ic} = \frac{K_{Ic}^2(1 - v^2)}{E} \qquad (13.55)$$

The unique meaning of CTOD testing is that a critical value, δ_c, can be measured in a test piece that has been subjected to extensive yielding, beyond the range of applicability of LEFM. As an example of calculation of the CTOD parameter from clip gage displacement, consider a CTOD test made on a steel having yield strength of 62 ksi, using a specimen thickness B equal to 1 in. and a width W of 2 in., as shown in Fig. 6.40. The specimen was precracked to a equal to 1.02 in. and a clip gage plastic displacement Δp was 0.013 in. The maximum load was P 11,240 lb and the knife edges z were 0.079 in. thick.

The elastic component from Eq. (6.24) was

$$\frac{K^2(1 - v^2)}{2ES_y}$$

where

$$K = \frac{PY_1}{BW^{1/2}} \qquad (13.56)$$

The value of the dimensionless polynomial Y_1 for $a/W = 0.52$ was given by Knott and Withey[4] as 11.33. Hence

$$K = \frac{11{,}240 \times 11.33}{1 \times (2)^{1/2}}$$

$$= 90{,}049 \text{ psi (in.)}^{1/2}$$

$$= 90 \text{ ksi (in.)}^{1/2} \quad (98.1 \text{ MPa m}^{1/2})$$

The elastic component, then, is

$$\frac{90^2 \times 0.91}{2 \times 30{,}000 \times 62} = 0.00198 \text{ in.} \quad (0.050 \text{ mm})$$

The plastic component from Eq. (6.24), also referred to as the geometrical term of the CTOD formula, is

$$\frac{0.4(2 - 1.02) \times 0.013}{0.4 \times 2 + 0.6 \times 1.02 + 0.079} = 0.00342 \text{ in.} \quad (0.09 \text{ mm})$$

The total CTOD, then, is

$$\text{CTOD} = \text{elastic part} + \text{plastic part}$$

$$= 0.00198 + 0.00342$$

$$= 0.0054 \text{ in.} \quad (0.14 \text{ mm})$$

Design Comment 13.7

In considering the application of fracture mechanics principles to the most likely production defects, the surface cracks are more common and are perhaps easier to analyze because of a better definition of shape and size.

The designer has selected the 4340 steel plate for a large structural component and it was found that there was a sharp surface crack 0.2 in. deep and 2 in. long. The material specification indicated 220 ksi as the ultimate strength and a yield strength of 185 ksi. The formula used for the calculation of the maximum stress at failure in this case,[23] was

$$\sigma = \frac{K_{\text{Ic}}(\Psi)^{1/2}}{\left[3.77a + 0.21 \left(\frac{K_{\text{Ic}}}{S_y} \right)^2 \right]^{1/2}} \tag{13.57}$$

The crack aspect ratio is defined for this case as depth divided by half-length of the crack, or (a/c). Hence

$$a/c = 0.2/1 = 0.2$$

and the crack shape parameter is found to be 1.1 from Fig. 13.6. The plane-strain fracture toughness follows from Fig. 2.15, using a tensile strength 220 ksi. This gives $K_{Ic} = 73$ ksi (in.)$^{1/2}$. The maximum stress at failure is calculated here from Eq. (13.57).

$$\sigma = \frac{73 \times (1.1)^{1/2}}{\left[3.77 \times 0.2 + 0.21\left(\frac{73}{185}\right)^2\right]^{1/2}}$$

$$= 86.5 \text{ ksi} \quad (596 \text{ MPa})$$

The conventional calculation for this case would be based on Eq. (13.48), which, however, has a slight complication since the stress term also enters the design chart in Fig. 3.21. Nevertheless a quick check can be made as follows. The crack aspect ratio for the chart is

$$a/2c = 0.2/2 = 0.1$$

For the first try, assume the following stress ratio

$$86.5/185 = 0.47$$

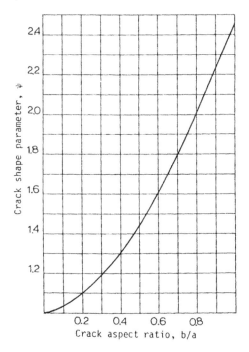

FIGURE 13.6 Crack shape parameter for surface flaws.

which should be sufficiently accurate for all practical purposes, as the calculation will show. Hence, from Fig. 3.21, the flaw shape parameter $Q = 1.09$. Next, assuming the front free-surface correction M_K to be close to unity, the stress can be estimated as

$$\sigma = 0.50 \times 73/(0.2/1.09)^{1/2}$$
$$= 85.2 \, \text{ksi} \quad (588 \, \text{MPa})$$

There is no need for the second iteration because the two results are quite close. The error is only 1.5%. The formula used here is essentially Eq. (13.48) with the constant factor of 0.50 ahead of the K_{Ic} term in order to be consistent with the behavior of a surface crack. When this factor is 0.56, the formula applies to embedded flaw.

Design Comment 13.8

It happens frequently during design calculations that the result is not what was expected. For instance, taking the maximum applied stress equal to the yield of the 4340 grade of steel (such as 220 ksi) and the ratio of (S_y/S_u) equal to 0.88, we are dealing with high-strength material (250 ksi) for which the limit of K_{Ic} is only of the order of 50 ksi $(\text{in.})^{1/2}$. What should be the level of K_{Ic} for a given crack length of $(2c)$ equal to 1.0 in. and a depth a of 0.1 in., for the applied stress approaching the materials yield? The pertinent formula here may be

$$K_{Ic} = \frac{1.94\sigma(a)^{1/2}}{\left[\psi - 0.21(\sigma/S_y)^2\right]^{1/2}} \tag{13.58}$$

The relevant crack aspect ratio is

$$0.1/0.5 = 0.2$$

and from Fig. 13.6, $\psi = 1.1$. Hence, substituting

$$K_{Ic} = \frac{1.94 \times 220(0.1)^{1/2}}{(1.1 - 0.21)^{1/2}}$$
$$= 143 \, \text{ksi} \, (\text{in.})^{1/2}$$

Obviously, there is a substantial difference between the two values of fracture toughness, requiring a major decision. In order to reduce the toughness requirement to come anywhere close to the available toughness, the applied stress must be reduced drastically, as can be seen in Fig. 13.7. If on the other hand, the stress must stay as is, it is highly unlikely that a material of a rather

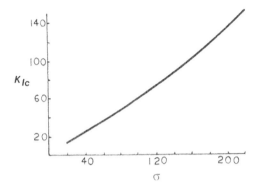

FIGURE 13.7   Toughness vs. stress (dimensions: ksi (in.)$^{1/2}$ and ksi, respectively).

high strength can be matched with high fracture toughness under plane-strain conditions. The other alternative, of course, is to repair the existing crack and to recertify the part. This, as is shown in Chapter 7, under certain conditions can become the problem rather than the solution. It should also be noted that when designing with thinner sections (when practical), the crack propagation may be governed by a combined plane stress–plane strain condition under which the parameter K_{Ic} should not be used. The crack propagation in a thinner section is likely to be slower, and if possible, the plane-stress parameter K_c should be selected for the job.

Design Comment 13.9

The normal practice in fracture mechanics is to determine either plane-strain or plane-stress fracture toughness depending on the degree of constraint. In this way, the information on K_{Ic} or K_c is available for design calculations. Because of the inherent complexity, these values are obtained empirically in terms of the stress and crack size. However, when a structural element of brittle nature is subjected to a combined loading and shows lack of symmetry, it may well be impossible to rely on the size and orientation of the flaw. This is certainly true in the case of a highly brittle material, such as glass, and it may become necessary to describe the fracture mechanics property in terms of the strain energy release rate, denoted by the G_c and G_{Ic} symbols for plane-stress and plane-strain conditions. These parameters can be defined as the quantities of energy released per unit area of crack surfaces as the cracks extend.[24] At the same time, the parameter G in general may be regarded as a measure of the force driving the crack. It is also fortunate that the G parameters can be correlated with the key quantities such as K_{Ic} and K_c. It can be shown how useful such a feature may be with reference to a simple calculation. It is of interest to estimate the approximate critical length of the crack in a glass panel subjected to tension on the premise that the conventional properties such as E and S_u are known.

Modulus of elasticity, $E = 10 \times 10^6$ psi
Ultimate strength in tension, $S_u = 40,000$ psi

The only known LEFM parameter consistent with this problem, G, could be taken, for instance, as follows.

Strain energy release rate, $G_{\text{Ic}} = 0.08$ in.-lb/in.^{2}

From Eq. (13.13), and $\nu = 0.3$, we have

$$K_{\text{Ic}} = 1.05(EG_{\text{Ic}})^{1/2} \tag{13.59}$$

Hence, substituting gives

$$K_{\text{Ic}} = 1.05(10 \times 10^6 \times 0.08)^{1/2}$$
$$= 938 \, \text{psi (in.)}^{1/2}$$

On the premise that a is one-half crack length for a through-thickness flaw in a wide plate in tension and σ is nominal applied stress, the critical crack length is

$$a_{\text{CR}} = 0.32 \left(\frac{K_{\text{Ic}}}{\sigma} \right)^2 \tag{13.60}$$

Therefore, on substitution, the half-length is

$$a_{\text{CR}} = 0.32 \left(\frac{938}{40,000} \right)^2 = 0.000175 \, \text{in.}$$

and the total critical length becomes

$$2 \times 0.000175 = 0.00035 \, \text{in.}$$

The foregoing result shows that the critical crack length in glass can be extremely small, consistent with the anticipated behavior of a highly brittle material. Certainly, even the best nondestructive techniques cannot pinpoint a minute discontinuity of this type.

Design Comment 13.10

Throughout the various sections of this book emphasis is maintained on the balanced approach to design issues involving the elements of fracture mechanics, materials evaluation, and conventional stress analysis. Even the most elementary formula, such as $K_{\text{Ic}} = \sigma(\pi a)^{1/2}$, requires knowledge of the stresses driving the crack, irrespective of the nature of loading and geometry. In other words, the LEFM process can come to a halt if a reasonable level of stress cannot be assigned to the problem at hand. The parametric studies are interesting and useful, but in the end the technical decision must hinge on a specific deterministic quantity that makes a practical sense.

The problem selected in support of the foregoing discussion is intended to illustrate how LEFM and practical stress analysis can work together. The task is to estimate the maximum allowable radial interference Δ in a press-fit assembly consisting of a solid shaft fitted into a cylinder of inner radius $R_i = 0.85$ in. and outer radius $R_o = 1.8$ in. on the premise that the parts are made of high-strength stainless steel with a minimum yield of 210 ksi and a K_{Ic} value of 70 ksi (in.)$^{1/2}$. The initial flaw depth of 0.05 in. can be assumed to extend in a radial direction from the inner surface of the cylinder. The aim is to make a comparison between the results based on the conventional hoop stress criterion (no initial flaw) and the design involving LEFM equations.

The tangential stress in a cylinder subjected to internal pressure[25] is given by

$$\sigma_t = P_i \frac{(R_o^2 + R_i^2)}{(R_o^2 - R_i^2)} \tag{13.61}$$

The conventional formula for a shrink-fit assembly, when the shaft and cylinder are made from the same material, is

$$\Delta = \frac{2P_i R_o^2 R_i}{(R_o^2 - R_i^2)E} \tag{13.62}$$

Eliminating the contact pressure term P_i between Eqs. (13.61) and (13.62) yields

$$\sigma_t = \frac{\Delta E}{2R_i} \left[1 + \left(\frac{R_i}{R_o} \right)^2 \right] \tag{13.63}$$

The classical expression for the case of a notch having depth a is

$$K_{Ic} = 2\sigma(a)^{1/2}$$

from which

$$\sigma = 0.5 K_{Ic}/(a)^{1/2} \tag{13.64}$$

Making $\sigma = \sigma_t$, Eqs. (13.61) and (13.64) give an expression for the radial interference Δ in terms of fracture toughness, notch size, and cylinder physical details. The term Δ_c represents the interference calculated on the basis of fracture mechanics criterion:

$$\Delta_c = \frac{K_{Ic} R_i}{E(a)^{1/2} \left[1 + (R_i/R_o)^2 \right]} \tag{13.65}$$

Substituting the relevant numerical values in Eq. (13.65) gives

$$\Delta_c = \frac{70 \times 0.85}{30{,}000(0.05)^{1/2}[1 + (0.85/1.8)^2]}$$
$$= 0.0073 \text{ in.} \quad (0.185 \text{ mm})$$

Solving next Eq. (13.63) for Δ, yields

$$\Delta = \frac{2R_i \sigma_t}{E[1 + (R_i/R_o)^2]}$$
$$= \frac{2 \times 0.85 \times 210}{30{,}000[1 + (0.85/1.8)^2]}$$
$$= 0.0097 \text{ in.} \quad (0.247 \text{ mm}) \tag{13.66}$$

In this particular example, the prediction based on fracture mechanics is conservative by about 25%. The degree of conservatism depends on the three parameters, as shown by the following ratio:

$$\frac{\Delta_c}{\Delta} = \frac{K_{Ic}}{2(a)^{1/2}\sigma_t} \tag{13.67}$$

This ratio is obtained directly from Eqs. (13.65) and (13.66).

Design Comment 13.11

There are two basic questions in the area of leak-before-break evaluation in relation to such components as pressure vessels and piping. For instance, if the crack were located at the inner surface of the wall, how long it would take to grow through the wall would be the first question. The second required piece of information would concern the crack behavior just after the break through the wall. Would this be a case of localized fracture with a detectable leak, or simply a large-scale catastrophic failure? It is also clear in this situation that a detectable leak, with an opportunity to make a repair, would be most desirable.

Starting with a known or a postulated defect, the initial K value will increase as the crack grows, and as long as the K parameter does not exceed the material's resistance to fracture before the crack grows to a very large size, a leak situation should prevail.

For example, it is required to make a quick estimate of the internal pressure consistent with the leak-before-break condition in a high-strength alloy steel cylinder having mean radius R equal to 32 in. and a wall thickness t of 2 in. The plane-strain fracture toughness is given as 85 ksi $(\text{in.})^{1/2}$. On the basis of a through-thickness crack growth the nominal applied stress comes from the standard expression of LEFM:

$$\sigma = \frac{K_{Ic}}{(\pi t)^{1/2}}$$

The premise here is that the crack has already gone through the entire thickness of the wall, so that t replaces a in the foregoing equation. At the same time the membrane theory defines the conventional hoop stress, as long as the vessel (or pipe) is relatively thin:

$$\sigma_t = PR/t$$

Next, making $\sigma = \sigma_t$, the foregoing expressions lead to the formula for internal pressure, in terms of fracture toughness:

$$P = 0.564 \frac{(t)^{1/2} K_{Ic}}{R} \tag{13.68}$$

and substituting the numerical values from the sample problem, the required pressure becomes

$$P = 0.564(2)^{1/2} \times 85{,}000/32$$
$$= 2120 \, \text{psi} \quad (14.6 \, \text{MPa})$$

In many cases the material's properties of yield strength and toughness are not specified ahead of time, because the design engineer starts with conventional stress analysis such as membrane theory (thin cylinder) or Lameí theory (thick cylinder). The dividing line between the two theories has not been rigidly adhered to,[25] but certainly when the (R/t) ratio is equal to or exceeds 10, the membrane theory is fully justified. The design factor of safety establishes the likely (σ/S_y) ratio and the yield strength of the material. The actual design stresses will, of course, depend on the cylinder dimensions and the required internal pressure based on the specification of the particular system. The final calculational step will involve the estimate of the required toughness, which may or may not change the original selection of the material.

Consider, for instance, the selection process for a high-strength steel for a pressure vessel, the maximum outer diameter of which should not exceed 36 in. The design pressure is 5000 psi and the mandated factor of safety on yield is 3. Hence, the design stress σ is

$$\sigma = S_y/3$$

For the first approximation, consider the membrane theory.

$$\sigma = \frac{S_y}{3}$$

$$= \frac{PR}{t}$$

or

$$S_y = \frac{3PR}{t}$$

$$= \frac{3 \times 5000 \times 18}{t}$$

$$= \frac{270,000}{t}$$

Assuming $S_y = 180,000$ psi, the first estimate of thickness is

$$t = \frac{270,000}{180,000}$$

$$= 1.5 \, \text{in.}$$

Because

$$R/t = 18/1.5$$

$$= 12$$

the membrane theory is applicable, and the design stress is 60 ksi. Since, however, there is a limitation on the external diameter of the vessel, and since the radius in the membrane stress calculation is normally defined as the average, a small correction is in order:

$$R - (t/2) = 18 - 0.75$$

$$= 17.25 \, \text{in.}$$

and

$$\sigma = 5000 \times 17.25/1.5$$

$$= 57,500 \, \text{psi}$$

Also, the factor of safety is

$$180,000/57,500 \cong 3.1$$

with a stress ratio (σ/S_y) of about 0.32. Assuming that weight and cost limitations are not specified, the foregoing calculations satisfy the preliminary estimate based on the traditional approach. The next step is to estimate the minimum level of the plane-strain fracture toughness K_{Ic} using Eq. (6.17)

$$\frac{\pi \sigma^2 B}{1 - \frac{1}{2}(\sigma/S_y)^2} = K_{Ic}^2 \left[1 + 1.4\left(\frac{K_{Ic}^4}{B^2 S_y^4}\right)\right]$$

The required input to this equation may be obtained from the preliminary calculations.

$$\sigma = 57,500\,\text{psi} = 57.5\,\text{ksi}$$
$$B = t = 1.5\,\text{in.}$$
$$S_y = 180,000\,\text{psi} = 180\,\text{ksi}$$
$$\frac{\pi \times 1.5 \times (57.5)^2}{1 - 0.5 \times (0.32)^2} = K_{Ic}^2 \left[1 + 1.4\left(\frac{K_{Ic}^4}{1.5^2 \times 180^4}\right)\right]$$
$$16,421 = K_{Ic}^2 [1 + 0.62(K_{Ic}/180)^4]$$

Let

$$F_1^* = \frac{16,421}{K_{Ic}^2}$$
$$F_2^* = 1 + 0.62(K_{Ic}/180)^4$$

and

$$F_1^* = F_2^*$$

This problem can be solved by plotting two auxiliary functions F_1^* and F_2^* for a few K_{Ic} values from 100 to 140, as shown in Fig. 13.8. The intersection of the two curves should give a sufficiently accurate solution for all practical purposes.

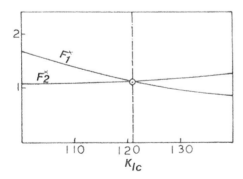

FIGURE 13.8 Graphical solution of a complex equation.

The graphical solution in Fig. 13.8 gives the K_{Ic} value of 121 ksi (in.)$^{1/2}$ as the minimum required plane-strain fracture toughness to assure the leak-before-break condition. The actual material selected for this case was 18Ni-8Co-3Mo (190 Grade) steel, vacuum induction melted, which had a yield strength $S_y = 187$ ksi and fracture toughness of $K_{Ic} = 160$ ksi (in.)$^{1/2}$. The corresponding critical crack length was calculated from the following equation:

$$a_{CR} = 0.32[1 - 0.5(\sigma/S_y)^2](K_{Ic}/\sigma)^2$$
$$= 0.32\left[1 - 0.5\left(\frac{57.5}{187}\right)\right](160/57.5)^2$$
$$= 2.36 \text{ in.} \quad (59.9 \text{ mm})$$

In this manner the major requirements of the traditional stress analysis and fracture mechanics have been satisfied.

Design Comment 13.12

It is well known that as the material thickness decreases, lateral constraint is diminished and the size of the plastic zone near the crack tip experiences a sudden growth. This process is often described as "crack tip blunting." The crack propagation velocity drops off, stresses drift beyond the yield level, and the state of "arrestable instability" develops, in line with the concept of plane stress represented by the parameter K_c. This factor can be correlated with the plane-strain fracture toughness K_{Ic} as indicated below:

$$K_c = K_{Ic}\left[1 + \frac{1.4}{B^2}\left(\frac{K_{Ic}}{S_y}\right)^4\right]^{1/2} \tag{13.70}$$

The concept of K_c may be adapted to the analysis of thin-walled components, although the fracture corresponding to K_c is of a mixed-mode type involving large amounts of plastic flow. It is also difficult to derive the K_c values experimentally, particularly for the thicker components made of low-strength materials. Nevertheless, some attempts have been made to apply the K_c concept to pressure vessels containing surface defects of longitudinal orientation.[26] This orientation is quite natural for the pressure vessel subjected to tensile hoop stresses.

The objective is to estimate the failure pressure when the flaw dimensions and the plane-stress parameter K_c are known, assuming a leak-before-break criterion. This should be pertinent to the plate proper and the welded regions.

Assuming the plane-stress conditions, the toughness parameter was found to be

$$K_c^2 = \frac{\pi(1 + 5v)a\sigma^2}{2(1 + v)\cos(\pi\sigma/2S_u)}\left[1 + \frac{1.7a^2}{Rt}(1 - v^2)^{1/2}\right] \tag{13.71}$$

On taking $v = 0.35$ (average for pure and wrought aluminum), Eq. (13.71) can be simplified:

$$K_c^2 = \frac{3.2a\sigma^2[1 + 1.6(a^2/Rt)]}{\cos(1.57\sigma/S_u)} \tag{13.72}$$

The corrections made in the derivation of Eq. (13.72) included the effect of going from a flat plate to cylinder, the plastic zone ahead of the crack tip, and the biaxial state of stress. The applied nominal stress comes from the conventional formula:

$$\sigma = PR/t$$

The half-length of the longitudinal crack is a, and the ultimate strength of the material is S_u. Also, when the nominal stress σ is in ksi, the parameter K_c is expressed in ksi (in.)$^{1/2}$. The foregoing derivation of Eq. (13.71) was the first part of the work in connection with burst tests of preflawed welded aluminum alloy pressure vessels.[26] The second part of this study assumed a part-through longitudinal flaw having a depth d with a rectangular shape and the area $A_f = 2ad$. From this, $a = A_f/2d$ can be substituted in Eq. (13.72) to give

$$K_c^2 = \frac{1.6A_f\sigma^2[1 + 0.4(A_f^2/Rtd^2)]}{d\cos(1.57\sigma/S_u)} \tag{13.73}$$

The membrane stress to failure was established experimentally as a function of the ultimate strength of the material S_u, calculated membrane stress σ, and the depth dimension of the part-through flaw d.

$$\sigma_f = \frac{(t - d)S_u^2}{tS_u - d\sigma} \tag{13.74}$$

The following sequence was established for the analysis. The term σ_f denotes the stress to failure. For a given area of a part-through flaw A_f and the maximum dimension d, Eq. (13.73) can be solved for σ provided the appropriate level for the parameter K_c can be assigned. Since the K_c values are difficult to measure and are seldom known a priori, an approximation can be made from Eq. (13.70) provided the relevant K_{Ic}, thickness, and yield strength of the material are known. This was certainly possible during the referenced study and it opened the way for a comparison of the two critical stress levels, σ_f and σ. The assumption was made that for $\sigma_f < \sigma$, the through-the-wall crack was likely to propagate, leading to a structural failure. However, for the opposite situation, that is, $\sigma_f > \sigma$, the vessel could be expected to leak before a catastrophic break.

Although the described design methodology in support of the burst tests and empirical studies is quite useful, this approach is still limited because of the difficulties in accounting for residual stresses during manufacture and service. In addition, highly stressed regions such as nozzle junctions and similar transitions pose separate problems of stress analysis, interpretation of test results, and inspection techniques. Further progress in this area will be affected by the balance between the forces of economics and potential consequences of failure.

ENGINEERING MATERIALS

In looking back at the trends in discretionary aspects of engineering design, we see that there was always a choice among wood, stone, brick, concrete, and then a variety of metals and exotic materials. In retrospect one would wonder whether the exotic fibers are really justified, knowing full well that nature, for instance, stumbled upon this long ago when she invented wood. And now, so recently, industry and technology has exploded with a growing list of new nonmetallics, metals, and steels that are developing a giant source of problems in themselves. Their inventories, transportation, and handling are no easy task. And among all this progress and complexity, the design engineer has to make a selection of the materials' properties, certification procedures, and fabrication techniques at a time of growing intolerance of engineering failures, and in light of the litigious character of modern society.

In connection with the topic of practical fracture mechanics, misinterpretation and misuse of materials data could be considered as a potential shortcoming of this branch of engineering science, although other sectors of industrial practice may also have a bearing on this problem. Our prime interest in this area is, of course, to assure the best sources and quality of fracture toughness data affecting the prediction of the permissible crack size, which can influence, for instance, the leak-before-break condition. The explosion of data banks, handbooks, and the general literature (including software libraries) is concerned with reduced data, which must have undergone some degree of interpretation and averaging that may not be insignificant. This is, of course, only natural because raw data are prone to scatter. Also, engineering judgment must come into play because data for the exact condition of interest are never available. The engineer has to base his technical decision on the similarity of alloys, operating environment, and design experience.

This brief section on engineering materials cannot compete with thousands of pages of details in materials handbooks and similar publications. Only a light sprinkling of typical numbers is employed here to make a special point related to design.

Since the plane-strain fracture toughness K_{Ic} is the material property of special value to design, there is wide interest in maintaining the importance of the K_{Ic} standard and excluding nonstandard data. However, the soundness of this argument is open to question when the standard laboratory conditions are compared with those of a structure in the field. In defense of the current practise, on the other hand, appears the similar argument, which says that fracture in the field and in the laboratory occurs at the same value of K. The reader interested in the various aspects of this argument is referred to a more detailed assessment.[13] In the meantime, in a thick material of a high constraint, the K_{Ic} parameter is the "bible."

Generally, all steels are divided into two main groups. The first is concerned with the so-called low-strength structural steels below the yield strength of 140 ksi, which are, however, temperature sensitive and dependent on the rate of loading. These steels become tougher with increasing temperature and decreasing loading rate. The criteria for rapid loading have been established by the ASTM practice.[20] The second group is concerned with high-strength materials, such as maraging grades, normally defined by the range of yield from about 180 to 300 ksi. No significant differences between static and dynamic K_{Ic} values were noted for the yield strength of 250 ksi and higher.

It should also be recalled that in comparing fracture resistance of steels, the best indication is obtained from the magnitudes of the (K_{Ic}/S_y) ratios. The key feature here is that the larger the (K_{Ic}/S_y) ratio, the better the resistance to fracture.

Another point to note is that the ASTM practise regulates the thickness requirement for plane-strain behavior according to the following criterion:

$$\frac{(K_{Ic}/S_y)^2}{B} \leq 0.4 \tag{13.75}$$

A few typical K_{Ic} values for a number of metals under room temperature conditions are given in Table 13.1.

Although the sampling of materials for Table 13.1 was rather arbitrary, steels have a higher (K_{Ic}/S_y) ratio than aluminum or titanium, on the basis of a simple average. The spread between the high and low numbers was shown to be significantly higher for steel than the other two metals. Although some of the characteristics based on Table 13.1 may be representative of the great many structural materials in the area of metallics, the information presented should not be used in any design without independent and thorough assessment of the selected materials on a case-by-case basis.

The problems with a number of aerospace components during the 1960s involved high-strength sheet metal alloys, susceptible to unstable fracture

TABLE 13.1   Yield and Toughness of Engineering Materials at Room
Temperature

Material	Condition	Min. yield (ksi)	Min. K_{Ic} [ksi (in.)$^{1/2}$]	K_{Ic}/S_y
Alloy steels				
A517-F	AM[a]	110	170	1.55
4147	AM	137	109	0.80
HY-130	AM	149	246	1.65
4130	AM	158	100	0.63
12Ni-5Cr-3Mo	AM	175	130	0.74
12Ni-5Cr-3Mo	VIM[b]	183	220	1.20
18Ni-8Co-3Mo	VIM	187	160	0.86
18Ni-8Co-3Mo	AM	190	112	0.59
4330 V	Tempered at 800°F	191	93	0.49
18Ni-8Co-3Mo	AM	193	105	0.54
4340	Tempered at 800°F	197	71	0.36
4330 V	Tempered at 525°F	203	77	0.38
PH13-8 Mo stainless	H1000	210	78	0.37
PH13-8 Mo stainless	H950	210	70	0.33
18Ni maraging	Aged 900°F, 6 h	210	100	0.48
4340	Tempered at 400°F	229	40	0.17
18Ni-8Co-3Mo	VIM	246	87	0.35
18Ni maraging	Aged 900°F, 6 h	259	78	0.30
Titanium alloys				
Ti-6Al-4V	Annealed	120	81	0.68
Ti-6Al-4V-2Sn	Annealed	144	45	0.31
Ti-6Al-4V-2Sn	Solution treated, aged	179	29	0.16
Aluminum				
7075	T7351	53	31	0.58
2014	T651	57	22	0.39
2024	T851	59	19	0.32
2021	T81	61	26	0.43
7049	T73	61	29	0.48
2124	T851	64	22	0.34
7075	T651	70	25	0.36
7049	T73	73	28	0.38

[a]AM indicates electric furnace air melted.
[b]VIM indicates vacuum induction melted.

emanating from small flaws. The well-known plane-stress relationship for an infinitely wide panel is stated as

$$K_c = \sigma_f(\pi a_c)^{1/2} \tag{13.76}$$

This formula is quite similar to that for the K_{Ic} parameter but has the following changes in symbols to be consistent with Ref. 27:

σ_f = fracture stress, ksi
a_c = one-half length of the critical crack, in.
K_c = plane-stress fracture toughness, ksi (in.)$^{1/2}$

The toughness parameter represents the resistance of a metal sheet to crack instability, and it can be used in design in two ways. (See also Design Comment 13.5 in the section on Special Problems, where the K_c formula is used in conjunction with the plasticity correction.) These design actions are:

- Comparison of fracture resistance of candidate alloys in a rational manner.
- Calculation of crack instability conditions from Eq. (13.76).

The materials selected for the particular investigation[27] included alloys of high-strength aluminum, titanium, and steel. One of the criteria was the high strength-to-weight ratio required in airborne vehicles. One of the practical difficulties of the test was that the specimen dimensions could not be a priori selected to assure a valid K_c experiment. In the end, the appropriate results were obtained, a sampling of which is given in Table 13.2.

The foregoing results were considered to be the initial step in the process of characterization of fracture resistance of high-strength sheet metals. Since that

TABLE 13.2 Plane-Stress Fracture Toughness Data for Selected Materials

Material	Thickness (in.)	Crack, length, $2a_c$ (in.)	Yield, S_y (ksi)	K_c [ksi (in.)$^{1/2}$]	K_c/S_y
4130 steel	0.063	2.00	170	157	0.92
4130 steel	0.063	5.00	178	159	0.89
Ti-13V-11Cr-3Al	0.063	2.00	207	45	0.22
Ti-13V-11Cr-3Al	0.125	4.00	217	33	0.15
Ti-6Al-4V	0.063	2.00	15	79	0.52
Ti-6Al-4V	0.125	2.00	146	97	0.66
7178-T6	0.090	2.00	79	55	0.70
7075-T6	0.090	2.00	77	65	0.84
7475-T61	0.090	3.00	62	94	1.52
7475-T761	0.090	2.00	59	92	1.56

time, the direction of research has changed in favor of the development of correlations between the CTOD and K_c parameters. Nevertheless it is instructive to retain Table 13.2, highlighting the order of magnitude of the plane-stress fracture toughness involved.

The majority of engineering applications (including conventional structures and pressure holding systems) require carbon and low-alloy steels, which are known to have pearlitic microstructures. As far as failure modes are concerned, these materials have either cleavage or microvoid coalescence characteristics, which can be recognized even without magnification, because cleavage has reflective and bright surfaces, whereas microvoid and dimple-type fracture traps the light and appears dull. A cleavage fracture is caused by a true tensile (or opening mode) type separation of material under the conditions of decreasing temperature, increased constraint, and high rates of loading. The result is that the fracture develops at a high rate of speed driven by elastic strain energy, and there is very limited plasticity ahead of the crack tip to slow down this process. Although microfracture phenomena belong to the scientific discipline of a metallurgist,[28] these comments should be a part of the blending process of at least some of the more elementary features of materials science, fracture mechanics, and stress analysis.

When mechanical and thermal conditions in pearlitic microstructure allow through-section yielding, the fracture mode is controlled by the ductility of the individual grains (and other effects) at which instability and separation occur.

The two microfracture modes represent the two extremes of fracture behavior. In one case, the elastic stresses can initiate the fracture process when cleavage effects cover at least 80% of the section. The other extreme, consistent with yielding, occurs when microvoid coalescence predominates. The mixed-mode area between the two extremes is governed by the "temperature transition" phenomenon, where the temperature and section thickness interact. This is an important area of knowledge directly related to the process of selection of engineering materials.

The other end of the metallurgical spectrum is the design of high-strength steels, where quench and temper (Q&T) or quench and age (Q&A) are pertinent to the characterization of fracture resistivity. The entire matter of microfracture modes is expanded because of complex alloying elements and temperature effects involved, and especially in the area of welded structures, where heat gradients can cause brittle conditions. Two additional microfracture modes, known as grain boundary separation and quasi-cleavage, can be of importance in the case of Q&T steels, where section thickness plays a part in developing fracture toughness gradients and contributing to the mechanical constraint. The latter element has the principal effect on the transition from plane-strain conditions at ambient temperature.

Metallurgical techniques, over the years, have managed to increase the yield strength of premium steels to levels approaching 300 ksi. Unfortunately, the same techniques have resulted in blocking the formation of dislocations and thereby limiting ductility. This process constituted the natural trend in steel of decreasing fracture toughness with the increase of yield strength. For premium steels, a sharp decrease in fracture resistance was observed between the yield strengths of 180 and 210 ksi.[29]

From the point of view of material selection in the area of nonferrous metals, special attention is normally placed on titanium and aluminum alloys. In general, fractures in nonferrous metals do not propagate as cleavage microfractures. The most common mode connected with a nonferrous metal fracture is a ductile microvoid coalescence. The plane-strain type of fracture is only possible in a high-strength nonferrous metal under conditions of a fast propagation rate.

Also, this metal can contain relatively large particles of phases that are essentially hard and brittle, and that can prevent the development of high levels of fracture resistance.

The absolute values of fracture toughness for aluminum alloys are generally low when compared with those for steels and titanium alloys. However, there is a well-defined strength transition from plastic–elastic to plane-strain elastic fracture similar to that for the high-strength steels and titanium alloys.[28]

It is now well established that the critical crack size is proportional to $(K_{Ic}/S_y)^2$ or $(K_{Id}/S_y)^2$; hence these ratios to the first power (K_{Ic}/S_y) or (K_{Id}/S_y) should be good indicators of the relative toughness of the materials under review. The only practical question here is how large such ratios should be to assure quality of the structure, particularly in those instances where continuous crack monitoring is not practical. The answer is not obvious and some conservative assumptions should be made in the selection of the materials in order to protect the structure against brittle fracture. And a ratio such as (K_{Ic}/S_y) becomes a material selection parameter. The problem can be started with an assumed value of K_{Ic} for the study using the entire theoretical range of yield strengths and some basic formula of fracture mechanics for the case, say, of a wide plate with a through-thickness central crack. The choice of the appropriate material would depend on either higher strength and lower toughness, or lower strength and higher toughness. The selection of the fracture-resistant design is still not obvious because of the economics. Hence the rather straightforward task of selecting a material with the required K_{Ic} and S_y values becomes a task of optimization of structural performance, safety, and economy.

This analysis of an engineering material indicates that the traditional methodology of designing a structure to a certain percentage of the yield strength does not give the same degree of reliability against fracture as that based on the LEFM criteria. There is, however, a problem with the lower strength steel because for a high ratio (K_{Ic}/S_y), the critical crack size can become unusually large, suggesting that fracture mechanics theory no longer applies. The problem here is not with the LEFM theory but with the application.

GLOSSARY OF TERMS AND FUNDAMENTALS

Individual technical terms are normally described in a text as the various topics appear. Since the science of fracture mechanics is relatively new in a conventional design office, a review of the basic notation and meaning of terms is essential. Fracture is a complex deformation process and it can be viewed from several angles requiring often lengthy definitions and interpretations. To avoid any potential problems with understanding the various relationships, it is hoped that the design engineer will assimilate a consistent set of terms and acronyms in order to survive the rigors of a new language during the transfer of fracture mechanics technology from the research laboratory to the design office. The following glossary of terms is selected for this purpose.

> *Brittle fracture*   A failure in structural materials that is catastrophic and occurs without warning at very high speeds and with virtually no plastic deformation. The fractured surface is flat and has a bright, granular appearance due to a cleavage of individual grains. The corresponding fracture stress is below that for a net section yield. The brittle-fractured surface can be recognized without magnification.

> *Charpy V-notch test*   A very popular test in industry, formally accepted (ASTM E370-88a) for assessing ductile-to-brittle transition of the material and establishing the reference nil-ductility transition temperature (NDTT). This is also referred to in the literature as the NDT temperature. The impact test is conducted with a small, blunt notch specimen. The test produces fracture energy, lateral expansion, and shear surface data for assessing the ductile-to-brittle transition.

> *Crack-opening displacement (COD)*   An elastic–plastic fracture mechanics (EPFM) method that is considered to be an alternative to the J integral, which is still the more generally accepted approach in the United States. COD or crack tip opening displacement (CTOD) denotes opening displacement of the crack face as a measure of the plastic strain at the crack tip.

> *Crack tip plasticity*   Plastic yielding can be localized or general. The "localized" plastic zone (small-scale yielding) is contained by the elastic stress field. The "general" plasticity extends across the ligament section (large-scale yielding), and it is the net section yielding.

> *Critical crack size*   When the applied stress intensity factor K reaches a critical level K_{Ic}, known as plane-strain fracture toughness (specific fracture resistance of the material), the corresponding crack reaches the criti-

cal level. At this point the crack becomes unstable and fast, brittle fracture takes place. The crack size a goes with the applied stress intensity K, while the critical crack size a_{CR} occurs in unison with K_{Ic}. K and K_{Ic} in fracture mechanics can be compared with σ and S_y in traditional stress analysis.

Deterministic fracture mechanics Method that assumes a single value of a parameter as the input. Such information is either considered to be known with certainty or it represents a conservative input to yield acceptable outcome of the analysis. Hence "conservative," or even "worst-case," input can at times lead to a highly pessimistic solution.

Ductile fracture When the use of linear elastic fracture mechanics (LEFM) is extended to a larger crack tip plasticity, the rules are changed because ductile rather than brittle fracture controls failure. This type of fracture is characterized by material tearing. Also, a significant amount of plastic deformation is developed at the expense of considerable energy. During this process the crack starts to extend in a ductile and stable fashion, until the structure fails by plastic overloading. The ductile behavior regime is applicable to ferritic steels, and the ductile fracture region is characterized by dimples with a dull appearance.

Dynamic yield strength A material parameter that is measured under conditions of high-speed tensile or impact loading. The test process is complicated because of adiabatic heating, nonuniformity of the applied strain rates, and dynamic wave effects. The effect of the rate of loading on the yield strength is more significant in low- and medium-strength steels (30 to 150 ksi yield).

Elastic modulus Conventional parameter indicating the ratio of stress to strain below the proportional limit. It is the slope of the stress–strain curve.

Elastic–plastic fracture mechanics (EPFM) The central issue of this development is the J integral concept, which represents a field parameter defining the plastic stress and strain intensity around the crack tip. The symbol J in EPFM corresponds to the symbol K in LEFM. The basic limitation of the EPFM approach is that the fracture zone must be relatively small in comparison with the surrounding zone and the planar dimensions of the cracked structural component, and the thickness of the material.

Fatigue crack growth threshold level (ΔK_{th}) Important fatigue parameter for characterizing crack growth rate behavior in terms of ΔK, which represents the cyclic range of the applied stress intensity factor. This term is similar to $\Delta\sigma$, or the stress range defining the fatigue cycling in traditional stress analysis. The ΔK range below which no crack growth occurs is denoted by ΔK_{th}.

Flow stress   Concept that defines the uniaxial true stress at the start of plastic deformation of a metal. It is equal to

$$\sigma_F = \frac{S_y + S_u}{2}$$

where S_y and S_u denote yield and ultimate strength of the material.

Fractography   The failure analysis part of fracture mechanics technology is conducted with the help of a fractographic examination of the failed part in order to check the mechanism of a subcritical crack growth and the nature of the basic microstructure. For instance, the stretch of the zone near the crack and the striation spacing can be determined using this branch of metallographical science.

Fracture mechanics   A branch of engineering analysis and testing con cerned with the assessment of the load-carrying capacity of a structural component containing a defect or a crack.

J integral (J)   The primary method of EPFM is given as a mathematical expression defining a line or surface integral that encloses the area con-tained by crack surfaces and represents the local stress–strain field in the vicinity of the crack.

J_{Ic} parameter   By analogy to the K and K_{Ic} parameters in LEFM, the J_{Ic} term defines the critical value J near the onset of stable crack extension.

Leak-before-break   A fracture control technique when a part-through crack in a thin-walled pressure vessel extends through the wall and either arrests or propagates at such a slow rate that the leakage is detected before the crack attains its critical size.

Linear elastic fracture mechanics (LEFM)   Currently a major tool of fracture control that determines the effects of flaws on the premise that the extent of local plasticity at the crack tip is very small and has no effect on load deflection characteristics.

Mechanical constraint   A geometric and section size system of con-ditions that promotes a triaxial state of stress referred to frequently in fracture mechanics as plane strain.

Membrane stress   The internal pressure in vessels and piping is often referred to as causing the membrane stress (hoop stress), associated usually with thin-cylinder conditions, having a diameter-to-thickness ratio of about 10 or more.

Microcracking   Frequently used in fractography for describing small cracks on a scale where the typical length may vary from atomic spacing up to the grain size.

Monotonic loading   In practical fracture mechanics, a term used rather infrequently. It describes a condition of loading that does not vary, or where the load is increasing in only one direction without any unloading.

Monte Carlo simulation When the input of information for the analysis cannot be determined with a high degree of certainty, it becomes necessary to fall back on the science of probability, where Monte Carlo simulation is a general numerical technique for determination of the distribution of the dependent variables. It is widely used in the probabilistic models of fracture mechanics.

Nil-ductility transition (NDT) temperature Described in referring to the concept of Charpy V-notch (CVN); it can also be obtained from the so-called drop-weight test (DWT). This technique of establishing NDT is formally recognized (ASTM E208-87a).

Nominal stress Unfortunately for the designer, there is enough discussion in the pressure vessel field as well as in fracture mechanics to muddy the waters. In the classical sense, nominal stress is calculated on the net cross-section (using simple elastic theory) without accounting for discontinuities in the form of cracks, grooves, fillets, and holes, to mention a few.

Plane strain A standard definition in stress analysis and LEFM of a stress condition in which there is zero strain in a direction normal to the axis of applied stress and to the direction of crack growth. This is consistent with maximum constraint in thick members.

Plane-strain fracture toughness (K_{Ic}) Defines the minimum value of fracture toughness for a condition corresponding to Mode I type loading and rapid crack propagation governed by the plane-strain fracture criterion.

Plane stress The classical definition of plane stress states that with two principal stresses always parallel to a given plane and constant in the normal direction, we have a state of plane stress. This is consistent with minimum constraint in thin members.

Plastic zone Refers normally to the region at the crack tip at which the tensile stresses are close to the yield strength of the material.

Probabilistic fracture mechanics Related to the comments on the Monte Carlo simulation technique. In short, probabilistic fracture mechanics allows input parameters to be randomly defined. This is quite opposite to deterministic fracture mechanics.

R-curve In this instance, reference is made to the EPFM technique, which defines a material's fracture resistance as a function of crack growth. The *R*-curve can also be given in terms of stress intensity *K* or CTOD, still within the range of small-scale plasticity.

Reference fracture toughness (K_{IR}) Sanctioned by the ASME code in which the K_{IR} parameter is treated as a function of temperature, the reference toughness represents the lower-bound critical stress intensity factors developed from a collection of static, dynamic, and crack arrest curves.

Strain energy release rate (G)   One of the earliest basic concepts is the so-called G parameter, which is a function of energy release per unit area, having the dimensions of in.-lb/in.^{2}. It is also directly proportional to the stress intensity factor in plane stress, known as K_c.

Stress-corrosion cracking (SCC)   It has been well established that in certain combinations of materials and environmental conditions, cracks can grow under constant loading. Below the value defined as the threshold parameter K_{Iscc}, the crack will not grow. This is certainly known in an aggressive corrosion environment. Also the crack growth rate depends on the stress intensity K, and it can increase rapidly as the K value increases. In the limiting case when the applied K reaches the critical level of K_{Ic}, the crack becomes unstable and rapid failure takes place.

Stress intensity factor   Well known LEFM parameter K, which denotes the stress intensity of the applied stress field at the crack tip.

Striation spacing   As the crack front moves to successive locations, a striation spacing is created for each fatigue cycle. On the microscale, these striations appear as the ripples on the fractured surface. Also, the striation spacing becomes a measure of the crack rate of propagation during a given cycle.

Subcritical crack growth   The useful life of a component depends on the rate of growth of a crack from a subcritical size under stress intensity of K to a critical crack size at K_{Ic}.

Unstable fracture   Interpreted as a rapid crack propagation without any increase in load.

Void nucleation   The advantage of utilizing the techniques of fractography is best illustrated in the study of the formation and coalescence of microvoids. For instance, when observed under an electron microscope, the coalesced voids will appear as minute dimples on the fractured surface.

The foregoing brief sampling of terms and principles touches on several engineering disciplines and emphasizes those elements of practical knowledge that may help the reader to appreciate the need for a common vocabulary in applying the rudimentary principles of fracture mechanics to design. It is more than likely that many critical questions will have to be answered in modern engineering concerning the structural integrity of the existing and new mechanical and structural systems, utilizing the technology of fracture mechanics. And probably the best way of enhancing an understanding of this task is by solving various practical problems, repeating numerous elementary calculations, and adopting a common language that fracture mechanicians, materials practitioners, and stress analysts can use.

SYMBOLS

a	Crack length (or depth), in. (mm)
a_c	Half-length of critical crack, in. (mm)
a_{CR}	Critical crack length (general symbol), in. (mm)
a_e	Effective crack length, in. (mm)
A_f	Cross-sectional area of part-through flaw, in. (mm)
B	Specimen thickness, in. (mm)
c	Half-length of crack, in. (mm)
d	Depth of flaw, in. (mm)
E	Elastic modulus, ksi (MPa)
F_1, F_2	Special factors (Design Comment 13.1)
F_1^*, F_2^*	Auxiliary functions
$f(a/w)_c$	Polynomial function
G	Strain energy release rate, lb/in. (N/mm)
G_c	Strain energy release rate (plane stress), lb/in. (N/mm)
G_{Ic}	Strain energy release rate (plane strain), lb/in. (N/mm)
J	J integral, lb/in. (N/mm)
J_c	J integral (plane stress), lb/in. (N/mm)
J_{Ic}	J integral (plane strain), lb/in. (N/mm)
K	Stress intensity factor, ksi (in.)$^{1/2}$ [MPa (m)$^{1/2}$]
K_c	Plane-stress fracture toughness, ksi (in.)$^{1/2}$ [MPa (m)$^{1/2}$]
K_I	Stress intensity factor (Mode I), ksi (in.)$^{1/2}$ [MPa (m)$^{1/2}$]
K_{Ic}	Plane strain fracture toughness, ksi (in.)$^{1/2}$ [MPa (m)$^{1/2}$]
K_{Id}	Dynamic fracture toughness, ksi (in.)$^{1/2}$ [MPa (m)$^{1/2}$]
K_{IR}	Reference fracture toughness, ksi (in.)$^{1/2}$ [MPa (m)$^{1/2}$]
K_{Iscc}	Stress-corrosion threshold toughness, ksi (in.)$^{1/2}$ [MPa (m)$^{1/2}$]
M_K	Back free-surface correction
P	Concentrated load, lb. (N)
P_i	Internal pressure, ksi (MPa)
Q	Flaw shape factor
R	Mean radius of vessel, in. (mm)
r	Radius from crack tip (or hole radius), in. (mm)
R_i	Inner radius, in. (mm)
R_K	Ratio of intensities (secant and tangent formulas)
R_L	Ratio of intensities (mixed trigonometric terms)
R_o	Outer radius, in. (mm)
R_P	Ratio of intensities (polynomials)
S_u	Ultimate strength, ksi (MPa)
S_y	Yield strength, ksi (MPa)
t	Wall thickness in vessels, in. (mm)
W	Finite width of plate, in. (mm)
w	Half-width of plate, in. (mm)

Y_1	Numerical value from polynomial
z	Thickness of knife edge, in. (mm)
β	Parameter in stress intensity formula
γ	One-half the surface energy, lb/in. (N/mm)
Δ	Press-fit interference (stress analysis), in. (mm)
Δ_c	Press-fit interference (LEFM), in. (mm)
ΔK_{th}	Threshold stress intensity, ksi (in.)$^{1/2}$ [MPa (m)$^{1/2}$]
Δ_p	Plastic displacement, in. (mm)
$\Delta\sigma$	Stress range, ksi (MPa)
δ	Crack-opening displacement, CTOD, in. (mm)
δ_c	Critical crack opening displacement, CTOD, in. (mm)
θ	Arbitrary angle in stress field, rad
σ	General symbol for stress, ksi (MPa)
σ_a	Asymptotic value of stress, ksi (MPa)
σ_F	Flow stress, ksi (MPa)
σ_f	Failure (or fracture) stress, ksi (MPa)
σ_t	Tangential stress, ksi (MPa)
ψ	Crack shape parameter

REFERENCES

1. Griffith, A.A. The phenomena of rupture and flow in solids. Phil. Trans. Royal Society **1920**, *221*, 163–198.
2. Irwin, G.R. Fracture Dynamics. In *Fracturing of Metals*; American Society of Metals: Cleveland, 1948.
3. Orowan, E. Fracture strength of solids. In *Report on Progress in Physics*, Vol. 12; Physical Society of London: London, 1949; 185–232.
4. Knott, J.F.; Withey, P.A. *Fracture Mechanics — Worked Examples*; Institute of Materials, The Bourne Press: Bournemouth, England, 1993.
5. Barsom, J.M.; Rolfe, S.T. *Fracture and Fatigue Control in Structures*, 2nd Ed.; Prentice-Hall: Englewood Cliffs, NJ, 1987.
6. *Standard Test method for J_{IC}, a Measure of Fracture Toughness*, ASTM E813-81, Vol. 03.01, Metals — Mechanical Testing; ASTM: West Conshohocken, PA, 1985.
7. Westergaard, H.M. Bearing pressures and cracks. Trans. ASME J. Appl. Mech. **1939**, *6*, 49–53.
8. Irwin, G.R. Analysis of stresses and strains near the end of a crack. *J. Appl. Mech.* **1957**, *24*.
9. Knott, J.F. *Fundamentals of Fracture Mechanics*; Butterworths: London, 1979.
10. Tada, H.; Paris, P.C.; Irwin, G.R. *The Stress Analysis of Cracks Handbook*; University of St. Louis: St. Louis, 1973.
11. Sih, G.C. *Handbook of Stress Intensity Factors*; Institute of Fracture and Solid Mechanics, Lehigh University: Bethlehem, PA, 1973.

12. Paris, C.P.; Sih, G.S. Stress analysis of cracks. in *Fracture Toughness Testing and Its Applications*, ASTM STP 381; American Society for Testing and Materials: Philadelphia, 1965.

13. Broek, D. *The Practical Use of Fracture Mechanics*; Kluwer Academic Publishers: Dordrecht, 1988.

14. Gran, R.J.; Orario, F.D.; Paris, P.C.; Irwin, G.R.; Hertzberg, R.W. *Investigation and Analysis Development of Early Life Aircraft Structural Failures*, AFFDL-TR-70-149, 1971.

15. Broek, D. *Cracks at Structural Holes*. Research Report MCIC-75-25; Battelle Columbus Laboratories: Columbus, OH, 1975.

16. Bowie, O.L. Analysis of an infinite plate containing radial cracks originating at the boundary of an internal circular hole. J. Mat. Phys. **1956**, *25*.

17. Pellini, W.S. *Principles of Structural Integrity Technology*; Department of the Navy, Office of Naval Research: Arlington, VA, 1976.

18. Rooke, D.P.; Cartwright, D.J. *Compendium of Stress Intensity Factors*; Her Majesty's Stationery Office: London, 1976.

19. Zahoor, A. *Ductile Fracture Handbook*, Vol. 1, *Circumferential Through Wall Cracks*, EPRI NP-6301-D, June 1989.

20. American Society for Testing and Materials. Standard method of test for plane-strain fracture toughness of metallic materials. in *ASTM Annual Standards*, ASTM Designation E-399-83, Vol. 03.01; ASTM: Philadelphia, 1987.

21. Wessel, E.T.; Server, W.L.; Kennedy, E.L. *Primer: Fracture Mechanics in the Nuclear Power Industry*; Electric Power Research Institute: Palo Alto, CA, 1991.

22. Murakami, Y. *Stress Intensity Factors Handbook*; Pergamon Press: New York, 1988.

23. Steigerwald, E. What you should know about fracture mechanics. Met. Progr. **1967**.

24. Sih, G.C. The role of fracture mechanics in design technology. Paper No. 76-DET-58. ASME J. Eng. Ind. **1976**.

25. Blake, A. *Practical Stress Analysis in Engineering Design*, 2nd Ed.; Marcel Dekker: New York, 1990.

26. Lake, R.L.; DeMoney, F.W.; Eiber, R.J. Burst tests of pre-flawed welded aluminum alloy pressure vessels at $-220°F$. In *Advances in Cryogenic Engineering*, Vol. 13; Plenum Press, 1968.

27. Freed, C.N.; Sullivan, A.M. Stoop, J. Influence of dimensions of the center-cracked tension specimen on K_c, fracture toughness. In *Proceedings of the 1971 National Symposium on Fracture Mechanics*, Part II, ASTM STP 514; American Society for Testing and Materials: Philadelphia, 1972.

28. Lange, E.A. *Fracture Toughness of Structural Metals*, NRL Report 7046; Department of the Navy, Naval Research Laboratory: Washington, DC, 1970.

29. Puzak, P.P.; Lange, E.A. *Fracture Toughness Characteristics of the New Weldable Steels of 180- to 210-Ksi Yield Strengths*, NRL Report 6951; 1969.

Index